S0-AVN-657

ACS SYMPOSIUM SERIES **508**

Pollution Prevention in Industrial Processes

The Role of Process Analytical Chemistry

Joseph J. Breen, EDITOR
U.S. Environmental Protection Agency

Michael J. Dellarco, EDITOR
U.S. Environmental Protection Agency

Developed from a symposium sponsored
by the Division of Environmental Chemistry
at the 201st National Meeting
of the American Chemical Society,
Atlanta, Georgia,
April 14–19, 1991

American Chemical Society, Washington, DC 1992

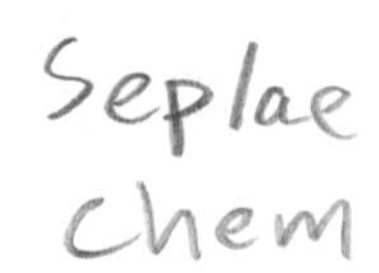

Library of Congress Cataloging-in-Publication Data

Pollution prevention in industrial processes: the role of process analytical chemistry / Joseph J. Breen, editor, Michael J. Dellarco, editor.

p. cm.—(ACS symposium series, ISSN 0097–6156; 508)

"Developed from a symposium sponsored by the Division of Environmental Chemistry at the 201st National Meeting of the American Chemical Society, Atlanta, Georgia, April 14–19, 1991."

Includes bibliographical references and index.

ISBN 0–8412–2478–1

1. Factory and trade waste—Analysis—Congresses. 2. Chemistry, Analytic—Congresses.

I. Breen, Joseph J. II. Dellarco, Michael J., 1951– . III. American Chemical Society. Division of Environmental Chemistry. IV. American Chemical Society. Meeting (201st: 1991: Atlanta, Ga.) V. Series.

TD897.5.P65 1992
628.5—dc20

92–30288
CIP

The paper used in this publication meets the minimum requirements of American National Standard for Information Sciences—Permanence of Paper for Printed Library Materials, ANSI Z39.48–1984. ♾

PRINTED IN THE UNITED STATES OF AMERICA

1992 Advisory Board

ACS Symposium Series

M. Joan Comstock, *Series Editor*

Foreword

THE ACS SYMPOSIUM SERIES was first published in 1974 to provide a mechanism for publishing symposia quickly in book form. The purpose of this series is to publish comprehensive books developed from symposia, which are usually "snapshots in time" of the current research being done on a topic, plus some review material on the topic. For this reason, it is necessary that the papers be published as quickly as possible.

Before a symposium-based book is put under contract, the proposed table of contents is reviewed for appropriateness to the topic and for comprehensiveness of the collection. Some papers are excluded at this point, and others are added to round out the scope of the volume. In addition, a draft of each paper is peer-reviewed prior to final acceptance or rejection. This anonymous review process is supervised by the organizer(s) of the symposium, who become the editor(s) of the book. The authors then revise their papers according the the recommendations of both the reviewers and the editors, prepare camera-ready copy, and submit the final papers to the editors, who check that all necessary revisions have been made.

As a rule, only original research papers and original review papers are included in the volumes. Verbatim reproductions of previously published papers are not accepted.

M. Joan Comstock
Series Editor

Contents

Preface

POLLUTION PREVENTION as a national environmental strategy has focused attention on technological disciplines that will make implementation of the strategy a reality. The two disciplines most discussed are industrial process engineering and alternative synthetic pathways for the design of "environmentally friendly" chemicals and formulations. Both are the center of attention as industry and environmental policymakers attempt to redesign processes to be more productive, while eliminating or minimizing pollutants.

A third area—process analytical chemistry (PAC)—is equally important. PAC provides the technological means to monitor, in real time, the redesigned industrial processes to verify productivity while documenting reduced or minimized levels of unwanted byproducts and pollutants.

Global competition and the need for improved quality, improved worker safety, higher productivity, and lower operating costs all couple with the new environmental paradigm—pollution prevention—to drive the demand within the chemical industry for more effective analytical techniques. To achieve success, instruments must be designed from basic principles to meet the particular process requirements. Continuous online and at-line analyzers must be rugged, maintain stability over time, and operate reliably and simply with minimal time for operation maintenance and service concerns.

The symposium on which this book is based, "Pollution Prevention and Process Analytical Chemistry", provided an exciting forum for industrial and academic researchers to be briefed on the new environmental ethic, pollution prevention, and the latest developments in community right-to-know legislation. It was equally exciting and informative for environmentalists and environmental regulators and policymakers to be apprised of the results and progress of industrial and academic efforts on the application of process analytical chemistry to the environmental issues of source reduction and waste minimization. The technical discussions represented a broad spectrum of interests from industry, academia, sensor research centers, and the Department of Energy (DOE) National Laboratories.

This volume includes real-world industrial applications from Monsanto, 3M, Du Pont, Dow, and Amoco. It includes contributions on sensor technology development—surface acoustic wave, chemical array, and electrochemical devices—from university and private sector programs;

contributions on coupling membrane sampling interfaces with mass spectrometers for industrial process and environmental applications by university and government research programs. Researchers from the DOE National Laboratories presented results on the use of fiber-optic immunosensors and on waste minimization via on-line and at-line process analytical chemistry.

The most-represented laboratory in the volume is the Center for Process Analytical Chemistry (CPAC), University of Washington; 5 of the 24 chapters are by CPAC authors. This fact properly reflects their preeminent role in U.S. industry–university research on process analytical chemistry. CPAC serves as a National Science Foundation success story in government–industry–university collaboration on seeking real-world solutions to industrial process and environmental problems.

This book, with contributions by university faculty and students and by government and industry representatives, suggests that pollution prevention may well serve as a framework for government, industry, and academia to work constructively together on the environmental issues of the 1990s.

DISCLAIMER

This book was edited by Joseph J. Breen and Michael J. Dellarco in their private capacity. No official support or endorsement of the U.S. Environmental Protection Agency is intended or should be inferred.

JOSEPH J. BREEN
Office of Toxic Substances
U.S. Environmental Protection Agency
Washington, DC 20460

MICHAEL J. DELLARCO
Office of Research and Development
U.S. Environmental Protection Agency
Washington, DC 20460

June 12, 1992

PERSPECTIVES ON POLLUTION PREVENTION

Chapter 1

Pollution Prevention

The New Environmental Ethic

Joseph J. Breen[1] and Michael J. Dellarco[2]

[1]Office of Toxic Substances and [2]Office of Research and Development, U.S. Environmental Protection Agency, Washington, DC 20460

Prosperity without pollution has become the fundamental environmental theme of the 1990s. Or at least, the consideration of how we will achieve this economic and environmental imperative. The new paradigm - **pollution prevention** - will serve as the keystone of federal, state and local environmental policy. Support for the new approach - **the new ethic** - is broad based and includes environmentalists, industrialists, law-makers, academicians, government regulators and policy-makers, and the general public. The challenge is to switch from two decades of environmental policy based on pollution controls and government mandated regulations, to a future environmental policy based on pollution prevention, source reduction, recycling, and waste minimization. It will require a new social compact amongst environmental, industrial, and regulatory interests. The roles and contributions of the chemical engineer, synthetic organic and inorganic chemist, and the process analytical chemist will be integral to the full articulation and implementation of the new vision.

Pollution prevention is the environmental ethic of the 1990s. It replaces two decades of national and state environmental policies based on pollution control (1-3). It represents the latest step in the evolution of environmental policy in industrialized nations, especially the United States. That policy over the past twenty years has progressed from a narrowly focused preoccupation with regulatory command and control of "end of pipe" releases; to a more practical "waste management" technology; to the more enlightened economics of "waste minimization."

The advent of "waste minimization" was a watershed in the evolutionary process. It prompted industry, regulators, and environmentalists alike, to lift their collective heads up from the seemingly interminable battles of special interests. It redirected their attention and energy away from "end of the pipe," and "fence-line" micro-environmental releases, back through the industrial facility or treatment process being controlled, right up to the front door and the planning table where the action begins. Added nagging problems of non-point source pollution provided further impetus to revisit the basic processes and systems polluting the environment.

Pollution prevention emerged as the theme around which to establish a framework to protect the environment, in part, to confront the economic realities of the enormous costs associated with hazardous waste treatment and disposal. The almost common good sense of the pollution prevention concept seemed obvious to all. Past improvements in one medium invariably resulted in contamination of another. Transferring pollutants between environmental compartments no longer was a viable solution. The successful control approaches of the 1970s and 1980s in dealing with macro-environmental pollution of air and surface waters no longer were sufficient to get the job done.

A new paradigm, a more flexible paradigm, that allowed creative solutions, jointly developed by industry, government and environmentalists, would have to be put in place. The new framework, with a defined hierarchy of possible responses - source reduction, recycling, treatment and disposal - provides industry, government and the environmental groups an array of options from which to seek acceptable solutions.

Pollution prevention is an attractive environmental strategy for industry for several reasons. If no pollution is generated, there are no pollutants to be controlled and managed. Future problems are avoided, such as those which occurred when land disposal methods were later found to be major sources of ground water contamination. The old policies and methods resulted in billion dollar site remediations. Preventing pollution before it occurs has the added feature of preventing exposures to the community at large and to the workers charged with the management of pollution.

A significant potential benefit of industrial pollution prevention is economic. When wastes are reduced or eliminated, savings in materials result - more product is produced from the same starting materials. Re-examination of manufacturing processes as part of a pollution prevention approach can produce a variety of unanticipated benefits such as conservation of energy and water and improved product quality. Given the escalating costs of waste handling in the 1980s, a program promoting source reduction can provide a major incentive to industrial firms. A dominant cost savings can be realized from significantly reduced future liability for future pollution. "Cradle to grave" responsibility for wastes generated has prompted industry to reassess old practices and long-established management techniques and perspectives on waste management and pollution controls.

On the environmental side, the advantages of pollution prevention include improving the effectiveness of managing reduced waste streams; minimizing the uncertainty associated with the environmental impact of released pollutants; avoiding cross-media transfers of released pollutants, and protecting natural resources. Finally, pollution prevention is consistent with the public's "right-to-know" and "right-to-know more" laws, and with increased public scrutiny of industrial practices.

Industrial Programs and Activities

Industry, working harder to be good neighbors and responsible stewards of their products and processes, is aggressively engaged in a variety of pollution prevention activities both as trade associations and as individual companies. An extended description of successful trade association and individual company programs is available from EPA's Office of Pollution Prevention (3). A snap-shot sampling of these programs would include:

Trade Association Programs (4). Three programs of merit are:

Chemical Manufacturers Association (CMA). CMA started its Responsible Care program in 1988 to improve the chemical industry's management of chemicals. All CMA members are required to participate and adhere to the 10 guiding principles of the program. The principles speak of protecting health, safety, and the environment, but do not address pollution prevention per se. The program does outline the framework for the reduction of waste and releases to the environment. To evaluate progress, CMA companies must submit an annual report to identify progress in implementation and to quantify facility-specific releases and wastes (5).

American Petroleum Institute (API). API also has a prescribed set of guiding environmental principles its members are encouraged to follow. API's 11 principles generically promote actions to protect health, safety, and the environment. One of API's principles directly speaks to pollution prevention by requiring its members to reduce overall emissions and waste generation (6). API articulates the 11 principles as goals to which members should aspire.

National Paints and Coatings Association (NPCA). NPCA has a bona fide Paint Pollution Prevention Program (April 1990). The goal of the program is "the promotion of pollution prevention in our environment through effective material utilization, toxics use and emissions reduction and product stewardship in the paint industry" (7). The statement recommends each NPCA member company establish a waste reduction program to include setting priorities, goals and plans for waste and release reduction with preference first to source reduction, second to recycling/reuse and third to treatment.

Company Programs (4).

A review of company pollution prevention activities reveals some companies have programs which they are willing to share with the public and other companies whose efforts are considered internal and proprietary. The more accessible programs are usually with large multi-facility companies. They are engaged in a wide range of operations, from specialty chemicals to high technology electronics. Some programs are well established with formal names and acronyms. Others are newer and more informal. The earliest dates back to 1975, with some following in the early and mid-1980s and others initiated in the 1990s.

The scope of programs varies considerably. Some are limited to one environmental medium, while others are multimedia. Some focus on certain types of pollutants, such as the Toxics Release Inventory (TRI) chemicals, others are more wide ranging. All include some form of pollution prevention, but vary in their emphasis. Most adopt EPA's environmental management hierarchy: source reduction first, followed by recycling, treatment, and disposal. Strict comparisons across companies is difficult because the same or similar terms mean different things to different programs. (See Table I. Selected Pollution Prevention Definitions.)

Program goals vary by amount pledged to be reduced, time frame for reduction, pollutant being reduced and environmental medium affected by release of the pollutant. Some goals are quite specific, others are more general. Some longstanding programs have an impressive list of accomplishments. A few programs have data collection and analysis programs to assess their pollution prevention progress. Others use a case study approach to illustrate and document successes.

A description of the pollution prevention program at the Monsanto Company is presented in Chapter 2 of this volume. It details one company's efforts and commitments to the new vision of environmental protection and industrial pollution prevention.

Public interest groups have raised questions as to the accountability and reliability of the accomplishments claimed. Legitimate questions remain to document whether the cited reductions are real and result from pollution prevention methods. Or whether they be artifacts of changes in reporting requirements or analytical methods, or from waste transfers between sites or between media, or from reductions in operations (8).

The concerns notwithstanding, a major change in industrial perspective on the way business is to be done has taken place. Most programs and activities are voluntary. The programs initiated by industry on pollution prevention are important because they raise expectations for future progress. If the successes are real and include financial gains, there is a legitimate expectation other firms will follow the leaders into this new era of environmental protection.

State and Local Programs (9).

States have provided the leadership role from the government side of the pollution prevention ledger. Some state programs have served as models for federal activities. The Toxics Release Inventory requirements of the Federal Emergency Planning and Community Right-to-Know Act were modeled after similar elements of New Jersey and Maryland programs. Early state programs promoted industrial pollution prevention by providing technical assistance and information. Local programs tend to be more limited than state efforts, but local governments realize the opportunities and advantages of fostering pollution prevention activities with companies in their communities.

State Legislation. State pollution prevention statutes have mushroomed in the last seven years. Prior to 1985, only one state law dealt with pollution

Table I. Selected Pollution Prevention Terms

Pollution/Pollutants. In context of pollution prevention refers to all nonproduct outputs, regardless of recycling or treatment that may prevent or mitigate releases to the environment.

Waste. Nonproduct outputs of processes and discarded products, irrespective of the environmental medium impacted. Most uses of "waste" refer to hazardous and solid wastes regulated under RCRA, and do not include air emissions or water discharges regulated by the Clean Air Act or the Clean Water Act. The Toxics Release Inventory, TRI, refers to wastes that are hazardous and nonhazardous.

Waste Minimization. An early initiative under pollution prevention paradigm. Initial focus on RCRA wastes, rather than on comprehensive evaluation of industrial emissions regulated under all environmental statutes. Term becomes source of controversy since some consider it designates approaches to treating waste to minimize volume or toxicity, rather than decreasing quantity of waste at source of generation. Current RCRA use of waste minimization refers to source reduction and recycling activities, and excludes treatment and energy recovery.

Industrial Source Generation. Defined under Federal Pollution Prevention Act of 1990 as "any practice which 1) reduces the amount of any hazardous substance, pollutant, or contaminant entering any waste (pollutant) stream or otherwise released into the environment (including fugitive emissions) prior to recycling, treatment, and disposal; and 2) reduces the hazards to public health and the environment associated with release of such substances, pollutants, or contaminants. The term includes equipment or technology modifications, process or procedure modifications, reformulation or redesign of products, substitution of raw materials, and improvements in housekeeping, maintenance, training, or inventory control" (19).

Does not entail any form of waste management (e.g., recycling and treatment). The Act excludes from definition of source reduction "any practice which alters the physical, chemical, or biological characteristics or volume of a hazardous substance, pollutant, or contaminant through a process or activity which itself is not integral to and necessary for the production of a product or the providing of a service"(20).

Waste Reduction. Conflicting usage: by Congressional Office of Technology Assessment means source reduction; by others means waste minimization.

Toxic Chemical Use Substitution. Describes replacing toxic chemicals with less harmful chemicals, although relative toxicities may not be fully known. Examples include substituting toxic solvent in an industrial process with a chemical with lower toxicity and reformulating product to decrease use of toxic raw materials or generation of toxic byproducts.

Toxics Use Reduction. refers to the activities grouped under "source reduction", where intent is to reduce, avoid, or eliminate the use of toxics in processes and/or products so as to reduce overall risks to the health of workers, consumers, and the environment without transferring risks between workers, consumers, or parts of the environment.

Industrial Pollution Prevention. "Industrial pollution prevention" and "pollution prevention" refer to combination of industrial source reduction and toxic chemical use substitution. It does not include recycling or treatment of pollutants, nor does it include substituting a nontoxic product made with nontoxic chemicals for a nontoxic product made with toxic chemicals.

SOURCE: Adapted from ref. 3, p.6-7.

prevention. By 1991, close to 50 state laws were in place, half of them enacted in 1990. More than half the 50 states have passed pollution prevention laws, some states passing more than one. Other states have legislation pending or on their agenda.

The state pollution prevention laws vary widely in their provisions. Some have detailed requirements. They target specific source reduction goals and provide measures to meet them. Other states have general statutes, declaring pollution prevention, as state policy, to be the preferred method of dealing with hazardous wastes. Some states have no formal laws, but have operationally included pollution prevention into their programs.

Facility Planning Requirements. A new important trend in state pollution prevention requirements is facility planning. These statutes require industrial facilities to submit pollution prevention plans and to update them periodically. Most plans cover facilities obliged to report federal Toxics Release Inventory (TRI) data. The chemicals covered by facility planning statutes vary, but focus on the TRI list of chemicals and hazardous wastes covered by RCRA.

Many of the facility planning laws require industry to consider only pollution prevention options. Others are broader in scope, but consider pollution prevention as the preferred approach when technically and economically practicable. Facilities required to prepare plans must either prepare and submit progress reports or annual reports. Facilities failing to complete adequate plans or submit progress reports may be subject to enforcement actions or to negative local publicity.

State Policies and Goals. State policies and goals for pollution prevention may be established by both legislative and non-legislative means. Seven state legislatures have captured their goals legislatively. The focus of their efforts is hazardous waste generation and how to reduce it. For other states, their agencies express their commitment to pollution prevention policies by setting specific hazardous waste disposal capacities and requiring industry to work within them.

State Pollution Prevention Programs. State programs are the best barometer of activity in pollution prevention. Programs vary along with their enabling statutes. Some programs are mature, well-established, and independent. Others consist of little more than a coordinator, who pulls together the pollution prevention aspects from other state environmental programs. Some states delegate their pollution prevention to third party groups at universities and research centers and provide state funding for their operation.

Program elements of state programs include: raise general awareness of benefits from pollution prevention; reduce informational/technological barriers; and create economic and regulatory incentives for pollution prevention. Some also attempt to foster changes in the use of toxic materials and the generation or release of toxic by-products.

Local Programs. A strong impetus exists for local governments to get involved. The effects of hazardous waste production are almost always felt first and foremost at the local level. Rather than relying solely on state and federal efforts, local governments are often in a better position to identify needs and limitations of local facilities. Local governments may also be more flexible in dealing with specific problems. One example is the potential offered by publicly owned treatment works (POTWs). POTWs receive and process domestic, commercial and industrial sewage. Under delegated federal authority, they may restrict industrial and commercial pollution from waste water they receive. Pollution prevention is becoming a recurrent theme in the operation of POTWs.

Federal Policies and Programs (10).

The federal government plays a major role in promoting industrial pollution prevention. It does so in three primary areas: first, as a pollution generator at federal facilities and public lands; second, as a consumer and large purchaser of products and services; and third, as a policy maker. By encouraging its own federal facilities to promote pollution prevention, it can serve as a leader and foster the ethic of prevention both inside and outside governmental institutions.

Major federal agencies and departments implementing industrial pollution prevention are the Departments of Defense and Energy (DOD and DOE) and the Environmental Protection Agency (EPA). EPA plays the lead role in developing with its sister federal agencies a pollution prevention strategy for the federal sector.

Federal Legislation. Congress began to develop a deliberate national policy of industrial pollution prevention under the Hazardous and Solid Waste Amendments to the Resource Conservation and Recovery Act (RCRA) of November 1984. It set as national policy "wherever feasible the generation of hazardous waste is to be reduced or eliminated as expeditiously as possible" (11). The 1990 Federal Pollution Prevention Act (12) expanded the scope of Congressional pollution prevention policy to cover pollutants in all media. In 1990, Congress also passed the Clean Air Act Amendments (13) with provisions supportive of reducing pollution at its source. As a package, they allow EPA not only to promote industrial pollution prevention, but also to refocus federal environmental policy from control to prevention.

DOD Pollution Prevention Policy. DOD programs exist on a department-wide and on a individual service basis. DOD has taken a staged approach to adopting pollution prevention, beginning with a hazardous waste minimization policy that identifies source reduction as the preferred waste management strategy. With 1985 disposal data as a bench-mark, DOD established a goal of a 50% reduction in the disposal of hazardous waste by 1992. A 1989 DOD Directive on Hazardous Material Pollution Prevention (14) establishes as policy that hazardous materials should be selected, used and managed over its life-cycle such that DOD incurs the lowest cost required to protect human health and the environment.

DOE Pollution Prevention Policy. DOE's 1990 Waste reduction Policy Statement identifies waste reduction as a prime consideration in research activities, process design, facility upgrade or modernization, new facility design, facility operations and facility decontamination and decommissioning. (15)

Waste reduction should be accomplished by a hierarchy of pollution prevention practices starting with source reduction. As DOE's waste minimization program evolves, the head of each DOE facility must develop a source reduction and recycling plan. Each plan is to be revised annually and updated every three years.

EPA Pollution Prevention Policy. Pollution prevention, while not new to EPA, has emerged as a priority in the 1990s. This represents a fundamental change from the historical interpretation of the Agency's mission as protecting human and environmental health through pollution control. EPA's pollution control emphasis eliminated options to release and transfer industrial pollution in the environment and to increase the cost of the remaining options of treatment and disposal. The net effect has been to encourage industry to limit their pollution through source reduction.

The formal shift in policies and priorities for EPA are reflected most recently in the 1990 passage of the Clean Air Act Amendments and Pollution Prevention Act. EPA issued a Pollution Prevention Strategy in 1991 to articulate its position and objectives. The strategy serves two purposes: 1) to provide guidance and direction to incorporate pollution prevention into EPA's existing regulatory and non-regulatory program; and 2) to specify a program with stated goals and a time table for their accomplishment (16). EPA's goal is to incorporate pollution prevention into every facet of its operations - including enforcement action, regulations, permits, and research.

This strategy confronts the institutional barriers that exist within EPA, which is divided along single environmental medium lines. The Agency has:

* established an Office of Pollution Prevention and Toxics which coordinates agency-wide pollution prevention policy;
* created a Waste Minimization Branch in the Office of Solid Waste to coordinate waste minimization and pollution prevention policy under RCRA;
* charged the EPA Risk Reduction Engineering Laboratory with conducting research on industrial pollution prevention and waste minimization technologies; and
* developed a Pollution Prevention Advisory Committee to ensure pollution prevention is incorporated throughout EPA's programs.

All areas of EPA are developing initiatives to promote a **pollution prevention ethic** across the Agency. They are characterized by the use of a wide range of tools including market incentives, public education and information, technical assistance, research and technology applications and the more traditional regulatory and enforcement actions. Examples include:

* establishing cash awards for EPA facilities and individuals who devise policies or actions to promote pollution prevention;
* publicly commending and publicizing industrial facility pollution prevention success stories;
* coordinating development and implementation of regulatory programs to promote pollution prevention; and,
* "clustering" rules in order to evaluate cumulative impact of standards on industry, which could encourage early investment in prevention technologies and approaches.

Further, EPA is implementing the 33/50 Program which calls for the voluntary cooperation of industry in developing pollution prevention strategies to reduce environmental releases of 17 selected chemicals over the next five years. A discussion of the Toxics Release Inventory and the 33/50 Program is presented in Chapter 3 by S. Newburg-Rinn from EPAs Toxics Release Inventory staff.

EPA's pollution prevention program is multifaced and expansive. A comprehensive presentation goes beyond the bounds of this brief overview. The reader is encouraged to access the Pollution Prevention Information Clearinghouse (PPIC) for up to date news and information on recent developments in this rapidly changing arena. The PPIC Technical Support Hotline is (703) 821-4800.

Nongovernmental Initiatives (17).

Colleges and universities play a vital role in developing a pollution prevention ethic among scientists, business people, and consumers. The efforts of academia assure environmental awareness among students who will design and manage society's institutions, and develop ties between industry and the campus. Public interest groups also have a role to play by raising community awareness to support pollution prevention goals.

Academia. Pollution prevention interests and coursework are newcomers to the campuses of the United States. Historically, few faculty members had developed the relevant background to make it an important element in the environmental, chemical engineering or business curricula. The 1990s find the prevention ethic being received with enthusiasm by the faculty and students not only of environmental and engineering schools, but also of business, economic, finance, and marketing programs.

University faculty have identified a broad range of research topics in pollution prevention. Under cooperative programs with state agencies, EPA has sponsored research on product substitutes and innovative wastestream reduction processes. An increasing number of industries are also beginning to support university research. The evolution of the pollution prevention perspective is reflected in the academic environmental programs. The progression starts with industry's initial control efforts of good housekeeping, inventory control and minor operating changes. In the waste minimization stage, technologies are used to modify processes and reduce effluents. The 1990s brings the introduction of highly selective separation and reaction technologies predicated on the precepts of **design for the environment** and **toxics use reduction** (18).

Chemical engineering's professional society, the American Institute of Chemical Engineers (AIChE), aggressively encourages industry sponsorship of university research. Targeted research areas include: identification and prioritization of wastestreams; source reduction and materials substitution; process synthesis and control; and separations and recovery technology.

The American Chemical Society's (ACS) efforts have been more modest. The Division of Environmental Chemistry's symposium that served as the basis for this volume is one such effort. Others are expected to follow as chemists attempt to sort out their roles in the new arena. Clearly contributions are needed from the synthetic organic and inorganic chemists to build more environmentally friendly molecules - molecules designed for the environment, while still fulfilling their intended function and use. The focus of this symposium is on the contributions to be made by the community of analytical chemists as they take the analysis out of the laboratory to incorporate it as an integral part of the process stream. The Center for Process Analytical Chemistry provides an insightful discussion of the role of process analytical chemistry in pollution prevention in Chapter 4 of this volume.

Community Action. The public, as consumers and disposers of toxic chemical-containing products, is a major source of toxic pollution. As such it must and has become involved in toxic pollution prevention. Public involvement has resulted from state-wide initiatives, the actions of interest groups, and individual initiatives. Environmentalists concerned with pollution control have advocated source reduction over waste treatment as the preferred environmental option. But lack of public information about industrial releases to the environment effectively blocked any concerted efforts on the part of community action groups to address the toxics release issue. The Toxics Release Inventory and the "right-to-know" laws changed that irreversibly (18).

Citizens, for the first time, have access to industrial release and transfer data which allows the evaluation of trends and the setting of community standards. Armed with TRI data educational and lobbying groups have assumed a more credible role in community and legislative pollution issues. TRI data are used by community action groups to highlight the need for pollution prevention. Many cases of pollution prevention initiatives have been prompted by the negative publicity surrounding the public release of TRI data.

Pollution prevention is an evolutionary concept and an environmental protection strategy. As a strategy, it is especially attractive by being parsimonious and robust; a simple yet rugged concept that also makes common good sense. Ordinary and environmentally sophisticated people alike understand the pollution prevention ethic: eliminate or minimize all waste and pollutants. Businesses and communities understand more than ever the positive linkage between economic prosperity and environmental protection. Preventing pollution rather than devising ever-more costly control methods is the key to industrial competitiveness and environmental health. Pollution prevention is the key to prosperity without pollution.

Disclaimer

This document has been reviewed and approved for publication by the Office of Toxic Substances, Office of Pesticides & Toxic Substances, U.S. Environmental Protection Agency. The use of trade names or commercial products does not constitute Agency endorsement or recommendations for use.

Literature Cited

1. Royston, M. G. Pollution Prevention Pays, Pergamon Press: New York, NY, 1979.
2. Hirschhorn, J. S.; Oldenburg, K. U. Prosperity Without Pollution, Van Nostrand Reinhold: New York, NY, 1991.
3. Pollution Prevention 1991: Progress On Reducing Industrial Pollutants, U.S. Environmental Protection Agency, 1991, EPA 21P-3003.
4. Pollution Prevention 1991: Progress On Reducing Industrial Pollutants, Chapter 3. U.S. Environmental Protection Agency, 1991, EPA 21P-3003.
5. CMA. Improving Performance in the Chemical Industry. September (1990): 9-16.
6. Chevron Corporation. 1990 Report on the Environment: A Commitment to Excellence. (1990): 27.
7. NPCA. Paint Pollution Prevention Policy Statement. Pollution Prevention Bulletin (April 1990).
8. Toxics in the Community: National and Local Perspectives. U.S. Printing Office: Washington, DC, Chapters 10 and 11, 1990.
9. Pollution Prevention 1991: Progress On Reducing Industrial Pollutants, Chapter 4. U.S. Environmental Protection Agency, 1991, EPA 21P-3003.
10. Pollution Prevention 1991: Progress On Reducing Industrial Pollutants, Chapter 5. U.S. Environmental Protection Agency, 1991, EPA 21P-3003.
11. The Hazardous and Solid Waste Amendments of 1984, modifying section 3002 of RCRA, 42USC 6922.
12. Pollution Prevention Act of 1990. 42USC 13101.
13. Clean Air Act as amended in 1990. P.L. 101-549, 104 Stat.2399 et seq., 1990.
14. DOD. Directive Number 4210.15. "Hazardous Material Pollution Prevention." July 27, 1989.
15. U.S. DOE. "Waste Reduction Policy Statement." June 27, 1990.
16. Pollution Prevention Strategy. U.S. EPA. Fed. Reg. 56:7849-7864. February 26, 1991.
17. Pollution Prevention 1991: Progress On Reducing Industrial Pollutants, Chapter 6. U.S. Environmental Protection Agency, 1991, EPA 21P-3003.
18. Hendrickson, C. and McMichael, F. C. Environ. Sci. Technol. 1992, 26(5), 844.
19. Pollution Prevention Act of 1990. Section 6603 (5)(A)(i) and (5)(A)(ii).
20. Pollution Prevention Act of 1990. Section 6603 (5)(B).

RECEIVED May 22, 1992

Chapter 2

Industrial Approaches to Pollution Prevention

A. M. Ford, R. A. Kimerle, A. F. Werner, E. R. Beaver, and C. W. Keffer

Monsanto Company, 800 North Lindbergh Boulevard, St. Louis, MO 63167

After a brief discussion of pollution prevention, data from major chemical companies' programs to reduce environmental releases are presented. Monsanto Company is among the first of the large chemical companies to compile and make public, worldwide, such data on both a media and a chemical basis. These data cover the years from 1987 through 1990, and quantify self-imposed efforts to reduce SARA Title III air emissions worldwide 90 percent by the end of 1992 versus a 1987 base. Further programs to reduce environmental releases worldwide to water and various modes of land disposal are described.

Through nearly all of history, mankind has viewed the earth's natural resources as limitless. Even scarce minerals have been viewed as being available, only difficult to find in a vast planet. Episodes of excessive use of the environment, such as the London Fog, were considered either isolated incidences or were unknown to the general public.

It was not until after the Second World War that public interest rose to the level necessary to pass the Water Pollution Control Act of 1948. Although social concern lagged governmental action somewhat, during the 1960s public interest in environmental issues increased markedly (1).

Public interest in the 1960s led to the passage of new environmental legislation in the 1970s and the creation of the Environmental Protection Agency. Public interest in environmental issues continued to increase through the 1980s and may, in fact, still be increasing in the 1990s. In response to this public interest, industrial firms, particularly the larger chemical companies, began looking at waste release and disposal practices.

Many big companies have energy use reduction programs in response to the oil crisis of the mid-1970s. The early phases of these programs became the model for the pollution prevention programs of the 1980s. Both energy

0097–6156/92/0508–0013$06.00/0

reduction and pollution prevention programs depended heavily upon employee involvement; both programs had strong upper management support; and, both led to significant development in technology. Technology development reduced operating costs but was not directly related to product or process innovation, fostering an expansion of the role of technology in supporting corporate goals which continues today.

Much has been gained through employee awareness programs including suggestions and participation in programs requiring changes in work habits such as recycling. In fact, Monsanto Company has an Environmental Pledge describing the corporation's environmental vision. While this pledge has been polished for external use, it may be that its greatest use comes from internal consumption. To support employee involvement the company, in 1990, introduced the Monsanto Pledge Awards program which recognizes employees with worthy environmental achievements. Award winners in four categories may designate up to $100,000 to environmental projects of their choice outside the company.

Employee involvement can do much to cost-effectively reduce discharges into the environment; in fact, the company today that isn't striving to enhance employee involvement in its environmental programs is missing an important source of improvement. Yet technology must play the major role in pollution prevention. During the early phases of pollution prevention, employee involvement may affect significant reductions in environmental discharges at minimal cost but at some point pollution prevention programs must enter a technology-intensive phase to meet public expectation. There have been major increases in expenditures for pollution control technology beginning in the 1970s and continuing today. Initially, much pollution control technology was available for adaptation to specific projects; but as discharge levels have fallen, the cost of technology normalized to production rates has risen sharply, requiring increased investment in research into new pollution control methodology.

The Approach of a Large Chemical Company

Recognizing that public interest in environmental performance was growing rapidly, Monsanto Company began a series of voluntary emission and waste reduction programs in the middle to late 1980s that extended beyond regulatory requirements. The programs involved its domestic and worldwide facilities encompassing regulated and nonregulated chemicals and are reported below.

These voluntary reduction programs were based upon 1987 baseline data with goals that disregarded subsequent production levels. If production levels increase, operating units are still required to meet goals set at a percentage of 1987 levels. If a manufacturing process is shut down, the emissions are considered to have been eliminated. If a manufacturing process is sold, both baseline data and eliminations are removed from the data base.

SARA Title III Data

Since 1988, manufacturing facilities in the United States are required to report the domestic emissions of some 300 chemical compounds under SARA Title III. Data are reported by manufacturing facilities to the states where they operate and to the EPA each July 1 for the previous calendar year. Monsanto data are shown in Table I. It is important to note the change in reporting rules for injection wells which took place in 1990. Only the ammonium content of ammonium sulfate was reported after 1989.

Table II shows a further breakdown of these data with respect to the most significant chemicals by weight for 1990.

Worldwide Air Emissions Reduction Program

Monsanto Company's efforts to reduce worldwide air emissions 90 percent from 1987 emission levels by the end of 1992 are shown in Table III. This Table includes emission of approximately 100 SARA Title III compounds in the U.S. and in addition, chemicals of local concern outside the U.S. For example, two locally designated chemicals at Monsanto Company's Newport, Wales facility, carbon monoxide and butane, account for 60 percent of the total worldwide air emissions under this program in 1990. Significant reductions at the Newport

Table I. SARA Title III Data Summary (Millions Of Pounds) For Monsanto Company

Releases:	1987	1988	1989	**1990**
Air	18.4	15.3	11.2	**7.8**
Water	5.0	5.5	5.1	**1.7**
Injection Wells	203.0	231.3	233.3	87.0[1]
Land	0.4	0.4	0.3	**0.1**
Total Releases	226.8	252.5	249.9	**96.6**
Transfers:				
Public Sewage Treatment	40.4	35.1	37.9	**34.3**
For Disposal	8.9	6.6	6.1	**6.7**
Total Transfers	49.3	41.7	44.0	**41.0**
Total (adjusted)[2]	**276.1**	**294.2**	**293.9**	**137.6**
Total (as reported)[3]	342.2	300.7	293.9	**137.6**

[1]The EPA no longer requires that ammonium sulfate be reported, only its ammonia content. In 1989, Monsanto reported 202 million pounds of ammonium sulfate released to injection wells, which contained approximately 55 million pounds of ammonia.

[2]Adjusted to account for chemicals that were added or delisted by the U.S. Environmental Protection Agency and for Monsanto plant locations or businesses that were purchased or sold.

[3]As reported to the U.S. Environmental Protection Agency prior to the aforementioned events.

Table II. SARA Title III Chemicals, 1990 (Millions Of Pounds)

	Releases				Transfers		
Chemical:	Air	Water	Land	Injection Wells	Public Sewage	For Disposal	Total
Ammonia	.70	1.58	<.01	59.7	8.37	.17	70.5
Hydrochloric Acid	.09	<.01	-	.10	15.0	<.01	15.2
Ammonium Nitrate (solution)	-	-	-	14.0	-	-	14.0
Formaldehyde	.09	.04	<.01	3.7	1.24	.04	5.1
Sulfuric Acid	.03	<.01	<.01	-	4.88	.18	5.1
Methanol	.51	.01	-	.13	1.45	2.55	4.7
N-Butyl Alcohol	.04	.01	-	2.4	.90	.18	3.5
Acrylonitrile	.81	<.01	-	1.5	<.01	.05	2.4
Phosphoric Acid	.02	-	<.01	1.5	.08	.12	1.7
Acetonitrile	<.01	-	-	1.4	-	-	1.4
Cyclohexane	1.24	-	-	.10	-	-	1.3
Xylene (mixed isomers)	.86	.01	-	-	.27	.16	1.3
Hydrogen Cyanide	.02	-	-	.96	-	-	1.0
Styrene	.71	<.01	-	-	<.01	.27	1.0
Toluene	.40	.03	<.01	-	<.01	.41	0.9
Maleic Anhydride	.04	<.01	-	-	-	.60	0.6
Phenol	.01	<.01	-	.37	.03	.09	0.5
Benzene	.22	-	<.01	.07	.15	.05	0.5
4-Nitrophenol	<.01	-	-	<.01	.40	.06	0.5
Butyraldehyde	.10	<.01	<.01	-	.35	<.01	0.5
Other	1.9	<.02	<.01	1.1	1.1	1.8	5.9
Total(96 chemicals)	**7.8**	**1.7**	**0.1**	**87.0**	**34.3**	**6.7**	**137.6**

Table III. Air Emissions Reduction Program (Millions)

	1987		1988		1989		1990	
United States	**lb.**	**kg.**	**lb.**	**kg.**	**lb.**	**kg.**	**lb.**	**kg.**
SARA Title III	18.4	8.3	15.3	7.0	11.2	5.1	**7.8**	**3.6**
Percent Reduction from 1987	-		17%		39%		**58%**	
Worldwide (including U.S.):								
SARA Title III plus Locally Designated Chemicals	61.3	27.9	59.6	27.1	56.6	25.7	**53.5**	**24.3**
Percent Reduction from 1987	-		3%		8%		**13%**	

Note: Actions are planned at the Newport plant in Wales, which will significantly reduce air emissions of carbon monoxide and butane. These two locally designated chemicals at Newport accounted for more than 60 percent of the total worldwide air emissions under this program in 1990.

facility are planned that will affect a significant reduction in worldwide air emissions.

The effect of carbon monoxide and butane on worldwide air emissions can be seen again in Table IV. It is important to note that these data are not normalized to manufacturing rates, at least in part due to the use of data derived from regulatory reports that do not account for rates of production. In addition, the company's reduction targets are absolute and do not factor in increasing production rates.

Waste Release Reduction Program

Monsanto Company is currently evaluating a corporate-wide goal of achieving a 70 percent reduction in overall high priority manufacturing waste releases by the end of 1995 based upon 1987 baseline data.

Most of Monsanto Company's high priority waste is generated by its major operating units: Monsanto Chemical Company and Monsanto Agricultural Company. Progress in reducing high priority manufacturing waste in these operating units is shown in Table V.

It is important to note that during the years 1987 through 1990, manufacturing wastes generally increased. The data in Table V suggest that waste release rates rose in 1988 and 1989, and then decreased in 1990. Data normalized to production rates would show a steady decrease, in that production rates have increased to the extent that normalized data would have shown a 40 percent increase if no action had been taken. This is shown in Figure 1, along with the end of 1995 target that is being evaluated. This will be a difficult target to meet.

Table IV. Air Emissions[1], By Chemicals (Millions)

Pollutant:	1987 lb.	1987 kg.	1988 lb.	1988 kg.	1989 lb.	1989 kg.	1990 lb.	1990 kg.
Carbon Monoxide[2]	17.40	7.91	18.56	8.44	19.79	9.00	**22.32**	**10.15**
Butane[2]	13.30	6.05	14.08	6.40	13.12	5.96	**13.05**	**5.93**
Ammonia	6.36	2.89	4.70	2.14	1.17	.53	**.55**	**.25**
Xylene (mixed isomers)	3.02	1.37	2.84	1.29	3.22	1.46	**3.79**	**1.72**
Acrylonitrile	1.80	.82	1.44	.65	1.85	.84	**1.04**	**.47**
Styrene	1.63	.74	1.42	.65	1.43	.65	**1.20**	**.55**
Hydrochloric Acid	1.40	.64	1.06	.48	1.18	.54	**.37**	**.17**
Ethyl Alcohol[2]	1.39	.63	1.48	.67	1.48	.67	**.44**	**.20**
Methanol	1.28	.58	.93	.42	.83	.38	**.71**	**.32**
Toluene	1.09	.50	1.10	.50	1.20	.55	**1.11**	**.50**
Acetone	1.09	.50	1.01	.46	1.15	.52	**.93**	**.42**
Cyclohexane	.87	.40	1.18	.54	1.32	.60	**1.24**	**.56**
1,4-Dichlorobenzene	.75	.34	1.00	.45	.85	.39	**.02**	**.01**
Benzene	.72	.33	.62	.28	.44	.20	**.34**	**.15**
Chlorine	.54	.25	.56	.25	.31	.14	**.25**	**.11**
Sulfur Dioxide[2]	.52	.24	.58	.26	.87	.40	**.92**	**.42**
Tetrachloroethylene	.52	.24	.55	.25	.42	.19	**.38**	**.17**
Trichloroethylene	.52	.24	.30	.14	.40	.18	**.16**	**.07**
Butyraldehyde	.49	.22	.26	.12	.26	.12	**.19**	**.09**
1,3-Butadiene	.42	.19	.27	.12	.24	.11	**.22**	**.10**
All Others	6.2	2.8	5.7	2.6	5.1	2.3	**4.3**	**1.9**
Total	61.3	27.9	59.6	27.1	56.6	25.7	**53.5**	**24.3**

[1] Based on U.S. EPA SARA Title III chemicals at Monsanto facilities worldwide plus chemicals designated locally in plant communities outside the United States.

[2] Chemicals designated locally in plant communities outside the United States.

Table V. Waste Release Reduction Program (Millions)

United States	1987 lb.	1987 kg.	1988 lb.	1988 kg.	1989 lb.	1989 kg.	1990 lb.	1990 kg.
Air	37.7	17.1	36.9	16.8	32.8	14.9	**28.4**	**12.9**
Water	27.3	12.4	26.3	12.0	27.3	12.4	**32.5**	**14.8**
Injection Wells	121.8	55.4	146.9	66.8	165.8	75.4	**173.6**	**78.9**
Land	81.8	37.2	81.1	36.9	77.6	35.3	**63.5**	**28.9**
Total U.S. Releases	268.6	122.1	291.2	132.4	303.5	138.0	**298.0**	**135.5**
Worldwide (including U.S.)								
Air	82.5	37.5	83.0	37.7	81.3	37.0	**77.0**	**35.0**
Water	44.8	20.4	44.9	20.4	49.8	22.6	**52.8**	**24.0**
Injection Wells	121.8	55.4	146.9	66.8	165.8	75.4	**173.6**	**78.9**
Land	99.4	45.2	99.9	45.4	100.0	45.5	**79.6**	**36.2**
Total Worldwide Releases	**348.5**	**158.4**	**374.7**	**170.3**	**396.9**	**180.4**	**383.0**	**174.1**

Note: The waste release reduction program targets the reduction of organic and toxic inorganic chemical releases (dry weight) to the environment from current manufacturing processes, using 1987 as a base. Data may not add to totals due to independent rounding.

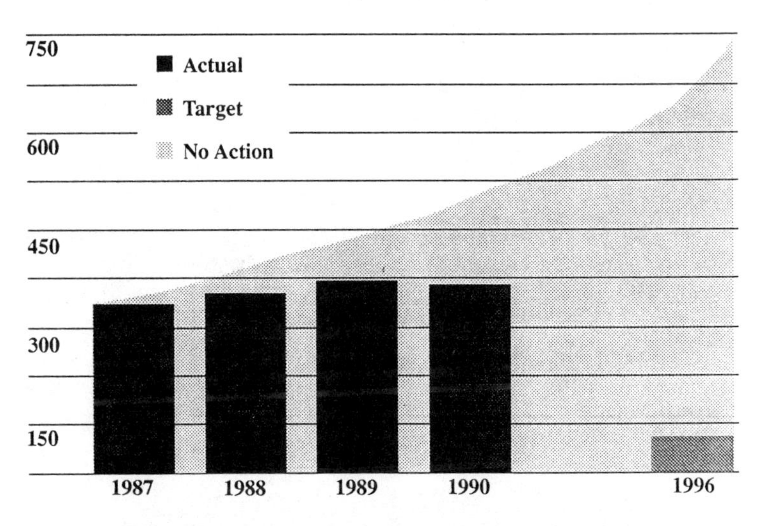

Figure 1. Waste release reduction program (millions of pounds).

Environmental Costs

Monsanto Company maintains an environmental, safety and health system that employs approximately 1200 people worldwide in its many programs. These programs require significant financial support in terms of capital. The data in Table VI reflect a continued increase in operating costs since 1987 with capital costs more than doubling.

Table VI. Environmental Expenditures (Dollars in Millions)

	1987	1988	1989	**1990**
Capital	36	36	60	**85**
Operation and Maintenance Expense	192	207	225	**265**
Remediation	18	19	27	**27**

Conclusions

While much of the data above speaks for itself, there are some important conclusions that can be drawn about the program as a whole.

- Environmental concerns are global in nature. Corporate pollution prevention programs need to have global focus to avoid shifting pollution across international boundaries. In addition, other countries have environmental concerns peculiar to their locale. These concerns must be taken into account in doing business in these countries.
- Energy programs of the 1970s have provided a useful model for the pollution prevention programs of today.
- Leadership from upper management in the form of demonstrated commitments is essential to gain employee support.

Acknowledgments

The data used in this paper were compiled by many people within Monsanto Company's environmental system. In addition, Mr. Steve Archer provided valuable editorial assistance. The authors wish to acknowledge the secretarial assistance of Ms. Debra Vollmer and Ms. Mary Schultz.

Note: Additional data may be obtained from Monsanto Company's Corporate Communications Department at 800 North Lindbergh Boulevard, St. Louis, Missouri 63167.

Literature Cited

1. Schaumburg, F. E. *Environmental Science Technology* 1990, 24 *(1), 17-22*

RECEIVED March 19, 1992

Chapter 3

Right-To-Know and Pollution-Prevention Legislation

Opportunities and Challenges for the Chemist

Steven D. Newburg-Rinn

Public Data Branch, Office of Pollution Prevention and Toxics, U.S. Environmental Protection Agency, Washington, DC 20460

New concepts in "citizens right-to-know" have significantly changed environmental management. Congress required EPA to give information concerning chemicals in the community directly to the public, hoping that the availability of the information cause change. This was particularly relevant to the releases and transfers of toxics. This program, together with additional legislation that builds upon it, will provide significant new opportunities and challenges to the chemist.

In 1986 Congress passed the Emergency Planning and Community Right-to-Know Act (EPCRA). It sought a significant change in environmental management. Frustrated by the lack of progress achieved through regulation, Congress embarked on what was really an experimental way to deal with toxic substances. When it passed EPCRA, Congress had EPA begin a program of giving information directly to the people, hoping that the public availability of the information would lead to change. This was especially true in the area of the releases and transfers of toxic substances, know as the Toxics Release Inventory ("TRI").

AND HAS CHANGE COME! Major chemical companies have promised drastic reductions in their releases. State right-to-know, toxics use reduction, and pollution prevention legislation has been passed. The Chemical Manufacturers Association has called for a "major shift in the way the chemical industry has approached its business."

Furthermore, additional national legislation has been adopted premised on what EPCRA started. Congress enacted the Pollution Prevention Act of 1990, and more additions to TRI are under consideration. In adding additional data gathering and other activities to the EPCRA TRI framework, the Pollution Prevention Act of 1990 states:

Published 1992 American Chemical Society

> . . .there are significant opportunities for industry to reduce or prevent pollution at the source through cost effective changes in production, operation, and raw materials use.

The Administrator of EPA has called for a voluntary 50% reduction in the releases of 17 major chemicals over the next several years, as measured in TRI. This program, called the 33/50 program, seeks to achieve a 33% reduction from 1988 levels in the releases and transfers of the 17 TRI chemicals by the end of 1992, and a 50% reduction by the end of 1995. To achieve these national goals and requirements the chemist will play an essential role. And don't expect this push to die down. Already there are calls for more information dissemination, more public access!

To understand the political and social environment that will be affecting your lives, this paper discusses briefly a) the Emergency Planning and Community Right-to-Know Act of 1986; b) the Pollution Prevention Act of 1990 ("PPA"); and c) the Clean Air Act Amendments of 1990. Furthermore, it is important for chemists to realize that TRI and PPA provide them with important new sources of information.

The Emergency Planning and Community Right-to-Know Act of 1986

In October, 1986, Congress passed the Emergency Planning and Community Right-to-Know Act of 1986. This is otherwise known as "Title III of SARA" (the Superfund Amendments and Reauthorization Act of 1986). EPCRA has four major sections: a) §§301-304 (Emergency Planning); b) §304 (Emergency Notification); c) §§311-312 (Community Right-to-Know Reporting Requirements); and §313 (The Toxics Chemical Release Inventory). It is this latter section, commonly referred to as the Toxics Release Inventory or TRI, on which this paper focusses.

The TOXICS RELEASE INVENTORY is embodied in a reporting rule which requires the annual reporting to EPA of direct release to all environmental media (air, water, and land, or off-site transfer to sewage treatment plants (POTW's) or other off-site facilities (such as commercial landfills). All facilities meeting the following tests must report:

- SIC codes 20-39 (from orange juice manufacturers to car companies to members of the chemical industry)
- with ten or more full-time employees
- which manufacture or process more than 25,000 pounds or use more than 10,000 pounds of any one of approximately 320 chemicals or chemical categories.

Industrial facilities meeting these tests submit, annually, information concerning facility information; off-site locations; chemical releases, transfers and treatment; and waste minimization.

What is TRI's purpose? Section 313(h) of EPCRA provides that "The release forms required under this section are intended to provide information to the Federal, State, and local governments and the public, including citizens of communities surrounding facilities. . . ." To accomplish this, Section 313(j) of EPCRA states that:

> EPA MANAGEMENT OF DATA. - The Administrator shall establish and maintain in a computer data base a national toxic chemical inventory based on the data submitted to the Administrator. . . . **The Administrator shall make these data accessible by computer telecommunications and other means to any person** on a cost reimbursable basis.

Once the data is collected, and has undergone rigorous QA/QC activities, we have made the data available on-line; on CD-ROM; on microprocessor diskettes; on microfiche; through an annual National Report; and through a reading room and a user support service.

All of this data is not collected for its own sake; it is collected to allow meaningful measuring and tracking of pollution. This can be done through nationwide analyses by geography (see Figure 1 and Table I); by mode of release or transfer (see Figure 2 and Figure 3); by industrial sector (see Table II and Figure 4); by chemical (see Table III and Figure 5); by toxicological profile (see Figure 6); and at an individual facility level, where the releases of particular facilities can be directly compared (see Table IV showing the top 50 TRI facilities.

The public availability and release of the TRI data is putting substantial pressures on companies to change their practices. Monsanto has pledged to cut its air emissions by 90%. For industry to meet their pledges will require substantial amounts of work by the chemist.

The Pollution Prevention Act of 1990

The PPA adds as much as 50% more data to TRI which must go through the same process, with little additional dollars and NO additional time. It will require EPA to change all aspects of its data management approach to insure the same level of data quality for all the additional data elements; and it will require submitters to learn the new reporting requirements, with the expectation that this new reporting will get better over time.

The PPA is designed to reduce the amount of industrial pollution in the United States by 1) establishing a source reduction program at EPA; 2) calling for increased technical assistance to industry by EPA and the states; and 3) requiring additional reporting on

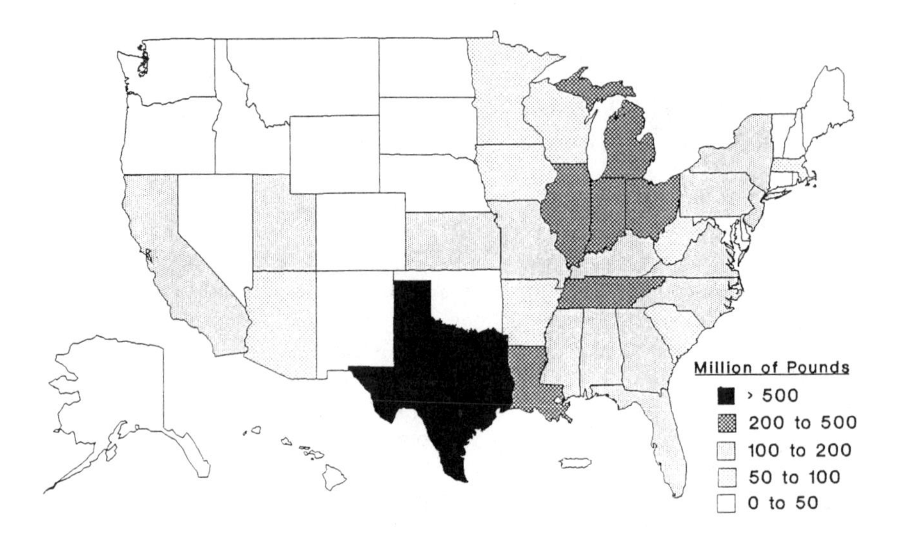

Figure 1. TRI total releases and transfers, 1989. (Reproduced from reference 3.)

- quantity of chemical (prior to recycling, treatment, or disposal) entering any wastestream or released to the environment;
- quantities of chemical recycled and treated at the facility and elsewhere;
- quantity of chemical released in one-time events not associated with production processes;
- information on source reduction activities and methods used to identify those activities;
- production ratio/activity index; and
- projections of future activities.

These reporting requirements, and the underlying challenge of pollution prevention and source reduction, will provide much meaningful work for the chemist.

The Clean Air Act Amendments

The Clean Air Act Amendments provide

> that facilities which demonstrate a 90% reduction "which shall be determined with respect to . . . emissions in a base year not earlier than calendar year 1987"

are able to avoid certain regulatory requirements. TRI is one means that will be used to track progress under these provisions.

Table I. The top 25 counties with the largest TRI total releases and transfers, 1989. (Reproduced from reference 3.)

TRI RANK	COUNTY	STATE	TOTAL RELEASES AND TRANSFERS Number	PERCENT OF TRI TOTAL Percent	PERCENT OF STATE TOTAL Percent	STATE TOTAL Pounds
1	Brazoria	TX	225,447,996	3.95	28.44	792,810,307
2	Jefferson	LA	196,108,289	3.44	41.41	473,546,487
3	Sedgwick	KS	154,109,629	2.70	83.24	185,131,051
4	Ascension	LA	141,752,206	2.48	29.93	473,546,487
5	Harris	TX	132,825,660	2.33	16.75	792,810,307
6	Tooele	UT	119,458,175	2.09	80.22	148,915,352
7	Jefferson	TX	111,046,679	1.95	14.01	792,810,307
8	Los Angeles	CA	79,856,790	1.40	47.30	168,825,335
9	Polk	FL	75,673,943	1.33	39.40	192,044,588
10	Lake	IN	72,486,534	1.27	28.42	255,023,626
11	Calhoun	TX	69,008,250	1.21	8.70	792,810,307
12	Cook	IL	66,559,508	1.17	26.86	247,813,608
13	St Louis	MO	65,188,780	1.14	39.97	163,105,846
14	Humphreys	TN	62,226,193	1.09	23.62	263,400,319
15	Mobile	AL	58,492,275	1.03	42.46	137,761,513
16	Allen	OH	58,001,380	1.02	16.17	358,677,545
17	Wayne	MI	57,535,048	1.01	26.14	220,137,364
18	Jefferson	KY	56,767,217	0.99	50.95	111,422,816
19	Galveston	TX	52,783,494	0.93	6.66	792,810,307
20	Sullivan	TN	49,181,587	0.86	18.67	263,400,319
21	Hopewell City	VA	46,748,824	0.82	34.73	134,592,526
22	St Clair	IL	46,470,224	0.81	18.75	247,813,608
23	Harrison	MS	42,896,317	0.75	35.56	120,617,983
24	Lewis And Clark	MT	36,632,896	0.64	93.68	39,103,261
25	Shelby	TN	33,442,929	0.59	12.70	263,400,319
	SUBTOTAL		2,110,700,823	36.99		
	TOTAL FOR ALL OTHERS		3,594,969,557	63.01		
	GRAND TOTAL		5,705,670,380	100.00		

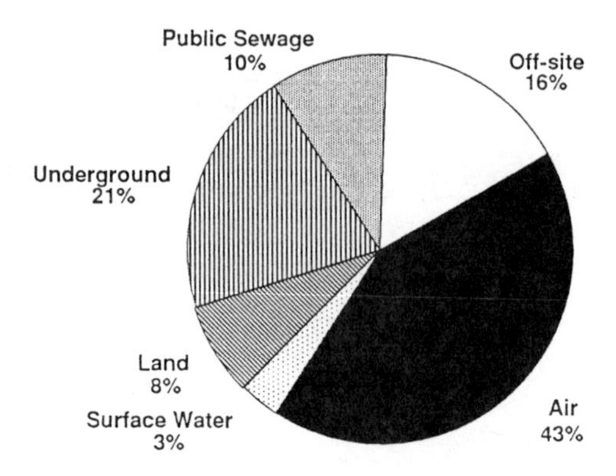

Figure 2. Environmental distribution of TRI releases and transfers, 1989. (Reproduced from reference 3.)

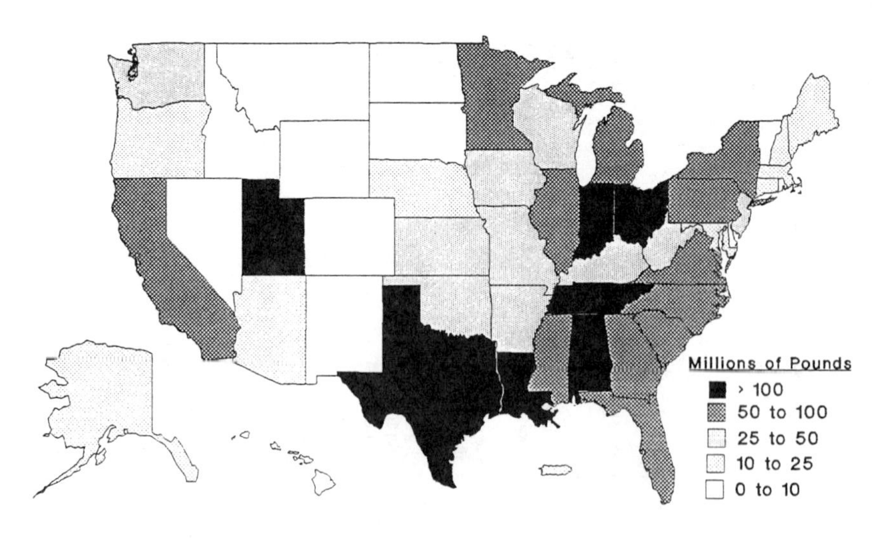

Figure 3. TRI total air emissions, 1989. (Reproduced from reference 3.)

Table II. TRI releases and transfers by industry, 1989. (Reproduced from reference 3.)

TRI RELEASE/ TRANSFER RANK	SIC CODE	INDUSTRY	TOTAL RELEASES AND TRANSFERS		FACILITIES (a)		AVERAGE RELEASES AND TRANSFERS PER FACILITY	NUMBER OF CHEMICALS PER FACILITY	
			Pounds	Percent	Number	Percent	Pounds	Average	Maximum
12	20	Food	67,803,337	1.19	1,779	7.88	38,113	2	32
22	21	Tobacco	1,485,626	0.03	17	0.08	87,390	2	4
16	22	Textiles	46,081,356	0.81	466	2.06	98,887	2	12
21	23	Apparel	2,059,373	0.04	47	0.21	43,816	2	6
18	24	Lumber	37,824,596	0.66	715	3.17	52,902	3	17
13	25	Furniture	65,369,507	1.15	484	2.14	135,061	4	21
4	26	Paper	313,254,241	5.49	624	2.76	502,010	4	17
14	27	Printing	60,923,661	1.07	374	1.66	162,897	2	16
1	28	Chemicals	2,745,768,071	48.12	4,259	18.87	644,698	5	85
9	29	Petroleum	103,136,599	1.81	412	1.83	250,332	8	47
7	30	Plastics	194,502,619	3.41	1,611	7.14	120,734	3	14
19	31	Leather	24,861,979	0.44	136	0.60	182,809	3	13
15	32	Stone/Clay	47,485,910	0.83	620	2.75	76,590	3	30
2	33	Primary Metals	756,808,577	13.26	1,649	7.31	458,950	4	33
6	34	Fabr. Metals	207,383,999	3.63	2,968	13.15	69,873	3	26
10	35	Machinery	74,922,470	1.31	1,026	4.55	73,024	3	25
8	36	Electrical	145,758,174	2.55	1,702	7.54	85,639	3	23
5	37	Transportation	245,316,145	4.30	1,194	5.29	205,457	4	28
11	38	Measure./Photo.	69,535,397	1.22	408	1.81	170,430	3	69
17	39	Miscellaneous	38,886,447	0.68	382	1.69	101,797	3	17
3		Multiple codes 20-39	437,278,275	7.66	1,448	6.42	301,988	4	44
20		No codes 20-39	19,224,021	0.34	248	1.10	77,516	3	55
		TOTAL	5,705,670,380	100.00	22,569	100.00	252,810	4	85

(a) Total number of facilities reporting to TRI, including those reporting no releases and transfers.

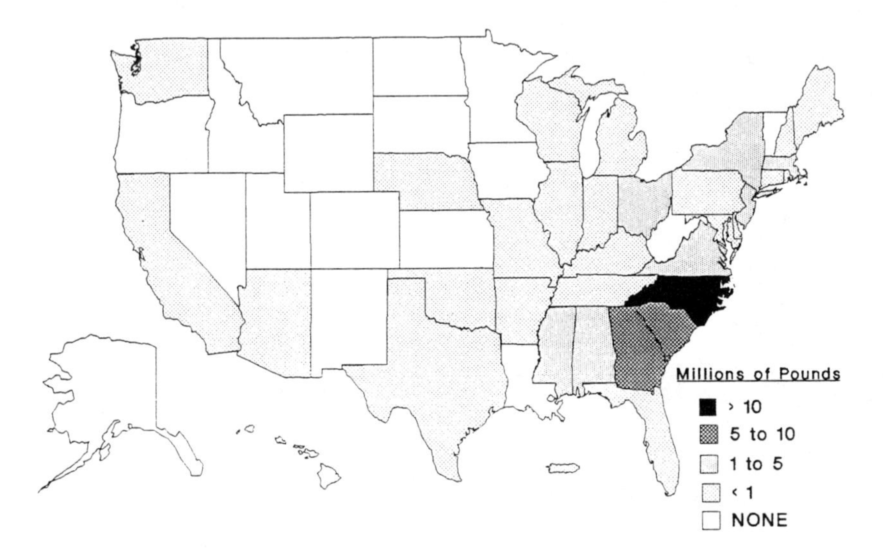

Figure 4. TRI total releases and transfers by the electrical industry, 1989. (Reproduced from reference 3.)

Table III. The 25 chemicals with the largest TRI releases and transfers, 1989 (Reproduced from reference 3.)

TRI RELEASE/ TRANSFER RANK	CHEMICAL	TOTAL RELEASES AND TRANSFERS		FORMS WITH RELEASES AND TRANSFERS(a)		TOTAL FORMS(b)		AVERAGE RELEASE/ TRANSFER PER FORM(c)
		Pounds	Percent	Forms	Percent	Number	Percent	Pounds
1	Ammonium sulfate (solution)	750,649,064	13.16	301	0.42	376	0.46	1,996,407
2	Hydrochloric acid	495,609,047	8.69	2,536	3.50	3,250	3.97	152,495
3	Methanol	408,119,093	7.15	2,418	3.34	2,521	3.08	161,888
4	Ammonia	377,248,848	6.61	2,846	3.93	3,120	3.81	120,913
5	Toluene	322,521,176	5.65	3,898	5.38	3,942	4.81	81,817
6	Sulfuric acid	318,395,014	5.58	3,391	4.68	5,547	6.77	57,399
7	Acetone	255,502,080	4.48	2,685	3.71	2,723	3.33	93,831
8	Xylene (mixed isomers)	185,442,035	3.25	3,472	4.79	3,525	4.30	52,608
9	1,1,1-Trichloroethane	185,026,191	3.24	3,826	5.28	3,893	4.75	47,528
10	Zinc compounds	164,799,357	2.89	1,746	2.41	1,935	2.36	85,168
11	Methyl ethyl ketone	156,992,642	2.75	2,441	3.37	2,464	3.01	63,715
12	Chlorine	141,428,470	2.48	1,445	2.00	1,779	2.17	79,499
13	Dichloromethane	130,355,581	2.28	1,523	2.10	1,552	1.90	83,992
14	Manganese compounds	119,825,790	2.10	550	0.76	638	0.78	187,815
15	Carbon disulfide	100,150,670	1.76	80	0.11	90	0.11	1,112,785
16	Phosphoric acid	98,660,456	1.73	1,507	2.08	2,638	3.22	37,400
17	Nitric acid	74,861,200	1.31	1,460	2.02	1,924	2.35	38,909
18	Ammonium nitrate (solution)	73,313,949	1.28	161	0.22	222	0.27	330,243
19	Freon 113	67,837,298	1.19	1,472	2.03	1,483	1.81	45,743
20	Glycol ethers	65,736,857	1.15	1,776	2.45	1,822	2.22	36,080
21	Ethylene glycol	57,792,359	1.01	1,280	1.77	1,469	1.79	39,341
22	Zinc (fume or dust)	57,487,663	1.01	571	0.79	625	0.76	91,980
23	Copper compounds	54,465,732	0.95	1,092	1.51	1,254	1.53	43,434
24	Chromium compounds	50,881,050	0.89	1,260	1.74	1,363	1.66	37,330
25	n-Butyl alcohol	50,095,319	0.88	1,162	1.60	1,177	1.44	42,562
	SUBTOTAL	4,763,196,941	83.48	44,899	61.99	51,332	62.66	92,792
	TOTAL FOR ALL OTHERS	942,473,439	16.52	27,529	38.01	30,559	37.32	30,841
	GRAND TOTAL	5,705,670,380	100.00	72,428	100.00	81,891	100.00	69,674

(a) Forms reporting releases and transfers greater than zero.
(b) Total number of forms reporting the chemical, including those reporting no releases and transfers.
(c) Calculated based upon total number of forms.

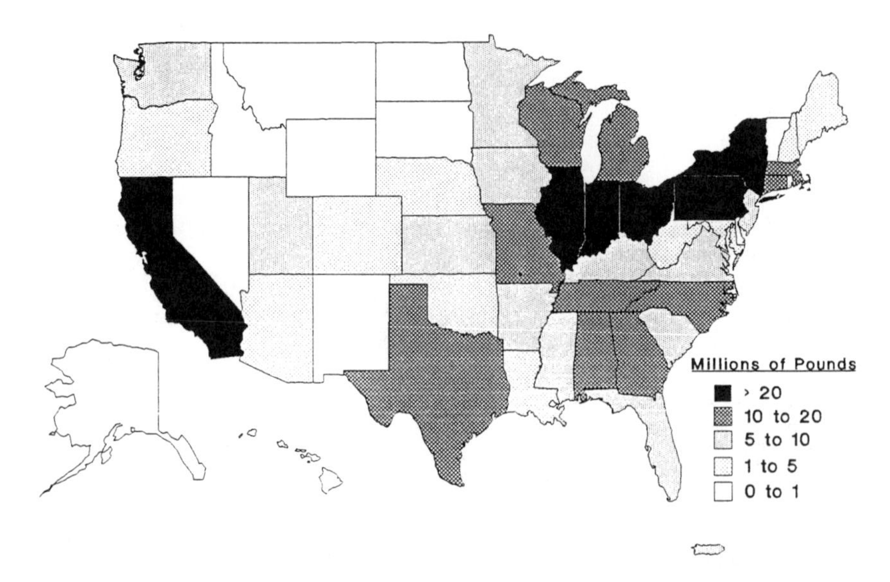

Figure 5. Total TRI releases and transfers by state for the 33/50 halo-organics. (Reproduced from reference 3.)

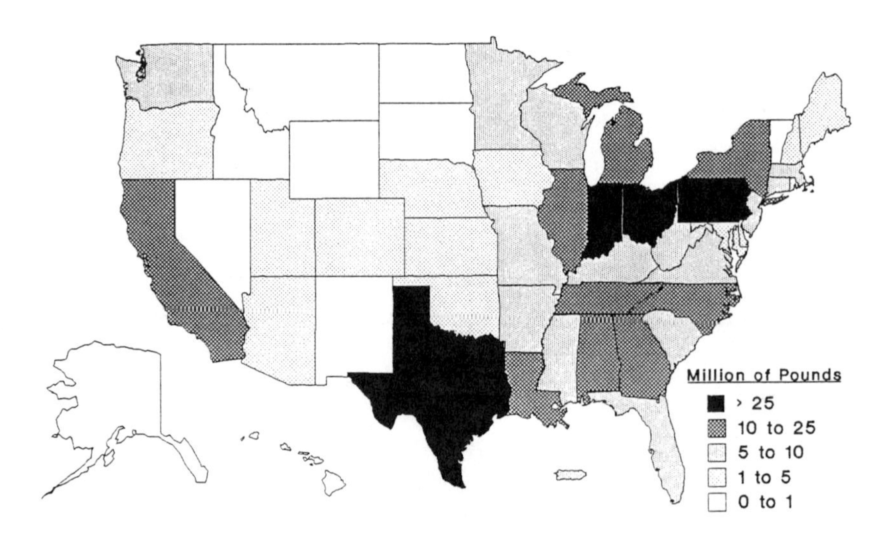

Figure 6. Total releases and transfers of carcinogens. (Reproduced from reference 3.)

Table IV. The 50 TRI facilities with the largest total releases and transfers, 1989 (Reproduced from reference 3.)

TRI RANK	SIC CODE	FACILITY NAME	CITY	STATE	TOTAL RELEASES AND TRANSFERS Pounds	Percent
1	28	Monsanto Co.	Alvin	TX	206,533,205	3.62
2	28	American Cyanamid Co.	Westwego	LA	192,350,800	3.37
3	33	Magnesium Corp. of America	Tooele	UT	119,060,425	2.09
4	28	Vulcan Chemicals	Wichita	KS	92,349,716	1.62
5	28	Du Pont Beaumont Works	Beaumont	TX	88,094,984	1.54
6	28	BP Chemicals	Port Lavaca	TX	65,543,672	1.15
7	Mult	Inland Steel Co.	East Chicago	IN	57,273,300	1.00
8	28	Du Pont	New Johnsonville	TN	57,022,345	1.00
9	28	BP Chemicals Inc.	Lima	OH	56,731,350	0.99
10	28	Atochem N.A.— Racon Facility	Wichita	KS	54,505,751	0.96
11	28	Columbian Chemicals Co.	Saint Louis	MO	52,461,020	0.92
12	28	Tennessee Eastman Co.	Kingsport	TN	45,253,659	0.79
13	28	Courtaulds Fibers Inc.	Lemoyne	AL	44,397,570	0.78
14	28	Du Pont Delisle Plant	Pass Christian	MS	42,517,217	0.75
15	28	BASF Corp.	Geismar	LA	40,802,088	0.72
16	28	Du Pont	Louisville	KY	39,829,058	0.70
17	33	Asarco Inc.	East Helena	MT	36,615,059	0.64
18	28	Allied Signal	Hopewell	VA	34,748,970	0.61
19	28	Du Pont Victoria Site	Victoria	TX	32,314,457	0.57
20	28	Kaiser Aluminum & Chemical Corp. (a)	Mulberry	FL	32,013,400	0.56
21	28	Filtrol Corp.	Vernon	CA	29,595,850	0.52
22	Mult	3M Consumer AV& Consumer Prod.	Hutchinson	MN	28,998,967	0.51
23	28	Triad Chemical	Donaldsonville	LA	26,152,046	0.46
24	28	BASF Corp.	Lowland	TN	25,624,950	0.45
25	28	Monsanto Co.	Cantonment	FL	25,622,958	0.45
26	28	Monsanto Co.	Cahokia	IL	25,261,353	0.44
27	28	Arcadian Corp.	Geismar	LA	24,967,500	0.44
28	33	National Steel Corp.	Portage	IN	24,586,030	0.43
29	28	The Upjohn Co.	Kalamazoo	MI	22,997,339	0.40
30	33	Phelps Dodge Mining Co.	Playas	NM	20,847,699	0.37
31	28	Sterling Chemicals Inc.	Texas City	TX	20,574,970	0.36
32	28	Union Oil Co.	Kenai	AK	18,558,521	0.33
33	38	Eastman Kodak Co.	Rochester	NY	18,123,187	0.32
34	28	Du Pont Sabine River Works	Orange	TX	18,010,133	0.32
35	28	Texasgulf Inc.	Aurora	NC	17,852,400	0.31
36	28	3M	Cordova	IL	17,631,532	0.31
37	33	National Steel	Ecorse	MI	17,562,820	0.31
38	28	Freeport McMoran,Agrico Chem. Div.	Donaldsonville	LA	17,254,750	0.30
39	33	Herculaneum Smelter	Herculaneum	MO	17,110,851	0.30
40	33	Asarco Inc.	Hayden	AZ	16,574,800	0.29
41	33	Copper Range Co.	White Pine	MI	16,330,870	0.29
42	28	Pfizer Pigments Inc.	East Saint Louis	IL	16,071,496	0.28
43	Mult	Elkem Metals Co.	Marietta	OH	15,771,470	0.28
44	33	Kennecott Utah Copper	Bingham Canyon	UT	15,473,300	0.27
45	28	Freeport McMoran Agrico Chemical Div.	Uncle Sam	LA	15,410,826	0.27
46	28	Air Products Mfg. Corp.	Pasadena	TX	15,190,634	0.27
47	28	Coastal Chem Inc.	Cheyenne	WY	15,133,055	0.27
48	29	Amoco Oil Co.	Texas City	TX	14,217,928	0.25
49	28	Mississippi Chemical Corp.	Yazoo City	MS	14,176,423	0.25
50	33	Cyprus Miami Mining Corp.	Claypool	AZ	13,660,904	0.24
		SUBTOTAL			1,975,763,608	34.63
		TOTAL FOR ALL OTHERS			3,729,906,772	65.37
		GRAND TOTAL			5,705,670,380	100.00

(a) Kaiser Aluminum & Chemical Corp. has submitted a revision to TRI regarding their 1989 releases and transfers. The revised amount for their total releases and transfers is 4,487,000 pounds.

Other Activities to Come

This year two other major EPA statutes are up for reauthorization - The Resource Conservation and Recovery Act (RCRA); and the Clean Water Act. Already, right-to-know aspects are being considered for both legislative activities. Also, explicitly, several versions of a "Right-to-Know More" Act have been introduced in the House and the Senate.

I do not believe that industry will be able to comply with all of these new legislative activities without the active participation of their chemists. Furthermore, new information will become available to the chemists through these various reporting requirements that will aid in these activities.

Fully expect additional demands to be placed on you as chemists to achieve the new goals set in these statutes!

LITERATURE CITED

1) The Emergency Planning and Community Right-to-Know Act of 1986 (otherwise known as Title III of the Superfund Amendments and Reauthorization Act of 1986), 42 USC §§ 11001-11050 (1986), or Public Law 99-499, 100 Stat 1729.

2) The Pollution Prevention Act of 1990 (otherwise known as Title VI of the Omnibus Budget Reconciliation Act of 1990), 42 USC §§13101 et seq., or Public Law 101-508, 104 Stat 1388-321.

3) United States Environmental Protection Agency, Office of Pesticides and Toxic Substances, Toxics in the Community: The 1989 Toxics Release Inventory National Report, EPA 560/4-91-014, September, 1991 (available through the United States Government Printing Office).

For additional information concerning this program, please contact the Emergency Planning and Community Right-to-Know Hotline at 1-800-535-0202.

RECEIVED May 18, 1992

Chapter 4

The Role of Process Analytical Chemistry in Pollution Prevention

Elizabeth A. McGrath, Deborah L. Illman, and Bruce R. Kowalski

Center for Process Analytical Chemistry, University of Washington, BG-10, Seattle, WA 98195

The Center for Process Analytical Chemistry (CPAC) was launched in 1984 as a partnership with the chemical and materials industry to address fundamental research questions at the frontier of process analysis. In collaboration with 46 corporate sponsors and 4 national laboratories, CPAC is developing new sensors and analyzers, integrated with chemometrics methods, to enhance process control. In addition to improving product quality and process efficiency, advances in this field will lead to improved process safety and accountability. By obtaining better chemical information about the process, it will be possible to achieve tighter process control, preventing process upsets and release of pollutants, and ultimately, to improve the design of new chemical plants in terms of safety and reliability. Programs at the Center focus on four aspects of process analytical chemistry for pollution prevention: (1) new sensors, such as optical waveguides for *in situ*, remote, automated and/or continuous environmental monitoring; (2) instrumentation coupled with chemometrics for tighter process control; (3) sensing methods for recycling (such as a remote technique for plastics identification); and (4) new graduate curriculum, team-taught by CPAC sponsors, with case studies from actual industrial processes.

The Center for Process Analytical Chemistry (CPAC) was first proposed by Bruce Kowalski and colleagues at the University of Washington in 1982. The original vision was a Center that could provide a common ground for scientists and engineers from academia and industry to pursue areas of interest and need in continuous chemical monitoring, remote sensing, and crisis alerting systems. "The initial idea behind the concept of process analytical chemistry came when Professor Kowalski matched these needs and interests with the possibility of using chemometrics to enable analytical instruments to monitor a chemical process, on-line, as an integral part of the process" (*1*). At this time, Dr. James Callis, now Co-Director of

0097–6156/92/0508–0033$06.00/0

CPAC, was asked to participate in the creation of the Center by contributing his expertise in the area of instrument development.

The development of in-line and noninvasive analytical instrumentation for process control and optimization, intelligent analytical instrumentation, remote sensing and continuous chemical monitoring was the specific focus of the Center. Since the industrial environment presents a number of challenging problems (e.g., noise, vibration, temperature extremes, challenging sample conditions, etc.), new sensors needed to be rugged, reliable, inexpensive, with fast-response, and incorporating automated calibration, error detection and correction features (*2*). They must also be able to detect the analyte of interest in the presence of interference and matrix effects and provide information that is relevant and useful to process control. Therefore, a concomitant purpose of the Center was the development of "the potential of multivariate data analysis and pattern recognition techniques to enhance instrument calibration, resolution, sensitivity and for data interpretation and feedback... for on-line and continuous monitoring systems" (*1*). Because manufacturing operations are not easily modeled, statistical methods that empirically determine correlations between process parameters and product performance are essential for feedback control.

In 1984, CPAC became a reality as a partnership between the University of Washington, the National Science Foundation, and 21 founding industrial sponsors. Researchers at CPAC are developing new sensing technologies for use as an integral part of chemical systems, including in-line sensors, rapid separation techniques, chemometrics, and methods for nondestructive spectral determinations (*1*). This approach is designed to move chemical sensing away from the off-line/at-line measurements and toward the ultimate goal of in-line and noninvasive methods that can be utilized in real-time feedback control of chemical processes.

Riebe and Eustace (*3*) have characterized process analytical chemistry (PAC) as a subdiscipline of analytical chemistry that, unlike traditional analytical chemistry, is performed on the "front lines" of the chemical process industry. The philosophy of PAC is quite different from traditional analytical chemistry, which is performed in sophisticated laboratories by highly trained specialists. The first and most obvious difference is *location* of the analysis (*3*). While a traditional analytical lab is centralized, PAC analyzers are located right with the process. Delay time associated with transporting, bookkeeping and analysis of samples is reduced or eliminated. Second, this reduction in the *time frame* for obtaining results is another benefit of PAC; here results are used to adjust the process not identify products that must be scrapped or reworked (*3*). Production schedules are shortened since PAC data can be used immediately for process control and optimization. Lastly, by its nature the process analytical world demands generalists and problem solvers. Unlike traditional analytical chemistry, PAC requires a team approach since it is *problem-driven*, not *technique-driven* (*3*). Only through the close interaction of process chemical engineers, process and analytical chemists and instrument and electronics engineers and technicians can this goal be achieved (*4*).

For discussion, PAC may be divided into 5 eras: off-line, at-line, on-line, in-line and noninvasive (*5*). These eras describe a progression from traditional analysis toward the ultimate goals of process analytical chemistry, in-line or noninvasive monitoring. Both the off-line and at-line eras require the manual removal and transport of the sample to the measuring instrument. Off-line samples are analyzed in a centralized facility where the benefits of time-shared equipment may be outweighed by delays in obtaining results, additional administrative costs, and competition among users for the resources. At-line analysis employs a dedicated instrument, which may be simpler since it has limited applications, located in close proximity to the process. These two processes are closer in nature to traditional analytical chemical analysis.

The distinction between traditional and process analytical chemistry begins to widen with the on-line era. Here an automated sampling system extracts, conditions, and presents the sample to an analytical instrument for analysis. On-line analysis may be divided into two categories: intermittent, which requires injection of a portion of the sample stream into the instrument, and continuous, which allows the sample to flow continuously through the instrument.

The in-line methods eliminate the need to construct a separate analytical line to properly sample the main stream and present it to the instrument at a suitable temperature and pressure. Using the appropriate chemical probe, analysis is performed *in situ* within the process line. These probes may be chemically specific or composed of a sensor array, a set of semi-selective sensors, each of which has a different response profile to the analytes of interest.

The fifth era, noninvasive, may represent the ultimate goal of PAC. By eliminating a physical probe it is also possible to eliminate sampling problems associated with sensor fouling or sample contamination. This technology has great potential for use in the pharmaceutical and biotechnology industry.

Process analytical chemistry (PAC) is a new paradigm for chemical analysis. PAC involves both fundamental research and applications development, in order to produce information with the degree of chemical specificity, limits of detection, and rejection of interferences, necessary for meaningful analytical results.

Advances in this field are important to the continued competitiveness of American industry, which depends on such factors as process optimization and control, energy efficiency, and automation (*3*). In the chemical industry, it is generally recognized that the greatest returns on investments in manufacturing or processing plants are obtained from investments in process control. It has been claimed that profits could be greatly increased by reducing 'quality costs'—the dollars lost when a product fails to meet specifications because the process could not be adequately controlled (*3*). In a recent survey conducted by the United States National Institute of Standards and Technology (NIST), the respondents knew of no existing sensors, instruments or methods to address about 25% of the described measurement problems (*6*).

The primary benefit of PAC, a reduction in quality costs, results from streamlining manufacturing process efficiency and subsequently, improving product quality. Another important benefit is an increase in worker safety, also resulting from the better understanding, hence, control of these processes (*7, 8*). But there is an additional benefit, that is becoming more important as time goes on. Process analytical chemistry will play an increasing role in environmental protection. By improving manufacturing efficiency, there is a reduction of resources lost to the waste stream. Employing the technological and methodological advances in PAC, it is possible for industries to simultaneously minimize plant effluent release and improve product quality (*5*). Therefore, the environmental perspective is no longer fear of out-of-compliance operation, but knowledge that waste and byproducts are minimized or possibly eliminated by a process that is running within acceptable boundaries (*3*).

The United States Environmental Protection Agency (EPA) now realizes that "end-of-the-pipe" regulations are ineffective in dealing with today's environmental discharges (*9, 10*). The EPA, established as the primary Federal agency responsible for implementing the nation's environmental laws, has seen its mission largely as managing the reduction of pollution, particularly that pollution defined in the laws that it administers. As a result, past U.S. environmental policy has been fragmentary and reactive rather than proactive. A report titled, "Reducing Risks," prepared by the EPA Science Advisory Board, recommends that U.S. environmental policy must "evolve to become more integrated and focused on opportunities for environmental improvement" (*10*). Because this country and the rest of the world are facing burgeoning environmental problems resulting from unprecedented population growth and worldwide industrial expansion, past practices of controlling the "end-of-the-pipe" where pollutants enter the environment are not sufficient. In addition, it has become apparent that the task of remediating the damage to the environment "after-the-fact" is too difficult and expensive for the government and industries involved. Once generated, the cost of cleaning up and disposing of pollutants can be enormous, as exemplified by the Superfund program, Department of Energy's cleanup of nuclear sites (*11, 12*), and cancellation and disposal of chemicals already in use. End-of-pipe controls and waste disposal should be the last line of environmental defense. Of primary concern should be preventing pollution at the front end, by ensuring that the processes run under tight control, by substituting less hazardous chemicals, and by recycling by-products or redesigning production processes.

One obvious solution is a "front-end" reduction in pollutants in the waste stream. The diligent application of PAC can help reduce or prevent the release of pollutants. By obtaining better chemical information about the process, it will be possible to achieve tighter control and to prevent process upsets and release of pollutants. Unlike end-of-pipe controls, front end control eliminates the transfer of pollutants from one media to another because pollutants are not generated in the first place (*9*). Businesses can gain economic benefits by saving resources and avoiding the costs and liabilities associated with waste disposal and clean-up. Therefore, the ultimate in environmental protection is pollution prevention—eliminating problems before they occur. In this regard, PAC represents a major adjunct to the other facets of environmental analytical chemistry.

In addition, pollution prevention might also encompass the management of existing contaminated sites, e.g., "superfund" and government nuclear facilities. There is a need for new methods of monitoring the environment that can provide information on the location and movement of existing contaminants. Ecosystems are also chemical processes for which in-line, real-time analytical capabilities are desired (*3*). A recent Congressional Office of Technology Assessment (OTA) document finds that clean-up efforts at DOE nuclear production sites have been "hampered by lack of ready technical solutions, reliable data and qualified personnel... (*4*)." New sensors and instruments for *in situ*, remote, automated, and/or continuous analysis with built-in calibration, error detection, and correction must be developed to operate in complex environmental matrices. To be effective, these sensors will need many of the qualities (e.g., rugged, reliable, self-calibrating, etc.) necessary for sensors being developed for use in industrial chemical processes.

These new technologies will enable the surveillance of underground contaminant plumes and provide post-closure verification of the success of remediation efforts. They will also provide an "early warning" of potential contamination of nonpolluted soil and groundwater. These monitoring systems will need to be able to work in some very challenging settings (e.g., remote areas, down wells, radioactive matrices) and much of this technology has yet to be developed.

Underway at CPAC is a major environmental initiative to conduct fundamental research on new sensing technologies with respect to the two main thrusts described above. As we look to the future, the needs are compelling. The time has come for government and industry to join forces and create "kinder" industrial practices that protect the environment and make good economic sense (*13*). To quote EPA's Administrator William Reilly, "...let anyone who doubts the wisdom of pollution control—or who believes there is a conflict between economic growth and environmental protection—let them go to Eastern Europe. Policies in Communist Europe designed to stimulate economic development by foregoing pollution controls ended by wrecking the economy and also ravaging the environment" (*9*).

Pollution prevention has become the slogan for all EPA programs (*9, 13, 14*). Among the ten recommendations for establishing the direction of federal environmental policy in the future, the Scientific Advisory Board (SAB) stresses "...EPA should emphasize pollution prevention as the preferred option for reducing risk" (*10*). Armed with the results of the SAB report, Administrator Reilly has placed pollution prevention high on his strategic planning agenda.

The EPA recognizes the role of PAC in the prevention of pollution and has invited CPAC to showcase the work of our sponsors at this symposium. Many CPAC sponsors have led the way in instituting pollution prevention policies in the past and have expressed their desire to continue to take the lead in this area. Plant safety and the prevention of pollution are of vital concern to industry today. The EPA is also concerned with the reduction of total releases and transfers of 17 high-priority chemicals targeted from the Toxic Release Inventory (*15*). This inventory, part of the Emergency Planning & Community Right-To-Know Act, requires all manufacturing companies that use certain levels of chemicals to detail how much they emit into the environment (*16*). The U.S. Chemical industry, petroleum refining, and instrument manufacturers are among the 600 companies being asked by the EPA to voluntarily cut emissions of routinely used toxic chemicals. Tallies of 1988 reports show that the chemical companies, which release the greatest amounts of toxic chemicals, are also the facilities making the greatest reductions in emissions and waste reduction (*17*). The chemical industry had the greatest amount of pollution reduction of any single industry group (*17*). Most of the reported reductions were achieved by equipment changes, on-site recycling and housekeeping improvements.

An equally important part of this mission is the education of the next generation of scientists and engineers who will carry out the fundamental work needed to advance sensing technology for pollution prevention. The SAB also recommends that [the] "EPA...train a professional workforce to help reduce [environmental risks]..." (*10*). The DOE is also aware of the growing shortage of appropriately trained scientists and engineers to address the assessment and restoration of its contaminated facilities (*11*). Solutions to environmental problems will require a team approach similar to that found in process analytical chemistry. Scientists working on these problems will have to be familiar with a variety of disciplines, and will need to be able to communicate with other scientists and engineers on the team. This new paradigm is taking hold by a variety of means. In addition to the contact with industrial scientists and engineers that our graduate students have at semiannual meetings and throughout the year, students participate in internships at the sponsor companies and government laboratories. In the past, a common complaint of CPAC sponsors about new analytical graduates, in general, is that they are theory- and technique-oriented, as opposed to problem-oriented, and they are unprepared for real-world processes. To better prepare our students, we have created a new graduate course in process analytical chemistry that incorporates actual case studies presented by sponsor representatives. Students are given a set of measurement objectives and operating constraints for an actual process, and then work in small groups of chemists and chemical engineers to propose an analytical solution. This course offers a unique opportunity for students to be exposed to "real" problems and compare their ideas with the solutions prescribed by working professionals. As part of the Environmental Initiative, CPAC is expanding its educational activities to include courses in environmental sensing and expanding its student intern program to environmental agencies.

Today, CPAC has successfully grown to 50 industrial and National Laboratory sponsors, including the Office of Modeling, Monitoring Systems and Quality Assurance (EPA) (*18*). The National Science foundation has identified CPAC as one of its top two Centers based on the number of sponsors and scientific merit of the research (*1*). Meanwhile, interest in the potential applications of process analytical chemistry continues to expand. As advances in PAC are revolutionizing industrial chemical processes, we feel that they will also make a significant contribution toward pro-active environmental protection. The time has arrived to move beyond the "postmortem" mode of analysis and institute "front-end" control for pollution prevention.

Literature Cited

1. Scott, C.S. "A Comprehensive Historical Profile of the Center for Process Analytical Chemistry (CPAC): The NSF University/Industry Cooperative Research Center Located at the University of Washington," *Div. Ind. Sci. & Tech. Innov.*, NSF, 1991.
2. Illman, D.L. *Trends in Anal. Chem.* **1986**, Vol. 5, no. 7.

3. Riebe, M.T.; Eustace, D.J. *Anal. Chem.* **1990**, *62*, 65A.
4. Jacobs, S.M.; Mehta, S.M. *Am. Lab.* **1987**, *20(11)*, 15.
5. Callis, J.B.; Illman, D.L.; Kowalski, B.R. *Anal. Chem.* **1987**, *59*, 624A.
6. "Survey of Measurement Needs in the Chemical and Related Industries," (TN 1087), *NIST*, Superintendent of Documents, U.S. Government Printing Office, Washington, DC.
7. Lepkowski, W. "Revolutionary New Regulations, Law On Tap for Plant Safety," *C&EN*, July 30, 1990, p. 15.
8. Ainsworth, S.J. "Plant Disasters Fuel Industry, Government Concern Over Safety," *C&EN*, October 29, 1990, p. 7.
9. Reilly, W.K. "Aiming Before We Shoot: The Quiet Revolution in Environmental Policy," *The National Press Club*, Washington, DC, September 26, 1990.
10. US EPA, "Reducing Risk: Setting Priorities and Strategies For Environmental Protection, The Report of The Science Advisory Board: Relative Risk Reduction Strategies Committee," (SAB-EC-90-021), September 1990.
11. US DOE, "Basic Research for Environmental Restoration," (DOE/ER-0482T), December 1990.
12. US OTA, "Complex Cleanup: The Environmental Legacy of Nuclear Weapons Production," (OTA-O-484), February 1991.
13. Ember, L.R. "Rising Pollution Control Costs May Alter EPA'S Regulatory Direction," *C&EN*, February 18, 1991, p. 25.
14. Ember, L.R. "Economic Incentives for Environmental Protection," *C&EN*, March 25, 1991, p. 15.
15. US EPA, Office of Toxic Substances, "Toxics in the Community: 1988 National and Local Perspectives," (EPA 560/4-90-017), September 1990.
16. US EPA, "Chemicals in Your Community: A Guide to the Emergency Planning and Community Right-to-Know Act," (OS-120), September 1988.
17. Hanson, D.J. "EPA'S Emissions Inventory Shows Some Pollution Reduction," *C&EN*, October 22, 1990, p. 15.
18. Newman, A.R. *Anal. Chem.* **1990**, *62*, 965A.

RECEIVED April 21, 1992

Chapter 5

Chemical and Biochemical Sensors for Pollution Prevention

Jess S. Eldridge, Keith M. Hoffmann, Jeff W. Stock, and Y. T. Shih

3M Environmental Engineering and Pollution Control and 3M Occupational Health and Environmental Safety, St. Paul, MN 55133

A successful pollution prevention program depends upon monitoring chemical and biochemical information in a timely manner before or at the point of emission or discharge. Industry calls for product and technology developments for chemical/biochemical sensors needed for air, liquid, and soil. Sensor technology should be viewed as complementary to laboratory instruments and human senses. Fast response, reliability and compact size are essential to perform in-situ analysis or real-time monitoring. A broad sensing principle or mechanism is often more desirable than a compound-specific device, however, compound speciation is frequently needed. Also, selectivity should be balanced by minimizing interferences associated with complex chemical matrices. Sensors should be rugged and able to operate under varying conditions. Calibration must be performed on a routine basis to insure that results obtained are indicative of the conditions present at the time the measurements are taken; therefore, the calibration technique must be simple to implement and all encompassing.

Current Practices

How do chemical and biochemical sensors relate to pollution prevention? Basically speaking most sensors and analytical instrumentation are assessment tools. The data generated from this equipment provides the information necessary to make decisions regarding the matrix being monitored. Ultimately the decisions made are pertinent to defining concentrations of what pollutants are being emitted relative to the project objectives. The key to pollution prevention is to reduce or eliminate pollution at it's source which tends to affect compliance issues as a side benefit. Many industries are going through a transition from a compliance only mentality toward a more progressive approach to pollution. The

0097–6156/92/0508–0040$06.00/0

outdated attitude of simply complying is giving way to a more integrated proactive philosophy about pollution. The new approach is to prevent or reduce pollution at the source for one major reason; it's good business! Raw materials that are utilized on one end of the plant become waste products on the other end. In order to dispose of these wastes costly treatment processes may be necessary which cut into profit margins. Also by factoring in pollution preventive design and management early in the planning of various processes many retrofit modifications can be minimized. The outcome of preventive measures can be dramatic and very effective. The current perspective for the application of chemical and biochemical sensors is mostly from the less inclusive view point of compliance monitoring. Fortunately, by broadening the view to include preventive techniques their uses can be even more varying. With these thoughts in mind the use of sensors, probes, and instrumentation have many potential uses. Hopefully, this will outline the intent of this paper.

Sample collection and flow measurements are the major thrust of the field services group. The standard practice for a majority of collected samples dictates that analyses be conducted in a laboratory setting. The laboratory utilizes the required instrumentation for sample analysis, and appropriate QA/QC routines are common practice. The normal turnaround time for laboratory results ranges from several days to several weeks depending on the complexity of the requested analysis. However, test results are often needed instantaneously. The driving force behind acquiring more timely data is necessitated by two main challenges: the dramatic increase in environmental regulations, and the environmentally conscious attitude adopted by 3M. Faster analytical turnaround is a current requirement, and real time data is essential in many environmental monitoring schemes. Unfortunately, technology has not advanced as fast as government regulations and corporate sentiment have, so an urgent need for real time data is apparent.

The goal of the field services group is to collect information that accurately represents the overall composition of the matrix under test. The overall composition of the test matrix is not always apparent, so monitoring is often performed to determine what elements and compounds are present. The data is used to assess the impact of the effluent stream on the environment and to determine compliance with applicable regulations, optimization criteria, or preventive techniques. The presence of undefined components makes the design of a sampling and analytical scheme difficult. Collecting representative information becomes a challenge if unforeseen constituents are present or expected constituents are not present. Things get even more complicated if transitional compounds exist due to conditions in the matrix during a monitoring event. In many cases, this predicament is unavoidable, but the affects of such an occurrence can be lessened by previous experiences, utilizing broadly inclusive methods, and some luck. Often, repeat sampling trips may be needed to collect the desired information.

Portable monitoring equipment is utilized to aid in the collection of representative data. Many monitoring programs employ on line sensors/instrumentation to obtain representative data. The use of portable gas chromatographs, transportable mass spectrometers, portable Fourier transform infrared analyzers, and in situ water quality probes has increased the overall integrity of sampling and analytical schemes. In terms of on site response they offer additional capabilities not possible with lab equipment and occasionally they facilitate technical advantages as well. Many times the major consideration in whether to utilize portable equipment is the need for reduced turn around time as compared to a laboratory setting. The use of field deployable analyzers supplements the existing resources that can be applied to any given environmental investigation or assessment. Portable monitors are primarily meant to aid us in our abilities to make value decisions about environmental matters based on additional information otherwise not attainable. Using portable equipment in the proper context, and in conjunction with laboratory resources may produce benefits that neither could offer exclusively where one does not necessarily replace the other.

Many monitoring programs require flow measurement as one of the defining parameters. This is a basic yet critical requirement in determining mass loadings of effluent streams. Mass loadings are determined by incorporating the analytical results with the flow measurement. Emission permits are frequently established based on mass rather than concentration, so flow monitoring is essential. Equipment manufacturers have made great strides in the development of flow measurement by exploiting various sensing techniques such as ultrasonic, pressure, capacitance, magnetic, and thermal sensors. Many of the sensors developed are intended for portable applications, so the sensors fit well into monitoring schemes. Current monitoring programs use battery operated flow meters for lengthy time periods in remote locations. The flow meters are capable of long term data storage, instantaneous and totalized flow measurement, and are capable of initializing samplers when the desired conditions exist. Flow monitor control and interrogation can be conducted directly or by remote access via cellular phone modem, conventional modem, infrared, radio, or satellite telemetry. The type of telemetry utilized depends on the monitoring application. There is a need for improving conduit flow measurement by integrating multiple velocity readings simultaneously with a sensor array and computing power to match. Currently, in the absence of such a device, manual traverses or a single discrete reading is made to determine average velocity.

Challenge of Real Time Monitoring

The challenges that face effective real time data acquisition are sensitivity, operational range, reliability, and portability. Also, it should be stated that the interferences associated with analysis of any kind must be minimized. The calibration of the instrumentation must be conducted such that adequate accuracy is attainable. Possibly the largest hurdle to overcome in implementing more wide spread use of real time data acquisition, is the process of validation and acceptance. The process of correlating new instrumentation to existing methodologies can be laborious and time consuming.

The importance of low level detection is fortified by industry and by regulators, so sensitivity of in situ instrumentation is significant to the evaluation of real time data. Instrument sensitivity is a widely varying concern. Sensitivity requirements range from part per million to per cent levels, yet part per trillion levels are frequently desired and often essential. Field instrumentation must have the ability to isolate the analyte of interest from other analytes that may be present. Cross-sensitivity potential must be eliminated. Field instrumentation is often deployed in situations where the total composition of the effluent streams are largely unknown. The instrumentation must have the ability to selectively monitor for the specific analyte(s) of interest, but the instrumentation must not be affected by unexpected biases that could be present in the effluent stream.

It is common to experience varying conditions during field operations, so the instrument must be capable of handling wide physical and analytical operating ranges. Elemental and compound specific detectors must be able to compensate for changes of the composition in the matrix stream and for changes in ambient conditions. For example, an exhaust duct from a combustion source can change in composition each time the input or the operation variables are affected. If hydrocarbons are present, the instrumentation must be able to respond to variations in ratio deflections in the hydrocarbon output. The moisture content and source temperature is also subject to instantaneous changes, so the monitoring instrumentation must be able to respond immediately to changes in the composition of the gas stream. The location and ambient conditions near the field monitor must also be considered. Frequently, testing is needed in remote areas where AC power is not available, so battery or solar power may be needed. The ambient temperature can easily vary from -30° F to 100° F, and temperature changes of 50° F in an hour are not uncommon. The ambient conditions can also be extremely corrosive or moist, so field instrumentation must be able to exist in and respond to varying conditions.

Real time field monitors will be used to document emission violations and to monitor process variations, so field monitors must exhibit a high level of reliability and durability. During the design stages, manufacturers should take steps to insure that all components of the field monitors are easily accessible, and a supply of spare parts and consumables must be readily available. In order to document the instruments reliability, the instrument must be easily calibrated and maintained. Before field monitors can be utilized in a monitoring scheme, the reliability of the instrumentation must be proven. Extended instrument downtimes cannot be tolerated. The reliability issue is further compounded by the application of the monitoring programs. Field monitors will frequently be deployed in remote areas, so the time that a competent technician can devote to the instrumentation will be limited. Operational reliability is a limiting factor in the application of leading edge technologies to field monitoring, but field monitors that are both reliable and durable must evolve from technological advancements.

The use of in situ instrumentation and real time analyzers requires that portability be included as a major factor in the design of field instrumentation. Monitoring programs vary in duration from several minutes to several months. Frequently, monitoring programs are process dependent, and field monitors must be moved rapidly from site to site. Sample introduction and sample conditioning systems will also be needed for most field monitors. The sample interface system must also be compact and appropriate for the anticipated effluent streams. The need for compact, transportable monitoring equipment is governed by both time and logistical limitations. A field monitor that is too large, or too hard to transport will suffer limited use.

Categories of Real Time Data Acquisition

In terms of real time data acquisition usage, the basic distinctions to be made are between sensors and on line instrumentation. The notion of sensors can be described as devices such as ion specific probes that essentially measure a single parameter or phenomena directly without additional chemical processing. These type of sensors lend themselves to in situ deployment because of their basic simplicity. There is a dramatic increase in the development of these devices. For example, water monitoring applications utilize an in situ water quality probe that can integrate, store, and transmit six (6) directly readable parameters: dissolved oxygen, pH, oxidation reduction potential, specific conductance, temperature, and depth. Various specific sensors, a pressure transducer, and a thermocouple are incorporated into a single probe. The probe integrates the effects of one parameter on others when relevant. For example, pH is temperature sensitive and is automatically adjusted to compensate for changes in temperature as measured by the probe. The probe is specifically programmed by the user to perform virtually any combination of measurements for data storage and transmission via remote telemetry or direct connection. The sensors are calibrated prior to deployment through direct computer interface with the instrument using the proper procedures. The probe is cylindrically configured in such a way to permit access through a four inch diameter pipe; thus, broadening it's application range to include down-hole groundwater monitor wells. The probe meets the criteria for real time monitoring discussed above.

On line instrumentation is generally more sophisticated. On line instrumentation requires a dynamic interactive process of sequential events leading to a final output, such as temperature adjustments, addition of reagents, and extractions. The process leading to the analysis of a sample under these circumstances would typically require automatic integration and control of the components of such an instrument. Historically, the human element was in control of this process. However, with the advent of computer processing, and automated control equipment, close involvement of the human element can be reduced and the instrument can be placed on site to facilitate practical real time data acquisition and dissemination. There seems to be a steady increase in the development of these type instruments. Various manufacturers of equipment such as total organic carbon analyzers, volatile organic analyzers, portable FTIR, X-Ray fluorometer, gas chromatograph with flame ionization, photoionization, electron capture, mass spectrophotometer detectors, etc. are on the market. On line analyzers and in situ sensors will probably never fully replace the need for the classical lab analyses but will augment the general study of our environment by providing more tools with unique capabilities. On site monitors will help to characterize different matrices by providing supplemental information preparatory to the laboratory setting. In addition, interactive control of process/waste treatment equipment will be possible. This would greatly reduce the time lapse between knowledge of a problem, quantifying it, locating it's source, and correcting it. In the extreme case, it may be possible to entirely avert problems that would otherwise go undetected.

General and Specific Sensing

In the environmental field, both regulators and industry require specific or generalized analyses of given emissions or discharges. An emission source or discharge is characterized by identifying and quantifying a set of specific and general parameters defined beforehand by parameter or afterwards by the methods used. It is common to perform a set of tests which define the project. Generally, when planning a project, one works backwards from the purpose for which monitoring is desired to design the appropriate steps to accomplish the objectives. For example, one may be required to analyze the matrix of interest for five specific compounds each in a specific detection range and each requiring different procedures. The methods used to perform this testing will look only at the five parameters in the specified ranges and ignore all else. This is an example where the discharge is defined by parameter. Alternately, the objective may be to perform a certain method in a given detection range, and to report quantifiable results. This method may only be sensitive to chlorinated compounds in the parts per million range and preclude all others. The results are reported as having conformed to the prescribed method. This is an example where the discharge was defined by method. Sometimes a certain parameter cannot be directly measured; it may be possible to establish a correlation between a different parameter and thereby provide a way to meet the monitoring requirements. In order for real time data acquisition to be a viable resource in environmental evaluations, there must be a broad base of specific and general testing to successfully utilize real-time data acquisition on a diverse practical basis.

Future

Many technologies are being developed including infrared, lasers, fiber optics, radioactive, X-ray, magnetic, ultrasonic, thermal, radar, microwave, ultraviolet, potentiometric, the list goes on and on. The trend in the environmental testing area is changing; mainly due to new regulatory guidelines and industry demands. The laboratory must adjust to these changes even though the differences may conflict with today's accepted technology. Sampling specific locations require an intensive plan to ensure that samples are representative and that integrity is maintained. Often, decisions regarding waste stream profiles are a result of one time sampling and analyses which may or may not be consistent over a specific time period.

Ideally, access to real-time information allows for appropriate decision making and problem solving. The current practice of taking samples and delivering them to the analytical lab takes a great amount of coordination in the sampling, preservation, and shipping within specified holding times. Once at the lab, data generation may take days or even weeks, depending upon the analytical methodologies required. This delay in information could result in decisions being made after the fact.

On site data gathering is increasingly important as waste streams become more complex. A waste treatment facility will benefit from the ability to rapidly identify a change in the waste profile. Multiple sensor and instrumentation systems would serve the need in generating real time data. On demand interrogation coupled with limit alarms will announce changing conditions, and facilitate a response action. As more systems become available the emphasis on sample collection may diminish, and a greater significance will be placed on calibration and maintenance. The skills and training for personnel utilizing these technologies will also change accordingly.

In conclusion, the need for real time data generation has become a priority in the decision making process. The objective of the future will be to answer the question before it is asked. Hopefully, changing technologies and awareness will meet the challenges of real-time monitoring for pollution prevention.

RECEIVED December 12, 1991

REAL-WORLD PROCESS ANALYTICAL CHEMISTRY

Chapter 6

On-Line Analyzer for Chlorocarbons in Wastewater

Sydney W. Fleming[1], Bertsil B. Baker, Jr.[1], and Bruce C. McIntosh[2]

[1]E. I. du Pont de Nemours and Company, Wilmington, DE 19880–0357
[2]KVB/Analect, Irvine, CA 92718

The control of volatile organic contaminants in plant wastewater is a major concern for the chemical industry. Continuous measurements of the actual levels present allows effective control measures to be implemented. This paper describes a monitoring system for CCl_4 and $CHCl_3$ based on continuous closed loop sparging and FTIR vapor measurement. The system theory of operation, design and operating performance are described.

A technique has been devised and instrumentation developed for continuously determining the concentration of carbon tetrachloride and chloroform in an industrial plant wastewater stream. This technique, based on continuous closed-loop sparging and FTIR spectroscopic vapor analysis, has a sensitivity of 1 and 10 ppb respectively for CCl_4 and $CHCl_3$ and a 90% response time of approximately 3 minutes. This paper presents the theoretical basis of the technique and the performance achieved on-line. This technique has applicability to a wide range of volatile contaminants in industrial wastewater streams and can contribute to tightened control of these pollutants.

Volatile organic contaminants in industrial waste streams are currently measured by periodic sampling using a purge-and-trap technique to remove the organics from the water sample. Gas chromatography (GC) or other analytical methods are then used to quantitatively measure the various components. Although this well-accepted technique has high sensitivity, it is unsuited to provide tightened control of wastewater pollution. The time interval between samples preclude the detection of sudden or short-term excursions and thus does not allow timely action to prevent the inadvertent release of contaminants. Another drawback is related to sample handling. A water sample containing volatile contaminants such as CCl_4 must be kept totally sealed between collection and measurements. Finally, sampling a large-volume stream such as a plant waste outflow is subject to severe statistical errors because of spatial or temporal inhomogeneity in the stream.

0097–6156/92/0508–0048$06.00/0

The work reported here was undertaken to provide a much tighter level of control on a plant wastewater stream than had been achieved by existing purge-and-trap measurement. The apparatus developed in this work contains a new type of sparging system that continuously, rapidly, and repeatably extracts an equilibrium sample of the volatiles. The volatiles are then directed into a process FTIR spectrometer equipped with a long path gas cell for quantitative multicomponent measurement. The wastewater sparging analyzer allows the diversion of out-of-spec flows for subsequent reprocessing. This strategy had been impossible with previous purge-and-trap methods.

Sparging Techniques

Sparging, or passing a carrier gas through a liquid to extract solvated materials from the liquid, is performed in many ways and on many scales in industry. Here we discuss four such techniques used in analytical measurements.

Grab Point Sample Purge-and-Trap. This technique is based on passing a volume of gas, sufficient to remove essentially all the volatile components, through a relatively small volume of liquid. The preconditioned gas stream is then directed through a carbon-filled tube or a cryogenic trap for concentration of the materials of interest. Upon completion of the purge-and-trap step, the sample is expelled from the trap and measured, usually in a gas chromatograph. This technique achieves high sensitivity, and the equipment is relatively simple. It is, however, intermittent and the time required is relatively long, typically 30 to 45 minutes.

Continuous Sparging. To make this measurement continuous we needed to replace the fixed sample with a continuous liquid flow; to run a continuous flow of gas through a controlled region in the liquid flow; and to measure the gas with a rapid-response multicomponent gauge. In our case, the gauge is an FTIR spectrometer equipped with a long-path gas cell.

Single-Pass, Open Loop. The simplest implementation of this concept uses a frit through which the gas stream is passed, creating numerous small bubbles in the liquid. The gas, enriched with volatile compounds, is collected, passed through a condenser to reduce the water vapor content, delivered to the gas cell, and finally sent to waste. Although this approach responds rapidly to concentration changes in the water, it has two drawbacks. Since the carrier gas passes through the liquid only once, the vapor concentration does not achieve the equilibrium concentration predicted by Henry's Law. This discrepancy ranges from negligible to threefold, depending on solubility and vapor pressure. The important point is that the vapor concentration depends both on gas and liquid flow rates, as well as on the concentration being measured.

Multiple-Pass, Closed Loop. Our solution to these problems is to recirculate the gas stream through the water-interaction region. If a constant concentration exists in the liquid, the corresponding concentration in the vapor state achieves its equilibrium value in a few passes. As the liquid concentration increases or decreases, the vapor concentration follows with a modest delay. Although the delay varies with gas flow rate, the liquid to gas concentration scale factor for a given chemical varies only with liquid temperature, which can easily be measured accurately.

Theoretical Basis of Closed-Loop Sparging Calibration

When a small amount of volatile liquid or gas (the solute) is dissolved in a much larger volume of a second liquid (solvent), the solute will exist, in gas phase, over the solvent to a degree dependent on the solubility and vapor pressure of the solute. This condition is described by Henry's Law:

$$P = kC \tag{1}$$

where C is the concentration of solute in the solvent, P is the partial pressure of solute vapor over the solute, and k is a proportionality constant. At saturation, where $C = C_o$ (the solubility of the solute in the solvent), the vapor pressure of solute over solution will equal the vapor pressure of the pure solute P_o. (Strictly speaking, the vapor pressure is in equilibrium with the solute, e.g. CCl_4 saturated with solvent, that is water). For weakly soluble materials, this distinction is probably negligible. Under these conditions, $k = P_o/C_o$, and equation 1 may be written:

$$P = P_o(C/C_o) \tag{2}$$

This equation states that the vapor pressure of the volatile component over its solution in water is simply the product of its vapor pressure and its fractional saturation in water. It may be rearranged to:

$$C = C_o(P/P_o) \tag{3}$$

This equation yields concentration as a function of a measured vapor concentration (P) and two calculable quantities, assuming linearity over the entire range of solubility. This assumption is reasonably valid for the weakly soluble non-polar materials of interest here. Significant departures from linearity would be expected for polar materials of higher solubility in water.

Determining Solubility and Vapor Pressure. The terms P_o and C_o are both functions of temperature and must be measured and included in the quantitative analysis.

Vapor Pressure. Vapor pressure vs. temperature is well characterized for most solutes of interest. Tables and equations are available in the literature [1,2]. The equations from references 1 and 2 are of the form:

$$\ln P_o = A + B/(T_c + C) \tag{4}$$

Where T_c is temperature in degrees Celsius.

A rearranged form of the vapor pressure equation was used in the actual computations for this work.

$$\ln P_o = K + L/T_k + MT_k + NT_k^2 \tag{5}$$

The coefficients K - N were obtained from internal du Pont tables. These were checked to be in good agreement with the results using reference 1 and 2 data. Figure 1 shows the vapor pressure curves used for CCl_4 and $CHCl_3$.

Solubility. Solubility is less well characterized but literature values for chlorinated solvents are published [3]. Again we used a polynomial to predict the temperature behavior.

$$C_o = Q + RT_c + S(T_c)^2 + U(T_c)^3 \tag{6}$$

Figure 2 shows the solubility curves obtained using internal Horvath tables.

Liquid Vapor Concentration "Enhancement". The results of equations 5 and 6 can be combined to calculate the liquid concentration from the measured vapor concentration.

$$\text{ppmLiquid} = \text{ppmVapor}*7.60*\text{Co}/\text{Po} \tag{7}$$

Table I shows the result of computing these factors for CCl_4 and $CHCl_3$.

Table I Liquid-Vapor "Enhancement" Factors

	Carbon Tetrachloride			*Chloroform*		
Temp [C]	*P_o [mmHg]*	*C_o [%]*	*Enhancement ppmVapor/ppmLiquid*	*P_o [mmHg]*	*C_o [%]*	*Enhancement ppmVapor/ppmLiquid*
20	89.6	0.081	146	157.5	0.822	25.2
25	112.4	0.079	186	195.6	0.792	32.5
30	139.7	0.079	232	241.8	0.769	41.2
35	172.4	0.080	285	294.7	0.753	51.5
40	210.9	0.081	343	357.8	0.744	63.3

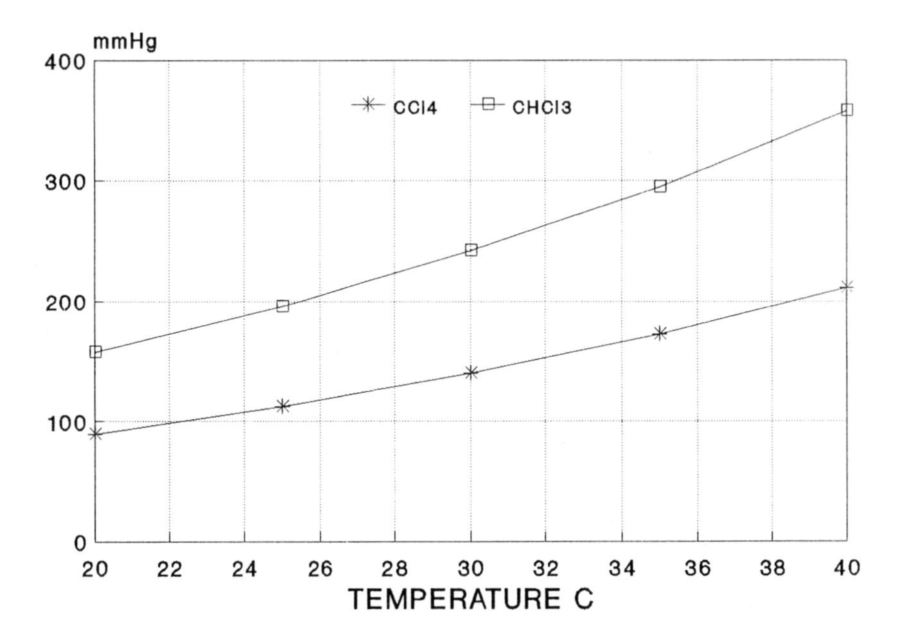

Figure 1 - CCl_4 and $CHCl_3$ vapor pressure

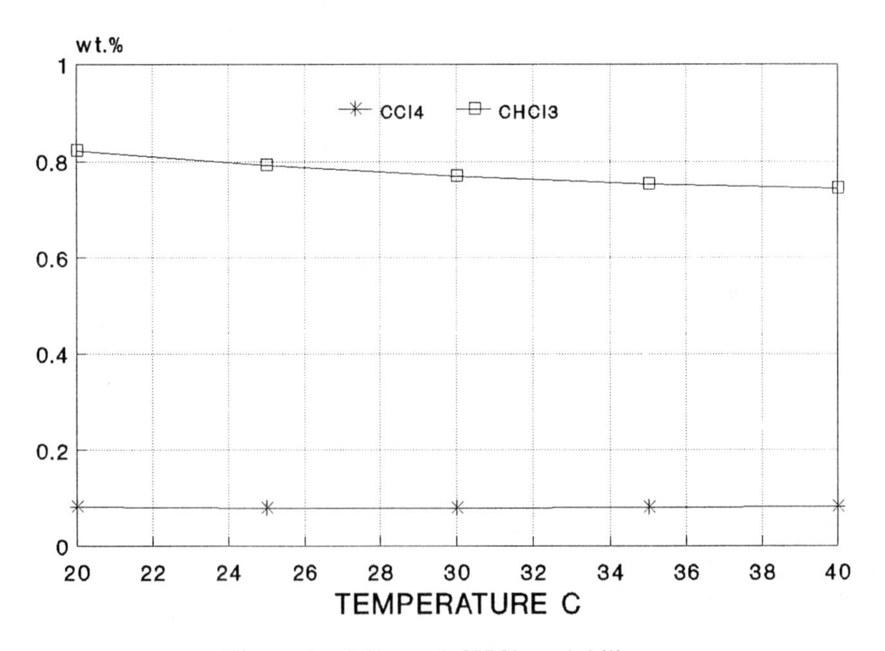

Figure 2 - CCl_4 and $CHCl_3$ solubility

The Enhancement factor may seem a bit misleading as it compares substantially different measures of concentration. It is however useful as it shows that, for these materials, the concentration is high enough for very sensitive detection. In fact for materials with factors of unity or less, it is still far easier to detect the contaminants by FTIR in water vapor than in liquid water.

Calibration Without *a priori* Solubility Data. In future work we will generally use references 1 and 2 to obtain data for equation 5; do a direct vapor calibration for spectroscopic sensitivity P; and then measure the complete instrument response to known water concentration samples at various temperatures over the range of interest. This procedure will allow us to develop a set of coefficients for equation 6 which expresses the effective solubility.

Spectroscopic Calibration

The final step to making a complete concentration measurement is to determine P in the sparged vapor stream. This involves the calibration of the FTIR spectrometer for the gases of interest. The initial calibrations for this system were done by injecting 0.25ml of saturated vapor into the evacuated gas cell which was then back-filled to 1 atmosphere. Knowing the vapor pressure of the material at ambient temperature and the volume of the gas cell, we calculated the concentration. The concentrations were about 15 ppm, yielding peak absorbance near 1.0A in the ten meter path, which was ideal for calibration. For the final instrument configuration, the path length was reduced to 1 meter to increase the full scale ranges.

The spectra used for the calibrations are shown in Figure 3. The solid line is for CCl_4, the short dashed trace is for $CHCl_3$ and the long dashed trace is for CO_2 which occasionally appears in high concentration in the water stream. There are no totally isolated frequencies for any of the three components, but the separation of the band centers allows a robust P-Matrix calibration to be created for this set of components.

On-Line Calibration Verification

Once the system was operating on-line, a check of its calibration was made by capturing a sample just as it exited the analyzer and subjecting it to laboratory purge-and-trap analysis. The results showed the analyzer to be 15% high for CCl_4 and 18% high for $CHCl_3$. With the number of separate calibration steps and the uncertainty of the original solubility data, we found this agreement excellent. The system's overall calibration was adjusted to remove these residual disagreements, since the purge-and-trap method is today's standard.

Sparger System Description

Figure 4 is a schematic diagram of the sparging equipment used in this work. It consists of two loops--a water loop and an air loop.

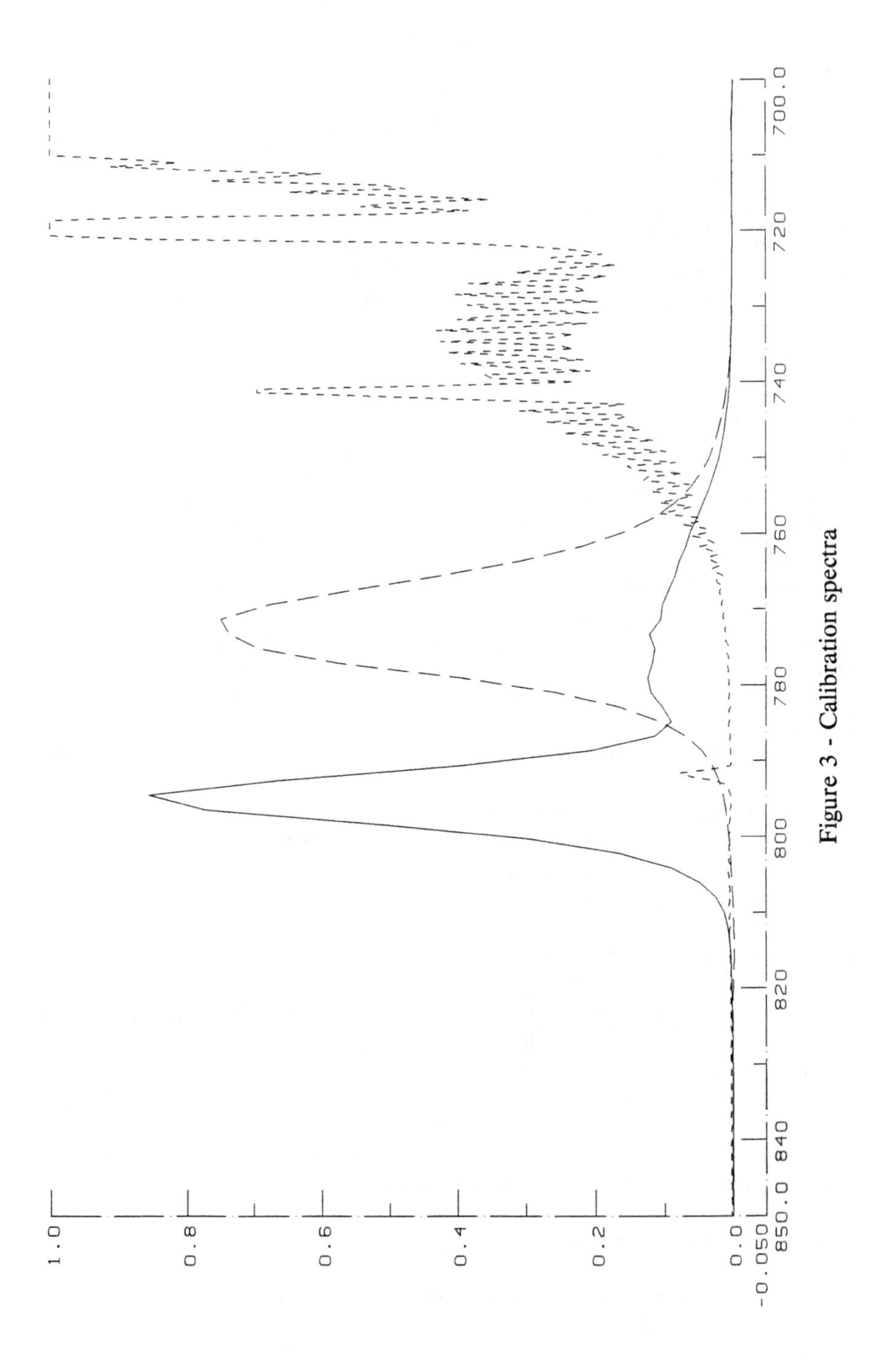

Figure 3 - Calibration spectra

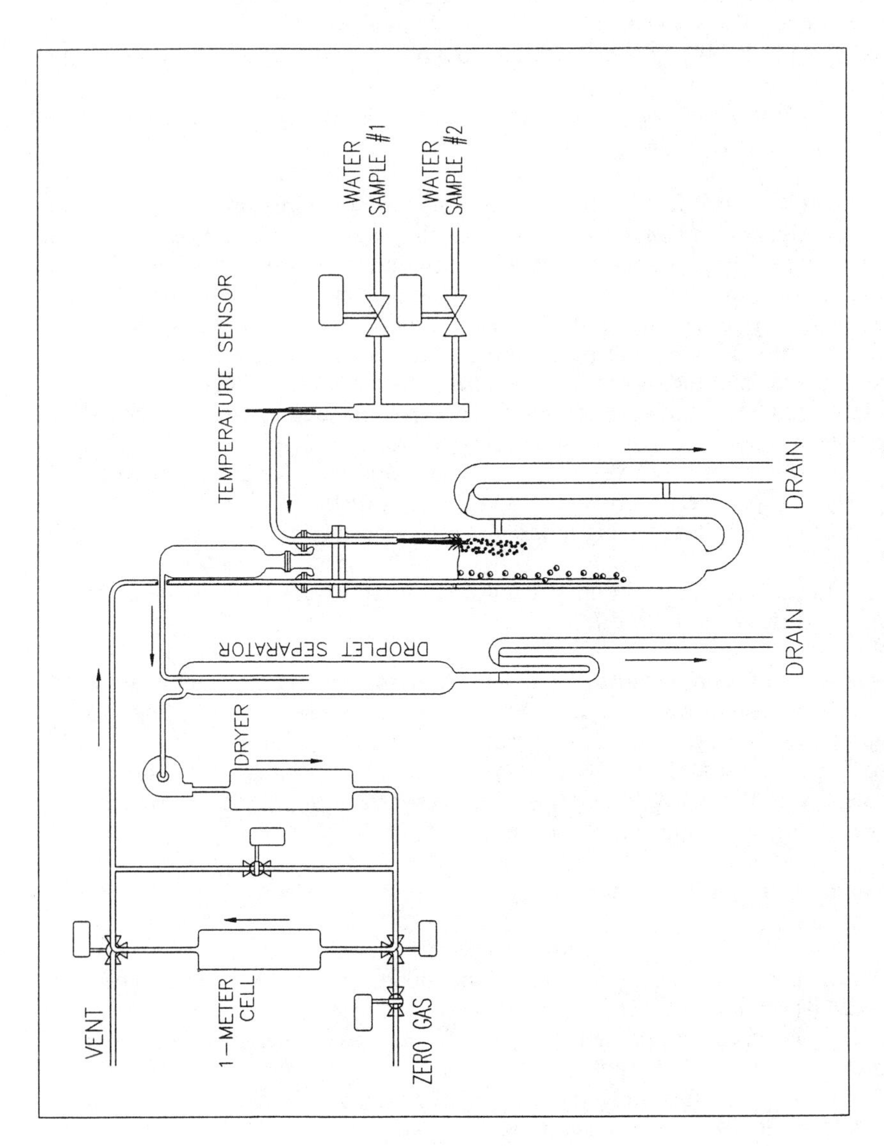

Figure 4 - System schematic diagram

Water Loop. The wastewater sample streams enter a manifold through computer controlled sample valves. A separate valve allows clean water to be injected directly into the top of the sparging vessel. At the outlet of the manifold, the water temperature is measured for use in the solubility and vapor pressure calculations. The water flows into the glass sparging vessel at a flow rate which causes many bubbles to form. These bubbles are the primary source of the sparging effect.

The water then flows out through a standard U-tube type trap assembly which establishes the liquid level in the vessel.

Air Loop. The air loop in the system is somewhat more complex. In sample measurement operation, air flows up from the liquid vapor interface, through a pair of expansion chambers which allow separation of water droplets from the stream, and into the input side of a diaphragm pump. From there it flows through a refrigeration-type drier to remove more water vapor and then into the cell. Finally it passes through the 1-meter path length cell in the FTIR spectrometer and through a dip tube back into the sparging region.

The FTIR also requires periodic background measurements. During these periods, the cell is connected to a source of gas (instrument air). The outlet of the cell is switched to a vent. At the same time the sparged vapors are sent through a bypass valve to maintain the sparging cycle.

The sparger operating in the laboratory during initial testing is shown in Figure 5.

The spectrometer with its gas cell is shown in Figure 6, while the complete sparger is shown in Figure 7.

Integration of Analyzer with Wastewater System. The normal sample point for the on-line system is at a collection sump before the plant outfall. This sump is equipped with large pumps which can divert the full plant water output into a holding tank whenever it is outside permit limits. The analyzer measurements are converted into 4-20mA signals which are transmitted to the process-control computer used to operate the water system.

System Operation Experiment

A comprehensive experiment was performed to demonstrate instrument operation. Please refer to Figure 8 for the following discussion. This trace is the standard output of the system, a trend chart of concentration and temperature vs. time. This appears on the system control screen and can be plotted out as shown here.

At the left of this trace, the system is operating normally with both species well below limits. The diversion pumps downstream of the sample point were turned on to prevent the experiment from causing a release.

The wastewater was then diverted around the normally used stripping column. This allowed the concentration to increase at about 16:55 hours to about 760 ppb.

At 16:53 the system was manually switched to the zero gas. The readings went down to near zero.

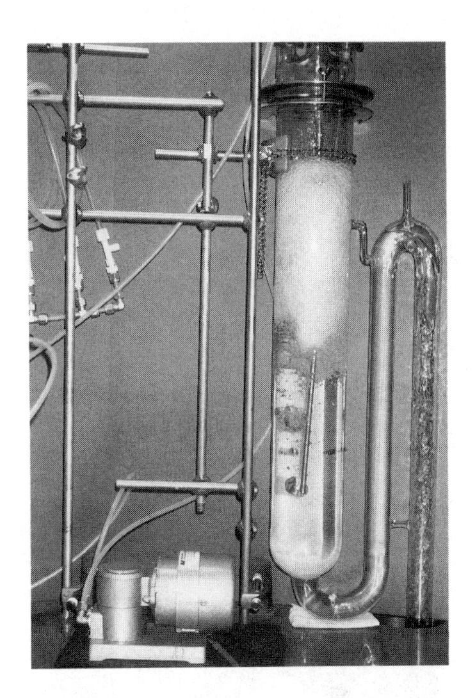

Figure 5 - Laboratory sparger test

Figure 6 - Process spectrometer

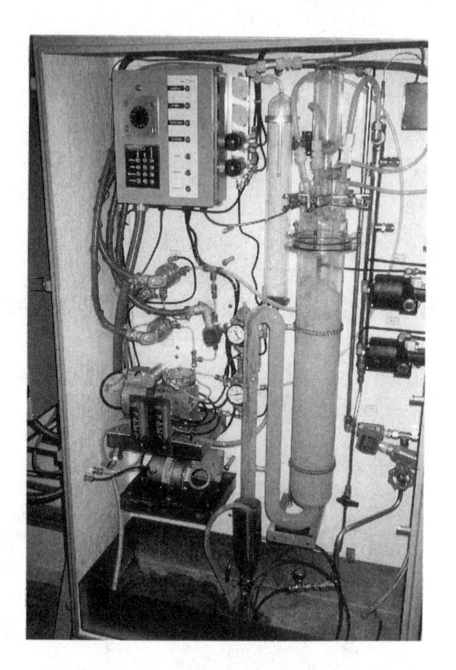

Figure 7 - Process sparging system

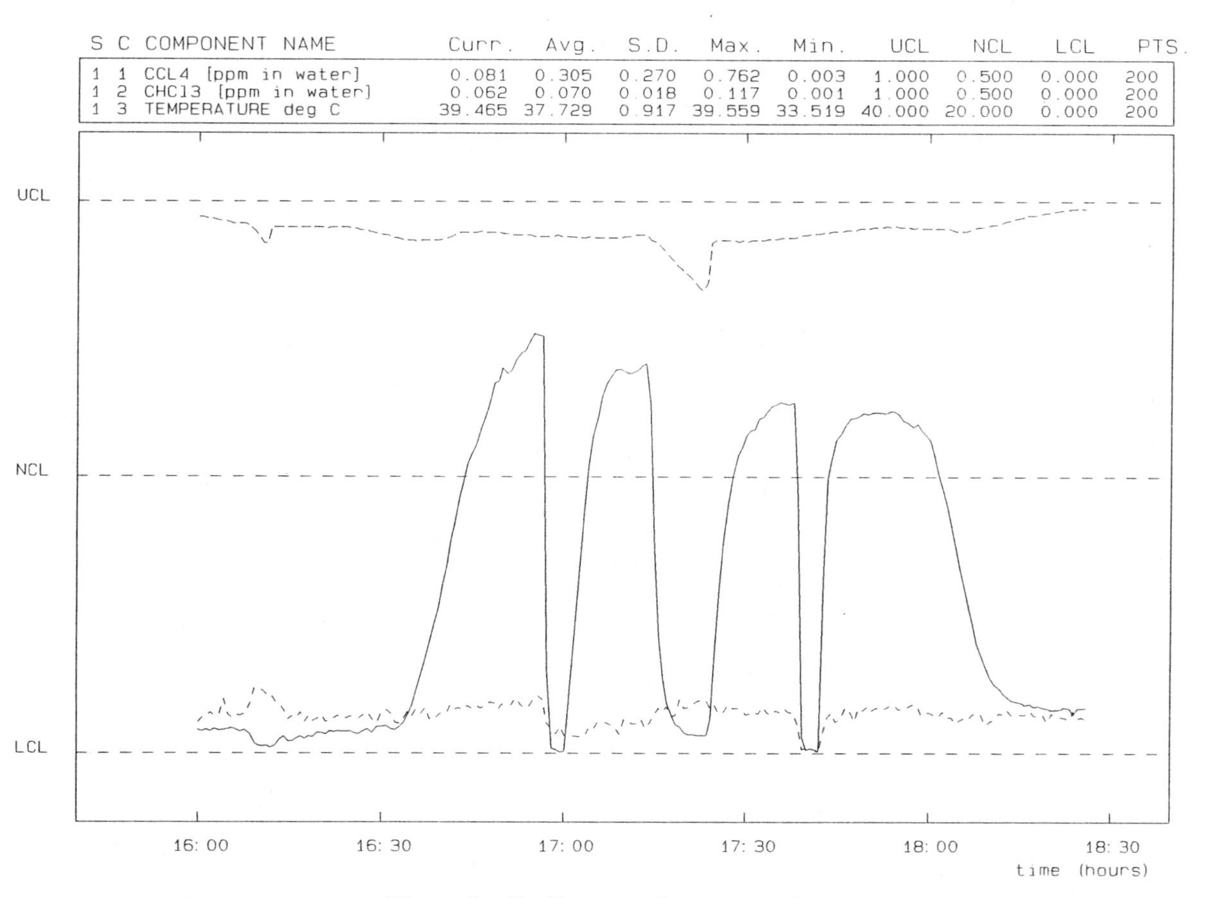

Figure 8 - On-line experiment record

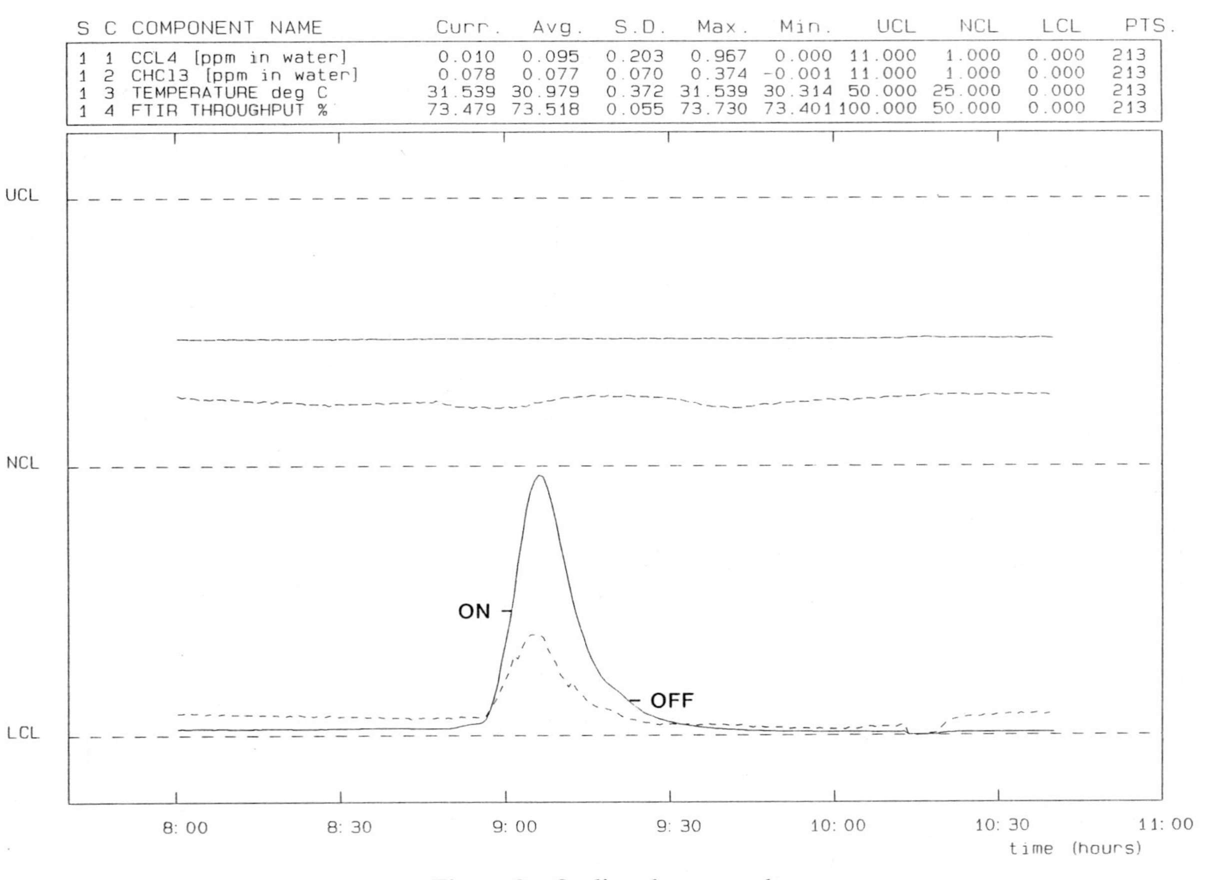

Figure 9 - On-line data record

At 17:00 the circulating sparged gas was switched back to the cell. The readings again went up.

At 17:14 the input manifold was switched from the waste stream to a clean water stream. The CCl_4 value went down but the $CHCl_3$ remained constant or even went up. This is because chloroform is normally present at low levels (100-200 ppb) due to the effects of chlorination of the water system. The temperature reading dropped because the clean water bypassed the temperature sensor.

Finally at about 17:55 the stripping system was reactivated, lowering the concentration level in the sump to normal over about 15 minutes.

Operating Event

A normal operating plot is shown in Figure 9. At just before 9 a.m. the CCl_4 went up to a peak value of about 970 ppb. As the level increased toward that value, the diversion pumps were automatically turned on. When the level fell again the pumps were turned off and the diverted material was recycled through the stripping system.

Future Work

This instrumentation technique will continue to be developed. We plan to perform calibrations for many other pollutants. Included in this work will be an examination of how highly soluble volatile materials such as alcohols can be measured. We also plan to examine the effects of cosolvation and pH on sparging efficiency.

Literature Cited

1 *Handbook of Chemistry and Physics*; Weast, Robert C.(Ph.D.), Ed.; CRC Press, Inc.: Cleveland, OH, 1975; 56th Edition

2 *Lange's Handbook of Chemistry*; Dean, John A., Ed.; McGraw Hill Book Company: New York, NY, 1979; 12th Edition

3 Horvath, A.L.; *Halogenated Hydrocarbons Solubility and Miscibility with Water*; Marcel Dekker Inc.: New York, NY, 1982

RECEIVED June 25, 1992

Chapter 7

Determination of Organic Compounds in Aqueous Waste Streams by On-Line Total Organic Carbon and Flow Injection Analysis

W. W. Henslee, S. Vien, and P. D. Swaim

Analytical and Engineering Sciences, Building A-915, Dow Chemical Company, Freeport, TX 77541

The determination of the soluble organic content of waste water streams is important for (1) efficient operation of bio-treatment facilities, (2) compliance with regulatory permits, (3) detection of process upsets and (4) material balances around manufacturing units. Grab samples and composites, followed by TOC measurements in the laboratory are often too slow to provide time for an appropriate response. Accuracy, precision, sensitivity and stability data are given for a new commercially available Total Organic Carbon (TOC) Analyzer. In some cases we have found that the time for a TOC measurement is too long for adequate response to a spill or process upset. Two cases are discussed in which flow injection analysis was used instead of TOC. Both involve the detection of amines in waste streams.

The determination of the soluble organic content of waste water streams is important for several reasons. Among these are (1) the efficient operation of bio-treatment facilities, (2) compliance with regulatory permits, (3) the detection of process upsets and (4) provision of data for material balances around manufacturing units. Grab samples and composites, followed by TOC measurements in the laboratory are often too slow to provide time for an appropriate response to a spill or upset. Our laboratory has been working with several of our site manufacturing units to evaluate commercially available equipment, and where necessary, to develop equipment and methods which provide accurate data in an on-line, continuous manner.

0097–6156/92/0508–0062$06.00/0

Laboratory and field evaluations have been made on a new commercially available Total Organic Carbon (TOC) Analyzer. Accuracy, precision, sensitivity and stability data were collected for the Shimadzu 5000. This instrument performed acceptably well, but opportunities for improvement were noted and communicated to the supplier.

In some cases we have found that the cycle time, even for an on-line TOC measurement is too long for an adequate response to a spill or process upset. Two cases are discussed in which flow injection analysis is used instead of TOC. Both involve the detection of amines in waste streams.

In all of the cases discussed here, the emphasis is on early detection of process upsets or spills so that the appropriate corrective actions can be taken quickly. This assures compliance with regulatory permits, provides data for process modelling and allows manufacturing personnel to operate their plants in a smooth and efficient manner.

Total Organic Carbon (TOC) Analysis.

Introduction. The analysis of organic carbon in waste water has been reported by three methods. These are (1) chemical oxidation TOC, (2) inductively coupled plasma (ICP) spectroscopy and (3) thermal oxidation TOC. A set of requirements and desirable features for such applications has been developed by our laboratory and is given in Table I. These may be accomplished by the analyzer or by sample conditioning prior to introduction into the analyzer.

TABLE I. TOC Analyzer System Requirements

Withstands a wide range of pH values (1 to 13)
Capable of oxidizing large organic molecules (e.g. humic acid)
Tolerates salt well (e.g. repeated 7% injections)
Handles suspended solids
Has good linear dynamic range (3 orders of magnitude)
Automatic range changing
Exhibits good sensitivity (detection limit 0.05 ppm)
Good reproducibility (<3 % rel. std. dev.)
Minimal drift
Automatic calibration
Acceptable maintenance (a few hrs. per month)
Rugged (maintenance predictable and infrequent)
Hardware and software are user friendly (easy to troubleshoot)
Rapid and programmable cycle times
Not susceptible to interferences
Acceptable long term cost of operation

It is very difficult for one technique to meet all of the criteria. A major problem with the UV induced chemical oxidation TOC analyzer is low TOC recovery when dealing with large molecular structures such as humic acid and fulvic acid (*1-2*). The kinetics of chemical oxidation are simply too slow for large organic molecules (*3*). A second problem is that, in the presence of a large amount of chloride salt, the efficiency of chemical oxidation is substantially diminished (*1*). The large amounts of chloride compete with organics to get oxidized, forming chlorine (*1*). Matrix matching therefore becomes important when large amounts of salts are present, or erroneous results will be obtained.

Inductively coupled plasma (ICP) has been investigated for TOC analysis (Vien, S. H. and D. Zacharkiw, D., Dow Chemical, unpublished data.). Good linear dynamic range and reproducibility were achieved for low salt samples. Plasma tube clogging and poor reproducibility are notable problems with the ICP method for high salt samples. Commercially available ICP's are set up only for total carbon (TC) analysis. Due to the wide pH range of many industrial waste streams, interference from carbonate can become a problem. An additional limitation is the high cost of this instrument.

The difficulties discussed above are magnified by the requirements of continuous, unattended operation. Thermal oxidation solves the large molecule recovery problem, but creates another. At traditional temperatures (950°C), salt bonds to and coats the catalyst. It forms a refractory and cannot be washed off. The newest thermal oxidation instrument to be commercialized for this demanding analysis is the Shimadzu 5000. This unit avoids the formation of refractory when salt is present through the use of a lower-temperature (680°C) catalyst. High recovery of large organic molecules is preserved.

We have recently evaluated and installed a Shimadzu process TOC analyzer which possesses many of the desirable features mentioned in Table I. The first part of this communication reports the analytical performance, interference studies, field trial results, and maintenance schedule for this instrument.

Experimental. This instrument is equipped with a non-dispersive IR (NDIR) detector, a 680°C oven which houses a quartz column containing platinum catalyst beads, a sample preparation unit which removes inorganic carbon interferences through acidification and sparging processes, and a sample introduction unit via a syringe. A programmable microprocessor handles data acquisition and operation parameters. Data can be output to a small thermal paper printer and to an external computer. This instrument is capable of storing two calibration curves at two different analytical ranges. The injection volume can be automatically changed as needed. In addition, it can be programmed to recalibrate with two standard solutions on a periodic basis.

The experimental procedure is as follows. Initially, the sample stream is filtered through a Collins SWIRLCLEAN (Trademark of Collins, Inc.) Model 8500 filter to remove particulate matter above 0.2 um. The filtrate fills up a 10 ml sample cell where it is acidified with HCl and sparged with zero nitrogen to remove the dissolved inorganic carbon as carbon dioxide. The sparged sample is then injected into the 680°C Pt catalyst column where the organic compounds are air oxidized to form carbon dioxide and water. Spectral interferents such as water, HCl, and chlorine are successively removed further downstream as they travel toward the IR detector via condensation and a halogen scrubbing system. The carbon dioxide is detected by the IR detector.

Results & Discussion. The following effects were investigated with the instrument in a laboratory setting: conditioning of fresh catalyst; the effect of salt on signal intensity; the effect of pH on carbonate interference; humic acid recovery. In addition, the calibration, reproducibility and dynamic range of the instrument were investigated. Last, the results of a successful field trial are presented.

Platinum Catalyst Conditioning. In order for the Pt catalyst to operate properly, it must be activated thermally under acidic conditions (See Figure 1). Data points on this figure were generated using a three minute cycle with 40 uL injections of 5% HCl for five hours. Unusually high and noisy TOC signals are observed for the first three and a half hours of the conditioning period. TOC signals then plateau and become more reproducible.

It must be emphasized that the successful conditioning of a fresh Pt catalyst for this thermal oxidation unit involves both heat and acid. The Pt catalyst that we used is Pt on an alumina support. The alumina is an excellent organic adsorber. Our data indicated that heat treatment alone (680°C for 16 hrs) was not sufficient to desorb these organics. The combination of acidic conditions and heat treatment oxidizes the adsorbed organics to form carbon dioxide and water.

Effect of NaCl Concentrations. Since NaCl is present in many process streams, its matrix effects on TOC analysis were investigated. Results at various NaCl concentrations (in 0.5% HCl) were obtained. Although the apparent TOC signal remained fairly constant at below 6.25% NaCl, tiny signals corresponding to 0.05 ppm and 0.08 ppm TOC were observed for 12.5% and 25% NaCl, respectively. This matrix effect is not significant compared to the TOC level in a typical waste water stream.

Effect of pH. Basic media are good solvents for atmospheric carbon dioxide. The dissolved carbon dioxide can not be removed by a simple sparging cycle under basic conditions. This causes interference by inorganic carbon. Tailing of the peak signal due to inefficient oxidation is another undesired effect of this method under basic conditions. Basic conditions also deactivate the Pt catalyst and provide an environment in the catalytic column for carbon dioxide adsorption.

In acidic media, inorganic carbon interferents such as carbonate and solvated carbon dioxide are converted to gaseous carbon dioxide. Some of the resulting carbon dioxide weathers off even without sparging, as suggested by a smaller, more symmetric signal. The benefit of acidifying sample is more apparent after the sparging step. The sparging step further removes the solvated carbon dioxide. However, highly acidic samples affect performance and result in higher maintenance frequency due to the corrosive nature of the acid. Volatile organics are also lost unless a special "purgeable organics" sytem is employed.

Humic Acid Recovery Studies. Humic acid is one of the major macro organic compounds found in our river water. The Shimadzu TOC analyzer achieved 106% TOC recovery relative to the Beckman TOC for humic acid. This recovery result suggests that the lower temperature (680°C) oxidation process utilized in the

Shimadzu TOC analyzer is just as efficient as the Beckman's high temperature (950°C) oxidation.

Reproducibility, Calibration, and Linear Dynamic Range. Figure 2 is one day of TOC data with 15 minute injections of a 500 mg/L TOC standard. The overall relative standard deviation is 1.7%. The calibration curve is generated with a pseudo two point calibration method. In this pseudo two point calibration method, only one standard is needed to produce a two point standard curve. The low standard (zero) signal is electronically produced whereas the high standard signal is generated by running a standard of desired concentration. This calibration method is adequate for TOC analysis of most waste waters containing relatively high TOC levels (>100 mg/L). For low levels of TOC, water blanks usually produce a signal equivalent to about 1.5 mg/L TOC. Figure 3 is a typical calibration curve for TOC standard from 0 to 1500 mg/L. The linear correlation coefficient is 0.9997.

Process Field Trial. Figure 4 is an example of one day of TOC data for water coming to a water treatment plant. Note that instrument was programmed to recalibrate with a TOC standard once every 8 hours. The curve has peak and valley features which reflect the dynamic nature of the TOC levels in water treatment plants serving several customers. The duration of the peak depends on the amount of water the plant is releasing whereas the magnitude of the peak depends on which plant is releasing the water and how much organic the discharge carries.

The advantage of a process TOC analysis over the conventional composite sampling method becomes obvious in a dynamic system such as this. The on-line TOC can provide close to real-time information about each stream. This timeliness becomes critical in the event of an excursion. For the same data shown in Figure 4, the composite sampling method gives an average of 1200 mg/L TOC which does not change when sampled once every eight hours. This may indicate a safe level for the microorganisms to digest the organics. On the other hand, the on-line TOC data may raise a concern about an unsafe high TOC level which may be harmful to the microorganisms in the water treatment plant.

Maintenance. The maintenance schedule for the Shimadzu TOC-5000 is summarized in Table II. This schedule was developed for a stream of 3 to 4% dissolved solid and a 15 minute (10 uL) injection cycle. This instrument operates at 680°C, which is below the melting point of salt. Salt accumulation in the catalytic column is a problem in that it reduces the oxidation efficiency and is reflected in a reduction of signal intensity. A control chart for standard TOC signal intensity is a good resource for making sound judgements on catalytic column maintenance.

Another problem with the accumulation of salt in a quartz column at high temperatures is that it vitrifies the quartz tube. Once vitrified, the column is subject to thermal shock due to the difference in thermal expansion between the inside and outside walls of the column. As a result, a crack may develop. Poor reproducibility is a good indication of a leak in the system. The maintenance frequency for the catalytic column is once every 6 weeks.

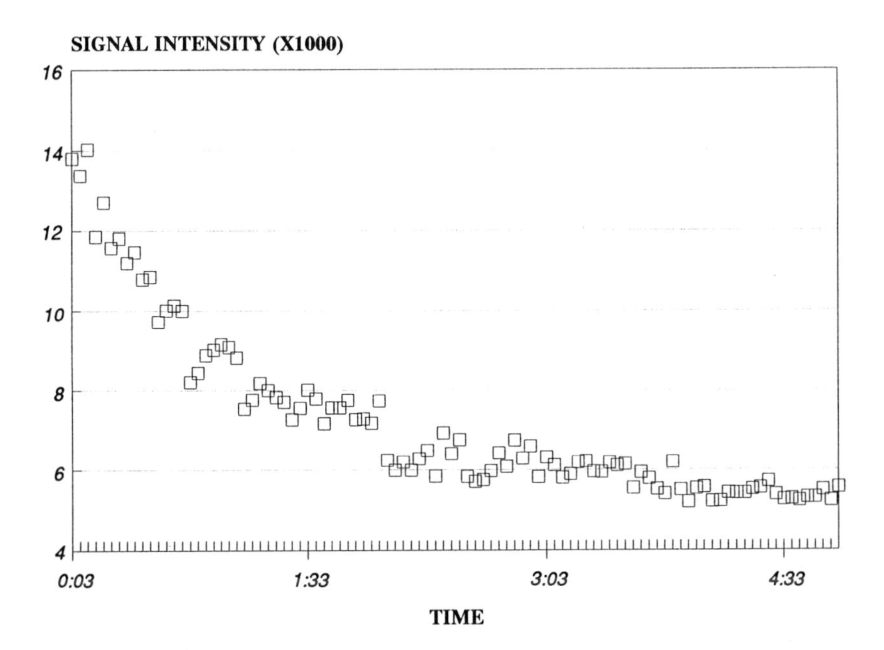

Figure 1. New Pt catalyst conditioning. Data were generated with 40 uL injections of 5% HCl solution under normal instrument operation condition.

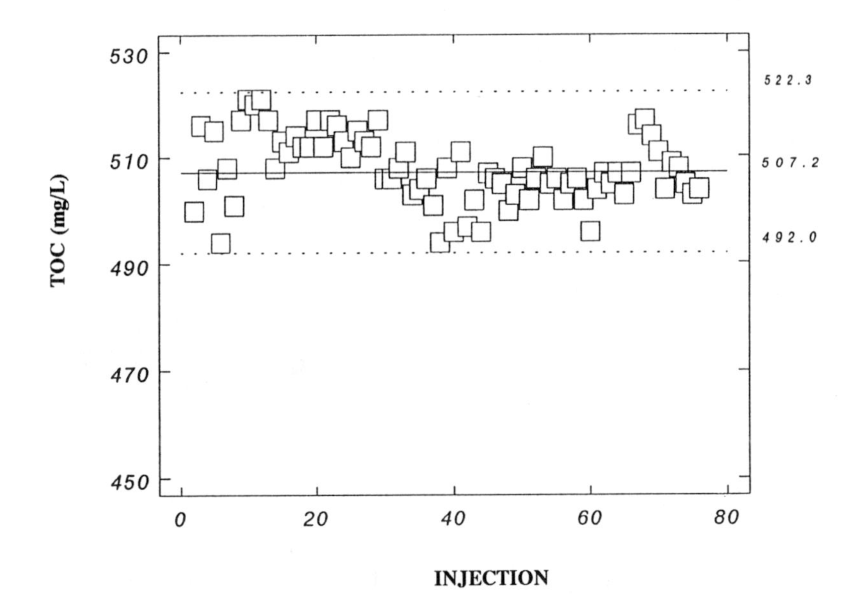

Figure 2. One day control chart run. Data were generated with 500 ug/mL TOC standard in 0.5% HCl. Injection Cycle = 3 min.

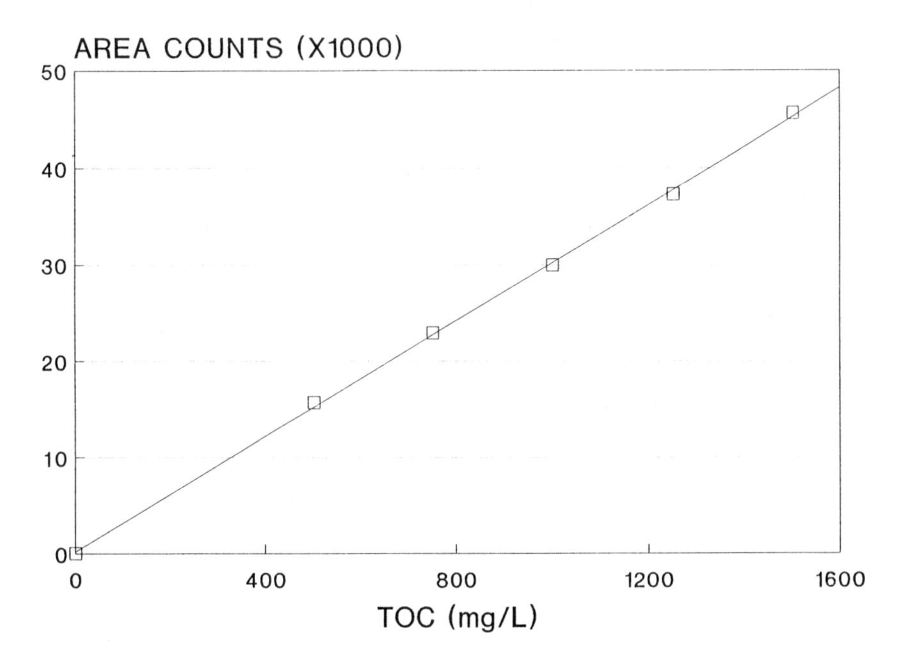

Figure 3. A typical TOC calibration curve. Standards made from potassium acid phthalate in 0.5% HCl. Correlation coefficient = 0.9997

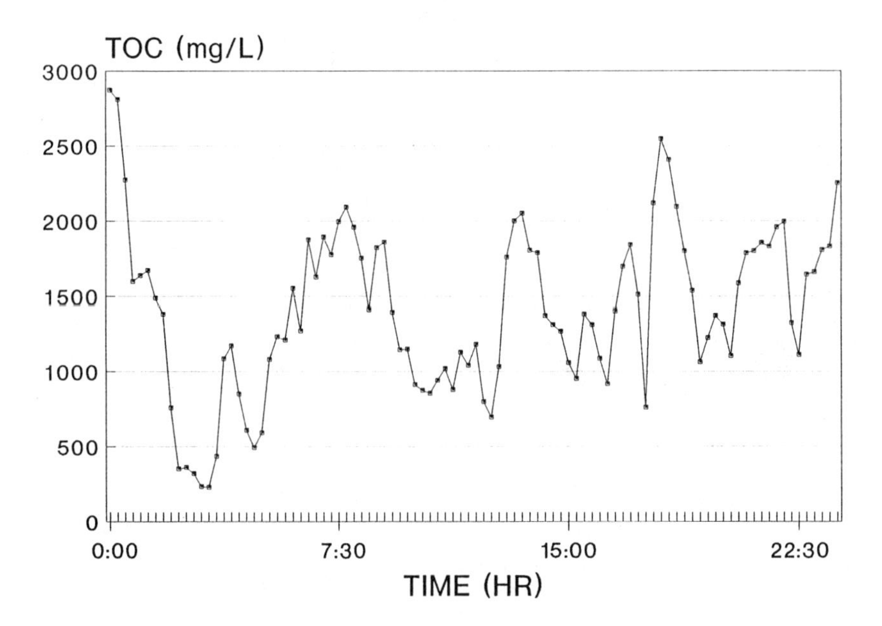

Figure 4. A typical one day on-line TOC monitoring of waste water in a water treatment plant. Injection Cycle = 15 min.

A buildup of a salt bridge at the interface between the injector block and the catalytic column can occur. This buildup is a result of smears of salt after drying. Poor reproducibility will result. TEFLON (Trademark of E.I. duPont de Nemours & Co., Inc.) plunger leaks develop at the end of the 3rd or 4th week of continuous use. This is a result of normal wear of the material and the required maintenance frequency is predictable. Poor reproducibility and traces of salt and liquid around the syringe barrel area are clues for this leak.

TABLE II. Typical Shimadzu TOC-5000 Maintenance Schedule for 15 Minutes Injection Cycle of a Stream Containing 3 to 4% Dissolved Solids

PARTS	FREQUENCY
Catalyst/Column	>6 Weeks
Injector Block	4 Weeks
Teflon Plunger	3 Weeks

Overall, the Shimadzu TOC 5000 unit performs acceptably well for long term, continuous service in waste streams. However, it must be housed properly and maintained adequately. Most importantly, the sample must be properly filtered before TOC analysis to ensure signal reproducibility and to prevent syringe clogging.

Detection of Simple Amines by Flow Injection Analysis (FIA)

Introduction. Aliphatic polyamines such as ethylenediamine (EDA), diethylenetriamine (DETA), triethylenetetraamine (TETA), tetraethylenepentaamine (TEPA), etc., are produced by the reaction of ethylene dichloride with ammonia to form a mixture of the corresponding amine salts. These salts are then converted to the free amines by neutralization with sodium hydroxide. This results in a slurry of solid salt in a concentrated amine solution. The salt is separated by centrifugation and ultimately discharged through the plant's waste system. Occasionally, these process centrifuges malfunction and allow some of the liquid amine to be discharged along with the salt.

In order to determine the onset of a spill, total organic carbon (TOC) analyzers were installed on-line to constantly monitor the salt stream. However, because of the 5 to 10 minutes required for these analyzers to respond, by the time that a spill was indicated, target limits had usually been exceeded. Flow injection analysis (FIA) is a versatile technique by which a wide variety of analytical methodology such as titrimetry, colorimetry, and other wet methods can be automated (*4*). Because of its mechanical simplicity, it has proven to be reliable for on-line applications. Response time is usually very fast so that as many as one or two determinations can be made per minute.

It has long been known that a number of aliphatic amines form complexes with cupric ion which are much more intensely colored than cupric ion itself. We found the amines associated with the EDA process to be no exception, and that the reaction appeared to occur instantaneously. Using this chemistry, a method was developed which could be incorporated into a FIA scheme and resulted in an on-line monitor which offered superior performance to TOC methods.

Experimental. FIA instruments have been in use for several years in locations throughout Dow USA and have proven quite reliable. The analyzer has a wet side which contains the reagent pump, sample valve, etc., and an electronics side which houses a microprocessor for controlling analysis parameters such as cycle time, sample injection, calculations, etc. The analyzer is housed in a Class A purged enclosure. A SPECTROCELL flow cell (Spectrocell Corporation, Oreland, PA) which has a path length of 1.0 cm and a volume of 40 microliters was used for this application. A Masterflex peristaltic pump equipped with three heads, in conjunction with a modified Balston model 95 filter was used for the sample filtration and dilution system. Silicone tubing was used for the pump tubes. The rollers in the pump heads were sprayed with silicone lubricant prior to use to extend the life of the pump tubes.

The reagent solution is prepared in the following manner: 17.6 g $CuSO_4{\cdot}5H_2O$ and 180 g sodium chloride are completely dissolved in 1500 mL of deionized water. Sixty grams of boric acid, 43 g citric acid monohydrate, and 64.5 g sodium hydroxide are then added. When dissolution is complete, the solution is diluted to 6 liters with deionized water. All reagents used are reagent grade or better.

Results and Discussion. Although these amines are relatively strong bases (pKb for EDA = 3.8), attempts to use pH to indicate their presence failed apparently because other acids and/or bases were present which had an even greater effect on pH. Therefore, it was important to determine the effect pH might have on the stability of the various cupric-amine complexes. We found that formation of these complexes occurred only when the amines were in their free state; no color change was observed upon addition of cupric ion to an acid solution of any of these amines. Thus, the pH of the reaction media needed to be greater than 9 - 9.5 to insure that any acid salts would be converted to the corresponding free amines.

Method Development. Since cupric ion is insoluble in basic media, a means was needed to render the cupric ion soluble that was not complexed by the amines. This could be accomplished by employing a different ligand to complex cupric ion provided (a) the resulting copper complex was soluble at pH values of 9-10, (b) it could easily be displaced by any of the amines, and (c) it yielded a solution whose color was significantly different from that of the copper-amine compounds. We found citrate ion to be such a ligand. As long as the pH of the solution was maintained below about 10.5, the copper ion remained soluble. At higher values, precipitation tended to occur. Figure 5 shows the absorption spectra of several solution in which equimolar amounts of EDA, DETA. TETA, and TEPA were added to an alkaline solution containing copper(II) citrate. It can be seen that the cupric-citrate complex absorbs relatively little in the range of 550 to 600 nm, an area where absorption is appreciable for the cupric amine complexes. From the above discussion it is obvious that the pH of the reaction solution must be maintained between about 9 and 10.5. This was done by buffering the cupric citrate reagent to a pH of 10 with a boric acid-borate buffer.

Sensitivity Study. It can be seen in Figure 5, that the wavelength of maximum absorption as well as the relative sensitivity differs for each of the amines. The result was that the method was not equally sensitive to the various amines. We found that on a TOC basis and with detection at 570 nm, the relative sensitivities by FIA were EDA-0.94, DETA-1.00, TETA-1.40, and TEPA-0.94. Therefore, exact results could be obtained only if the amine make-up of the standard(s) was similar to that of the sample, and if the sample composition remained constant. For the sake of convenience, however, and since absolute accuracy was not necessary, solutions of DETA alone were used for standardization. If all the amines are present in approximately equal amounts, then using DETA for standardization would result in calculated values differing from the actual ones by no more than 10 percent.

Effects of Salt. Initial attempts to analyze the filtered sample were not entirely successful because the high sodium chloride content (20-25 percent) generated significant peaks, even when no amine was present. This was caused by differences in refractive index between the reagent and sample. Also, plugging tended to be a problem since the sample was often saturated with salt and minor changes in temperature sometimes caused salt to precipitate from solution. Both problems were minimized by diluting the sample. The method of analysis was sufficiently sensitive that the sample could be diluted and the normal background level of amine could still be easily detected with a sample volume of 30 microliters. On-line dilution was accomplished by continuously pumping sample and deionized water at a rate ratio of 1 to 3.5 using a Masterflex multichannel peristaltic pump, and then mixing in a common line prior to filtration. Any residual effects due to differences in refractive index still remaining after dilution were eliminated by adding NaCl to the reagent solution to about 3 percent.

Filtration. The most difficult aspect of this application was finding a filtration system which required little time or effort to maintain. Sample filtration was needed in order to remove suspended solids which would otherwise have caused plugging of the analyzer. Much of this insoluble material was due to polymeric amines which were by-products of the amine process. Although it had the appearance of a rather course, gelatinous precipitate not unlike aluminum hydroxide, it was much easier to filter and could be removed from the surface of filter paper simply by washing with a stream of water from a wash bottle.

Initially, attempts were made to filter the stream prior to dilution using self-cleaning type filters such as the Collins SWIRLKLEAN. These filters use the high velocity of the sample itself to sweep solids from the filter's surface. We found, however, that even at flow rates of several gallons per minute the filter became coated and plugged. Although some types of filter materials were more effective than others, none could be found which did not require frequent replacement.

The system which was eventually adopted utilized a Balston filter, modified so that it could be used in a bypass configuration, and an automatic block valve controlled by a timer was added to allow a high flow of water to be used to flush the filter at regular intervals. Thus, a good portion of solids build-up on the filter element could be removed each time the filter was flushed, thereby extending the time period between filter changes. With a flush time of 45 sec every 4 hr at the a flow rate of 1/2-1 gal/min, the filter element needed replacing only once every two weeks. We also found that the filter functioned best when the removable housing pointed down and the port at the end of this housing served as the exit for the bypass stream. This was the most effective configuration for removal of solids during the flush cycle.

While the Model 95 filter has provisions to be used in the bypass mode, passages in the top of the unit are quite small and would not have allowed a high enough flow for the flush to have been effective. We modified the filter by machining a new center support (the part which actually holds the filter element) so that a screw-on cap could be added on top of the support to hold the element in place. This resulted in a greatly enlarged passage. It should also be mentioned that the holes in the support through which the filtrate passed, as well as the center channel, were greatly reduced in size in order to reduce dead volume and lag time. The time required for sample to flow from the process, through the filter, and into the sample valve on the analyzer was about 2 minutes.

Flow Scheme. The relatively low flow rate of diluted sample past the bypass side of the filter element made it necessary to use a pump to pull sample through the filter. This pump was actually a separate head attached to the pump drive used for pumping raw sample and diluent water. By flowing diluted sample into the unit at a much greater rate than filtrate was being taken out, and by keeping the end of the exit tube above the level of the filter to avoid syphoning, a slight positive pressure on the bypass side of the filter element was always assured. This prevented any tendency of air being drawn into the system which, if injected into the flowing reagent stream would produce very large peaks and therefore erroneously high results.

Figure 6 is an illustration of the flow scheme used to adapt the analysis for amines to FIA. The conditions were optimized for as high an analysis rate as practical. Thus, by employing 0.5 mm ID tubing for the mixing coil and transfer lines from the sample valve to the flow cell, dispersion of the sample zone was minimized. Peak width was minimized by using a sample volume of only 30 microliters. In conjunction with reagent flow rate of 1.0 mL/min, these conditions enable an analysis to be run every 60 seconds.

Conclusion. As stated previously, the object of developing this analysis was to improve response time so that the onset of an amine spill could be detected as quickly as possible. Figure 7 is a print-out from the plant's data system on which the output of both the flow injection analyzer and the TOC analyzer is shown. Zero minutes represents the current time while the previous hour is given in five minute intervals. It can be seen that the flow injection analyzer reached the 1500 ppm alarm limit almost 9 minutes before the TOC analyzer. In the event of a

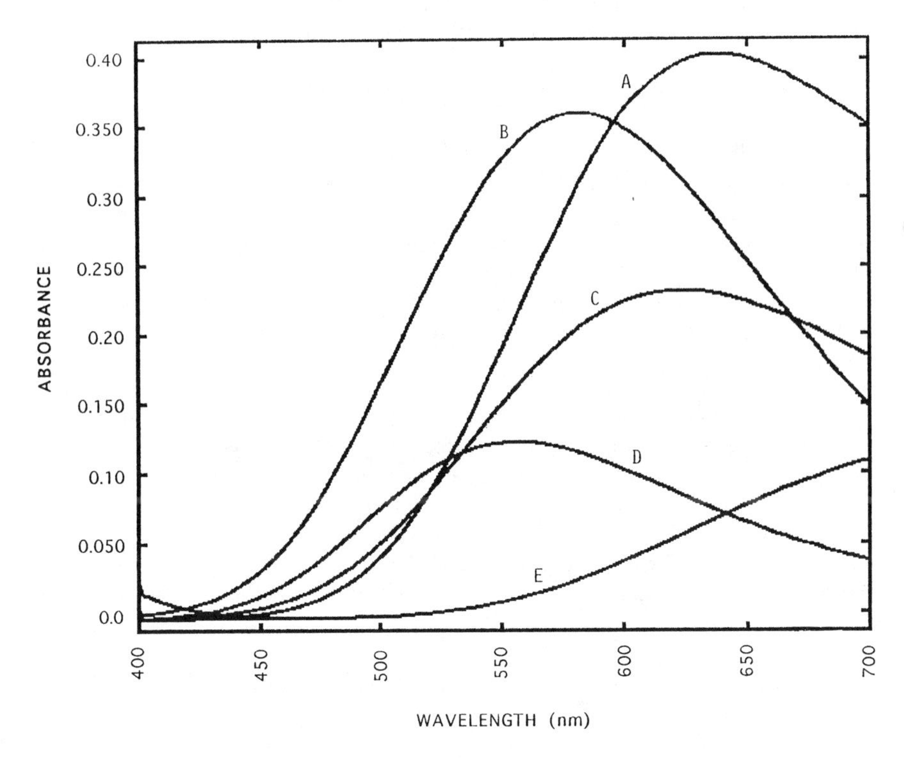

Figure 5. Absorption spectra of cupric ion complexes with selected organic ligands. Equimolar concentration of amines. A = tetraethylpentaamine (TEPA), B = triethyltetraamine (TETA), C = diethylenetriamine (DETA), D = ethylenediamine (EDA), and E = citrate.

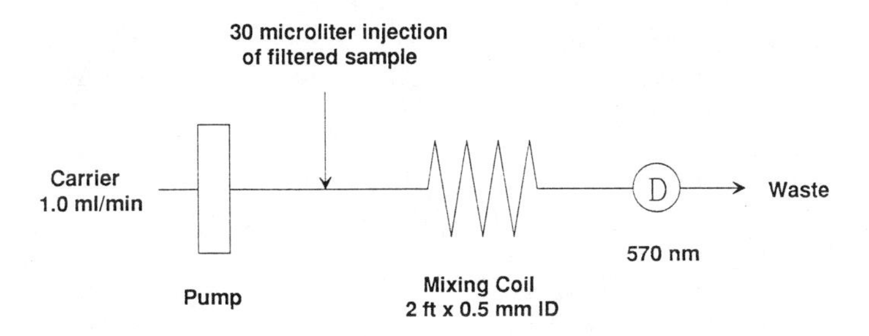

Figure 6. Flow scheme for the analysis of amines by FIA. Carrier solution = 3% sodium chloride, 0.2% cupric chloride, 1% sodium citrate, 1% boric acid, solution adjusted to pH of 10 with sodium hydroxide.

larger spill, it would be expected that the response from the TOC analyzer would be somewhat quicker and more positive. Even under these conditions, however, the response by FIA was always faster by at least 4-5 minutes.

Detection of Complex Amines by Flow Injection Analysis (FIA)

Introduction. One step of the methylene diphenyl diisocyanate (MDI) process at Dow's La Porte facility involves the conversion of aniline to methylenedianiline (MDA). A byproduct stream of dilute brine containing unreacted aniline is generated along with the MDA. This stream is processed in a tower called a brine stripper where any aniline which might be present is distilled overhead and eventually recycled back into the process. The net result is a waste stream whose volume and aniline concentration has been reduced considerably, but whose salt content has been increased to about 10 percent. If the stripping column malfunctions, aniline is left in the bottoms and can overload the biotreatment facility. An on-line analyzer was desired which could detect aniline in the exit stream from the bottom of this tower. Because of its very distinctive absorption band in the ultraviolet, spectrophotometry was considered to be the best approach. However, it would have been difficult to use an on-line UV spectrophotometer because the stream tended to deposit an opaque coating on any surface it contacted. It was decided that flow injection analysis (FIA) could offer a way to introduce a small amount of the sample into a UV detector in a way that would not foul the detector windows.

Experimental. Initial studies revealed that aniline in neutral or alkaline solutions has a strong and distinctive absorbance centered at 280 nm that is totally absent in acid media. Since the sample stream was quite alkaline, this would not be a problem. Because the absorbance of aniline is quite large at 280 nm and plant personnel were interested in levels less than 500 ppm, it was necessary to monitor the absorbance at some wavelength removed from 280 nm where the absorbance was less, or dilute the sample. Since diluting the sample might have other benefits, and since it could be easily done through FIA, this approach was taken.

Figure 8 schematically represents the FIA system used for the determination. The LDC detector was designed to be used for liquid chromatography, and was thus equipped with a flow-through cell. The cell used for this application had a pathlength of 3 mm and a cell volume of 47 microliters. All connections to the cell were via 1/8" tubing fittings, and all internal passages were at least 1/16". A Fluid Metering Inc. Model GRP 50 laboratory pump was used as the carrier pump. The sample valve was a Rheodyne Type 50 TEFLON 4-way rotary valve with 1.5 mm bore and 1/8" tubing connections. The mixing coil and all transfer tubing were 1/8" TEFLON tubing.

The carrier solution used was bottled drinking water, a source of clean water that was readily available at the site. Triethanolamine was added at about 30 ppm to prevent the accumulation of ferric hydroxide on the windows of the detector. At a carrier flow rate of 1.5 ml/min, about 1 1/2 min was required for a peak to be completely resolved at the detector. To minimize wear on the sample valve, the cycle time of the analyzer was extended to three minutes.

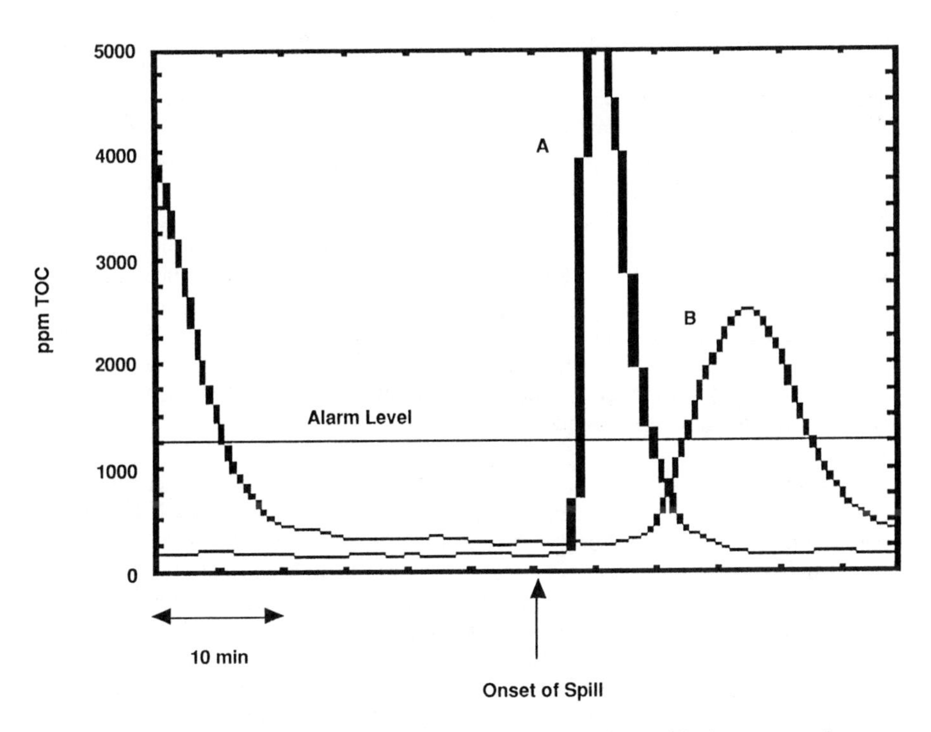

Figure 7. Responses of FIA and TOC analyzers to amine spill. A = output from flow injection analyzer. B = output from total organic carbon analyzer.

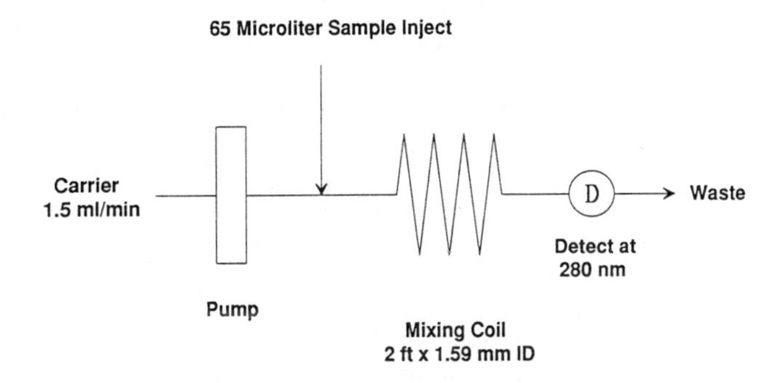

Figure 8. Flow injection analysis of aniline in waste brine. Carrier solution = 30 ppm triethanolamine in water.

The analyzer was calibrated using standards prepared by diluting reagent-grade aniline in distilled, deionized water. It was found that matrix matching of the standards to the sample stream composition was not needed. The analyzer was calibrated over the range of 0 - 1000 ppm aniline using standards of 0, 100, 500, and 1000 ppm aniline. At the end of each analyzer cycle, the amount of aniline in the sample was calculated based on peak height. A signal scaled to between 4 and 20 milliamps, corresponding to the amount of aniline found, was generated by the analyzer and sent to the control room. A signal above a certain level activated an alarm.

A Collins SWIRLKLEAN filter was used to provide particulate-free sample to the analyzer. A fast loop from the process was used to reduce lag time and to provide the necessary flow for proper functioning of the self-cleaning filter. A precautionary second filter was used downstream from the SWIRLKLEAN filter to remove solids that were originally dissolved, but tended to precipitate from solution with time. Both filters used 0.5 micron TEFLON elements. They were wetted with isopropyl alcohol prior to use.

Results and Discussion. Once the analyzer had been shown to operate properly on the lab bench using grab samples from the process, it was installed on-line between the plant but ahead of the treatment facility. Figure 9 shows the data collected over a period of one week. Note the overall high concentration and the brief excursions above the 500 ppm alarm level. Once the analyzer was on-line, we found analyzer response such as this was typical. We also found that almost all of what was detected was due to MDA rather than aniline. Both the absorptivity and wavelength of maximum absorption for MDA are almost identical to that of aniline. Since MDA is also toxic to the bacteria, albeit less so than aniline, having an indication of high levels of MDA in the waste stream is also useful information to plant personnel.

Although there were no adverse *analytical* effects created from the presence of higher-than-expected levels of MDA, other problems were encountered. As with most substances, the solubility of MDA in aqueous solution decreases with decreasing temperature. The temperature of the sample stream as it emerged from the process was 60-80 deg C, a temperature at which MDA is relatively soluble. Once the stream passed through the filter and was injected into the flowing carrier stream, however, the temperature of the sample had decreased to 20-25 deg C. When the concentration was high, the MDA tended to precipitate upon injection into the flowing water carrier stream. The larger tubing and valve sizes cited in the experimental section were installed to minimize the frequency of plugging due to high levels of MDA.

There was one other problem encountered with this application, and it began to surface almost immediately after the analyzer was put into service. We began to observe a steady increase in the background absorbance of the detector with time as if the windows in the sample cell were becoming coated. Over the course of about a week, the baseline increased so much that there was not enough span in the course and fine adjust potentiometers to return the baseline to a value of zero absorbance. At the same time the tubing downstream from the sample valve became increasingly discolored with a reddish-brown substance on its inside surface. When flushing the system with 1-1 HCl removed the coloration inside the

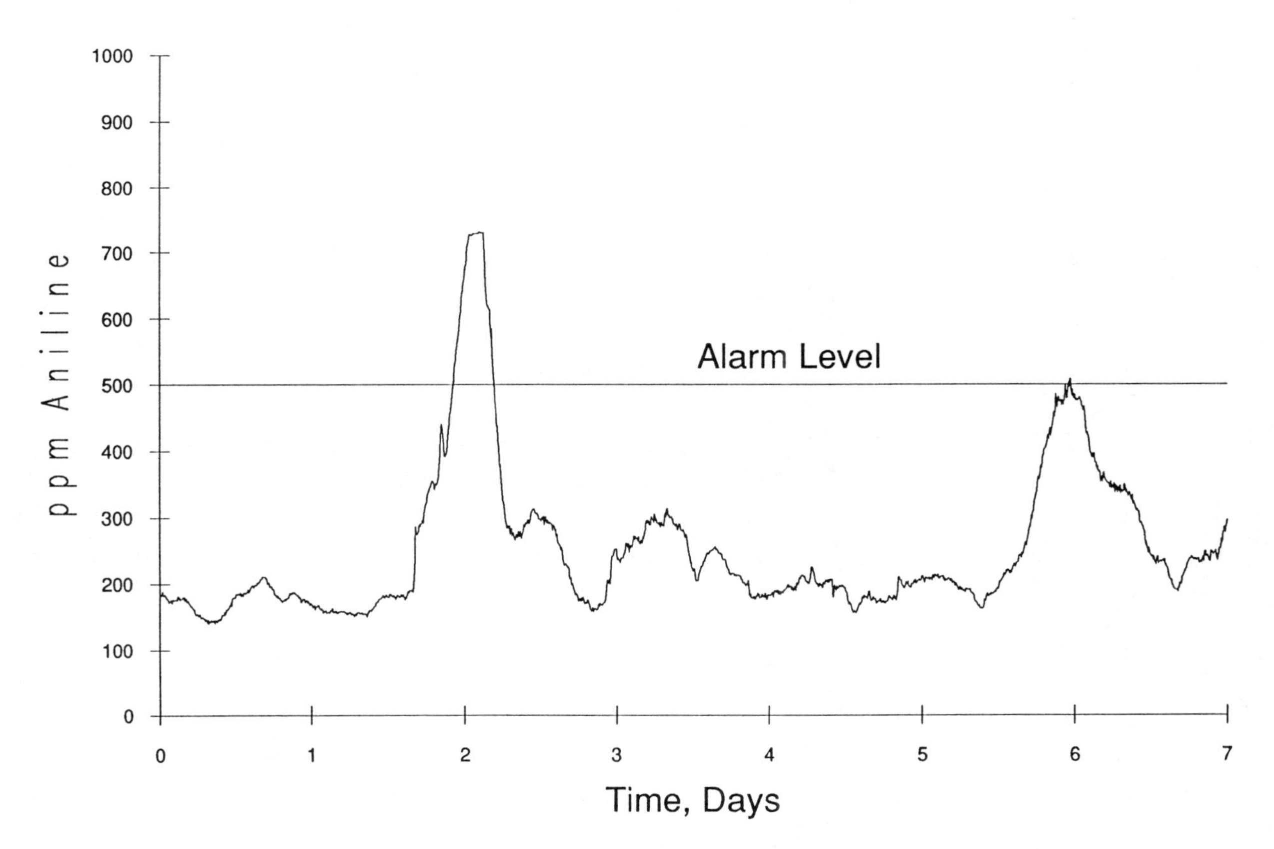

Figure 9. Aniline in brine stripper bottoms by flow injection analysis. Analysis frequency = 20 per hour.

tubing and lowered the baseline to its original level, it became obvious that these problems were caused by ferric hydroxide precipitating from solution after it had passed through the filter. The problem was solved by adding triethanolamine (TEA) to the water carrier to complex any iron (III) present and thus prevent its precipitation.

Conclusion An analyzer has been developed for the on-stream determination of aniline in a brine stream. The analyzer is based on UV absorbance as a means of detection, and flow injection analysis to both dilute the sample and to prevent mechanical interferences that would otherwise be encountered if the analysis were to be attempted directly.

Literature Cited

1. Sakamoto, T.; Miyasaka, T. *Ultrapure Water*, December 1987.
2. Sugimura, Y.; Suzuki, Y. *Marine Chemistry*, 1988, *24*, 105-131.
3. Small, J. W.; Linton, A. T.; Veroshco, C. E.; Burchett, R.; 17th Intersociety Conference on Environmental System, July 13-15, 1987, 871447.
4. Ruzicka, J.; Hansen E. H. *Flow Injection Analysis*; John Wiley & Sons: New York, 1981.

RECEIVED April 15, 1992

Chapter 8

Sulfur Recovery To Reduce SO_2 Pollution

E. H. Baughman

Amoco Corporation, P.O. Box 3011, MC B-6, Naperville, IL 60566

As hydrocarbons are extracted from the ground, they contain significant amounts of sulfur. Since most hydrocarbons are used for fuel, sulfur which remains with the product, will eventually appear in the environment as SO_2. While the term "Pollution Prevention" has been recently popularized, the oil industry has been active in recovering sulfur from the hydrocarbons for over forty years. In this paper we will look at the overall sulfur recovery process and mention an analyzer which allows for optimization of one particular part of that process.

As hydrocarbons are extracted from the ground, the sulfur content varies from the ppm to high percentage levels. Before the hydrocarbons can be used, the sulfur level must be reduced to the low ppm level. Generally the sulfur compounds are converted to H_2S and the H_2S is then carefully oxidized to elemental sulfur. A simplified flow diagram of a "gas sweeting" plant is shown on the figure. The "sour" gas contains H_2S and/or CO_2 which must be removed. An aqueous ethanol amine flows down through the contactor and extracts the H_2S and CO_2 from the gas. The "sweet" gas can then be used either by our customers or for internal fuel. The ethanol amine stream goes to the regenerator and the H_2S and CO_2 are removed. The amine stream is recycled to the contactor, the H_2S is converted to sulfur, and the CO_2 is vented to the atmosphere or compressed for reinjection into wells. Safe control of this sulfur recovery process requires that the H_2S and CO_2 content of both the "rich" and "lean" amine streams be known. Too much H_2S and/or CO_2 can lead to corrosion problems and unsafe operation. Excess circulation of the amine stream wastes energy and increases the hydrocarbon content of the H_2S/CO_2 stream going to the oxidizing unit. This leads to improper oxygen to H_2S ratio which causes SO_2 emission. Our H_2S-CO_2-amine analyzer is based on injecting a

0097–6156/92/0508–0079$06.00/0

reproducible sample into an acid bath and stripping the H_2S and CO_2 out of the acid bath past detectors which measure the H_2S and CO_2 in the vapor stream. With proper calibration, the H_2S and CO_2 content of the vapor stream are reproducibly correlated to the H_2S and CO_2 content of the original liquid stream. The first analyzer has successfully been on stream for over a year, following about two years of laboratory testing.

Magnitude of the Problem

Natural gas contains from 0-90% H_2S as it comes from the ground. Generally gas with very high H_2S levels is not brought to the surface because of the H_2S content. However, gas containing up to 35% H_2S is frequently processed. If this sulfur is not removed, the sulfur will be converted to SO_2 when the gas is burned for its heating value. To limit the SO_2 produced by burning natural gas, the upper limit on H_2S for pipeline gas is 4 ppm.

With crude oil, the sulfur content can vary from about 0-5% by weight. In crude oil, the sulfur is in various forms but must be removed before the refined products are suitable for use.

Table I shows the magnitude of the problem in a different way.

Table I

Crude	Weight % Sulfur	Long Ton Sulfur/100,000 BBL
B_S	0.4	51
Y	2.5	352
Trinidad	0.3	41
Light Arabian	1.6	213
		658

Suppose a 400,000 barrel/day refinery was running a crude slate of these four crudes, total sulfur per day coming in would be 658 long tons. If this is converted to SO_2 at the refinery or in the customers use of the refined product, ~1300 long tons/day of SO_2 would be generated. To show that these numbers are realistic, Amoco has the capacity to recover 1400 long tons/day of sulfur at one of our refineries. The sulfur recovery units at the natural gas plants range from 5-2000 tons/day. Clearly large amounts of SO_2 are prevented from entering the atmosphere by converting the naturally occurring sulfur compounds into elemental sulfur before the hydrocarbons are burned.

How is Sulfur Removed from the Product?

With crude oil, the sulfur is first converted to H_2S. With natural gas most of the sulfur is already in the form of H_2S. Figure 1 shows the separation of the hydrocarbon product from the H_2S. The sour gas, significant H_2S

and/or CO_2 level, enters the bottom of a tower which contains a base flowing down through the tower. Common bases are water solutions of ethanol amines. The base reacts with the acidic gases H_2S and/or CO_2 and pulls them to the bottom of tower while the hydrocarbon gases continue out the top of the tower. The amine stream from the bottom of the tower is referred to as "rich" amines. The "sweet" gas must contain less than 4 ppm H_2S. CO_2 should be minimized to prevent corrosion in the delivery system but is not as critical as the H_2S level.

The amine must be recovered or all that has been accomplished is to trade an air pollution problem, SO_2, for a solid or liquid waste problem. To recover the amine, the stream pressure is dropped and heat is applied. The H_2S and CO_2 is released and the "lean" amine can be recycled. We have now separated the H_2S and CO_2 from the desired product, the hydrocarbons.

Figure 2 shows a very simplified flow diagram of a Claus plant. The H_2S is reacted with oxygen to produce elemental sulfur and water. The CO_2 passes through the Claus plant and is released to the atmosphere or recovered for reinjection into the formation for enhanced oil and gas recovery. The details of the Claus plant are complex and are beyond the scope of this paper. The reader is referred to a paper by D. Parnell, "Look at Claus Unit Design", *Hydrocarbon Processing*, September 1985.

Control of Amine Scrubbing Units

The major goal of the sulfur recovery unit is to produce acceptable hydrocarbon product. A secondary goal is to produce sulfur in an economical fashion. The economic operation of the amine scrubbing unit requires a balancing act between wasting energy and corrosion problems. How is this control related to pollution prevention? Wasted energy is by definition pollution. Corrosion, if not caught in time, can lead to a major accident which results in significant localized pollution. Figure 3 is taken from a presentation by Don Ballard at the Laurance Reid Gas Conditioning Conference 3/86. It shows that in "rich" amine, total acid gas, H_2S plus CO_2, loadings of over 0.4 moles/mole amine result in corrosion. The "lean" amine loading should be less than 0.1 moles/mole amine. The reason the "rich" amine can have higher acid levels without causing corrosion problems is that the temperature of the "rich" stream is lower than that of the "lean" stream. Excess stripping of the "lean" amine or underloading of the "rich" amine results in wasted energy. Both the H_2S and CO_2 gas contribute to the total acid gas loading. Therefore, to properly control the loading level requires measurement of the amine concentration and the H_2S and CO_2 loading. Fortunately the amine concentration varies slowly which allows for laboratory measurement. However, the CO_2 and H_2S levels can change rapidly which calls for an on-line analyzer.

The schematic for an on-line H_2S-CO_2 analyzer for either the "rich" or "lean" stream is shown in Figure 4. The basic concept is very simple. A fixed quantity of the stream is injected into an acid bath. The strong acid,

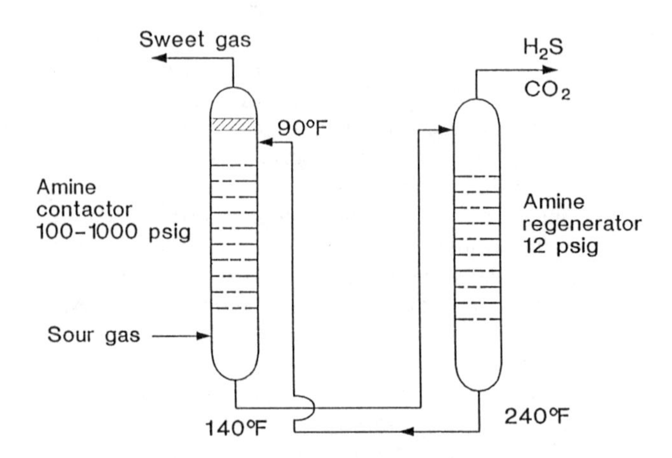

Figure 1. Simplified Amine Sweeting Unit.

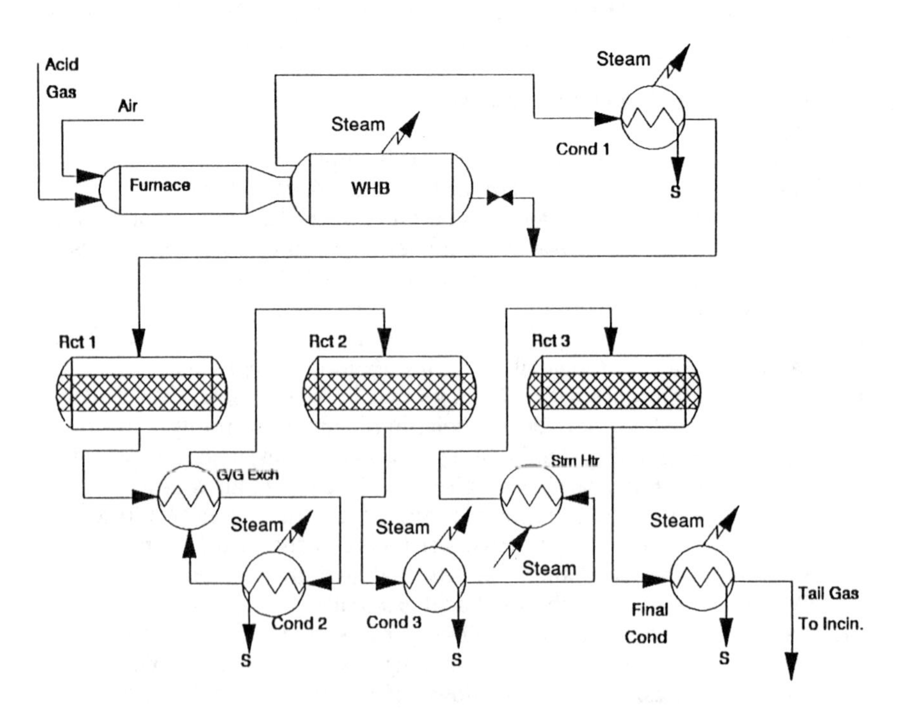

Figure 2. Typical Straight Through Claus Plant.

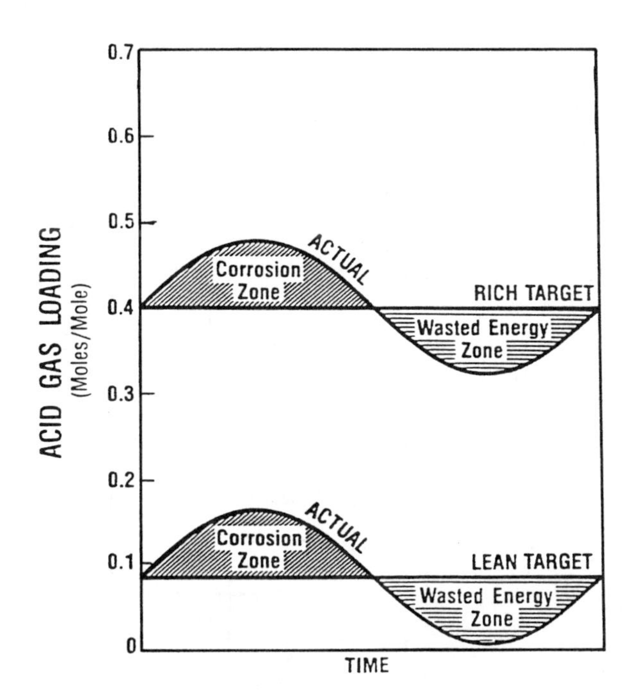

Figure 3. Target Loading Values.

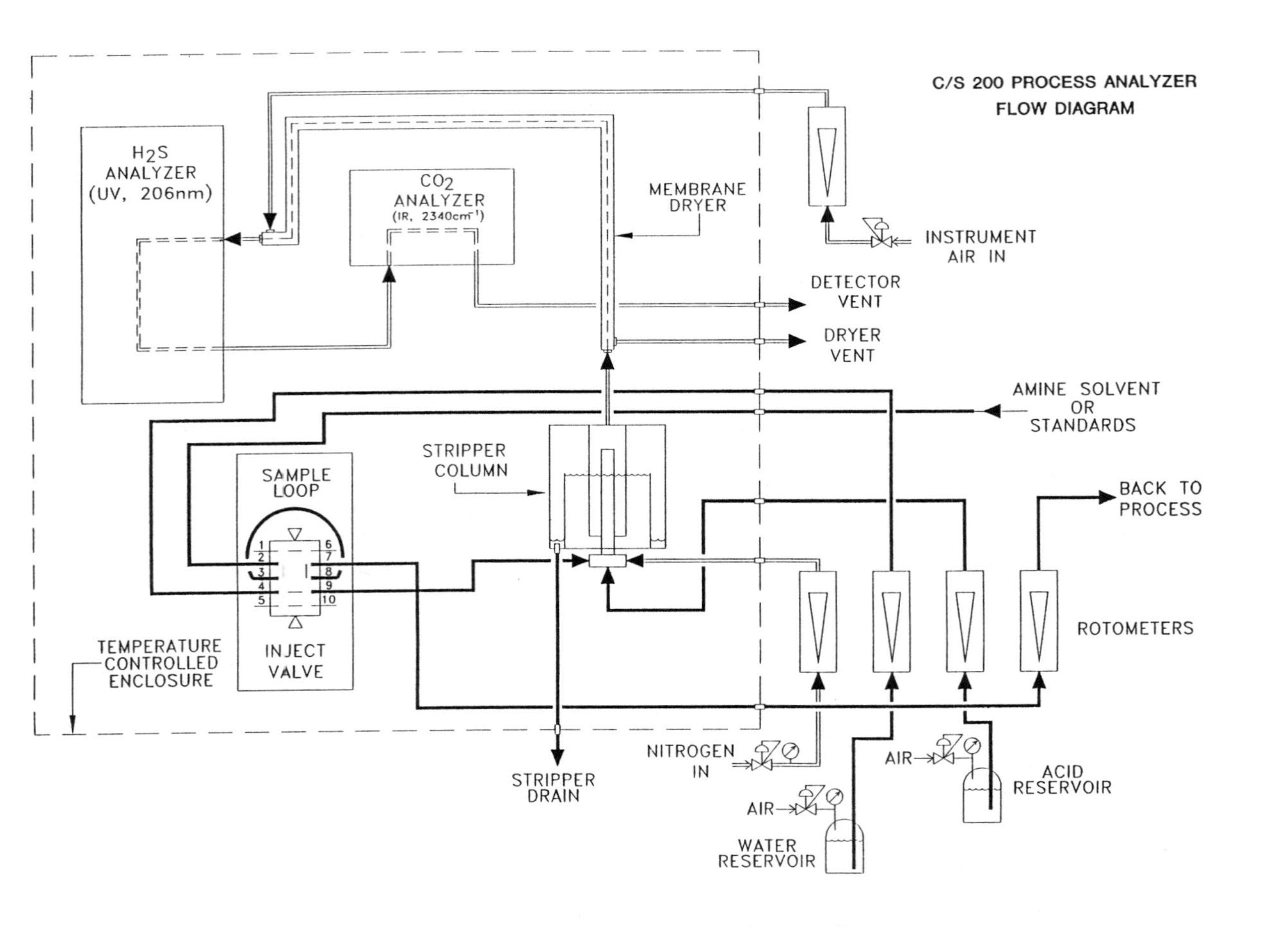

Figure 4. Eppendorf H_2S-CO_2-Amine Analyzer.

H_2SO_4, releases the weak acids, H_2S and CO_2, from the basic stream. N_2 transports the H_2S and CO_2 from the acid bath past a UV detector which determines the H_2S content of the gas and on IR detector which measures the CO_2 content. With proper calibration the H_2S and CO_2 concentration in the gas phase can be correlated with the H_2S and CO_2 content of the original liquid stream. While the overall concept is very simple, reducing the system to practice requires attention to many details and involves some nonobvious tricks which are the subject of a patent filing and will not be discussed further here.

We have had the analyzer on line for over two years in one location and ~1 year at a second location, both on "lean" amine streams. At both locations, the operating personnel were surprised at how rapidly the stream could change and how sensitive the loadings were to energy input. The analyzer has allowed for better control for the units.

RECEIVED April 30, 1992

Chapter 9

Surface Acoustic Wave Chemical Microsensors and Sensor Arrays for Industrial Process Control and Pollution Prevention

H. Wohltjen, N. L. Jarvis, and J. R. Lint

Microsensor Systems, Inc., 6800 Versar Center, Springfield, VA 22151

Surface Acoustic Wave (SAW) microsensors have many unique characteristics that make them ideal for monitoring gases and vapors associated with industrial chemical processes. Individual SAW devices are very small, rugged, inexpensive, and reliable. They can operate over a range of temperatures up to 50°C and provide electronic signals that can be readily integrated into computer networks for real time process control. They are sensitive to ppm levels of many chemical vapors, respond in seconds to changes in vapor concentration, and can be made selective for many different compounds or classes of compounds. The state of development of SAW technology and its application to process control for pollution prevention will be discussed. The use of arrays of SAW sensors for industrial monitoring will be emphasized.

It is accepted that the best approach to environmental protection is to prevent pollution at it's source, rather than repair environmental damage after the fact. Thus the Environmental Protection Agency is encouraging industry and the scientific community to explore new approaches for the monitoring and control of industrial processes. A new technology that shows considerable potential for this application is that of chemical microsensors. Chemical microsensors include a variety of chemically responsive devices, including surface acoustic wave (SAW) devices, organic and inorganic semiconductors, micro-electrochemical cells, chemfets, and other devices based on the use of solid-state electronic microfabrication technology. Even

0097–6156/92/0508–0086$06.00/0

though each type of microsensor has it's unique advantages and disadvantages, they all are very small, inexpensive, rugged, and have low power consumption. In addition to offering new analytical capabilities, chemical microsensors are also in the vanguard of attempts to reduce the cost of chemical information. Thus chemical microsensors should find increasing application as monitors of industrial processes where the excessive size, cost and power consumption of traditional sensors preclude their use.

One of the most promising of the chemical microsensors for chemical vapor monitoring is the Surface Acoustic Wave (SAW) device, which has been the focus of growing interest since the first studies reported in 1979 (1-14). The SAW device offers a simple, direct, and sensitive method for probing the composition of organic vapors.

SAW Sensor Technology

SAW Sensor Operating Principles. SAW devices are mechanically resonant structures whose resonance frequency is perturbed by the mass or elastic properties of materials in contact with the device surface. Rayleigh surface waves can be generated on very small polished chips of piezoelectric materials (e.g. quartz) on which an inter-digital electrode array is lithographically patterned (Figure 1). When the electrode is excited with a radio frequency voltage, a wave is generated that travels across the device surface until it is "received" by a second electrode. This Rayleigh wave has most of its energy constrained to the surface of the device and thus interacts very strongly with any material that is in contact with the surface. Changes in mass or mechanical modulus of a surface coating applied to the device produce corresponding changes in wave velocity.

The most common configuration for a SAW vapor/gas sensor is that of a delay line oscillator. In this configuration the device resonates at a frequency determined by the wave velocity and the electrode spacing. If the mass of the coating is altered, the resulting change in wave velocity can be measured as a shift in resonant frequency. SAW vapor sensors are similar to bulk wave piezoelectric crystal sensors, except they have the distinct advantages of substantially higher sensitivity, smaller size, greater ease of coating, uniform surface mass sensitivity, and improved ruggedness. Practical SAW sensors currently have active surface areas of a few square millimeters and resonance frequencies in the range of hundreds of MHz. However, SAW devices having total surface areas significantly less than a square millimeter are possible using modern microlithography.

SAW Sensor and Support Electronics. Most of the SAW vapor sensors reported in the literature employ two delay line oscillators fabricated side by side on the same chip, as shown in Figure 1, with one delay line used to monitor the selected chemical vapor and the other to act as a reference to compensate for changes in ambient temperature and pressure. The dual SAW devices can be mounted in a variety of electronic packages, depending upon the intended application. The 158 MHz SAW device shown in Figure 2 is bonded to a standard 1.5 cm diameter circular package that can be "plugged in" to a receiving fixture on a circuit board and is therefore easily replaceable. The size of the sensor package is determined by the size of the SAW sensor, thus a much smaller 300 MHz dual SAW sensor could be mounted in a package only half the diameter.

The SAW sensor, as shown in Figure 2, can be exposed directly to the industrial process environment, relying on diffusion to transport the various vapors to the sensor surface. Alternatively, the sensor packages can be covered with closely fitted lids, with vapor inlet and outlet fittings, so that vapor samples can be actively moved passed the sensor surfaces by a small pump, for more accurate, controlled sampling. A SAW sensor package with a sealed lid is an extremely rugged device and can easily withstand the shock and vibration associated with most field or industrial monitoring applications.

A typical SAW Vapor Sensor Module will generally contain a dual SAW sensor and all necessary support electronics. All components of a typical 158 MHz SAW Vapor Sensor will easily fit on a card approx. 3" x 2" x 0.5". This card is essentially a complete chemical vapor monitor. It requires only power to operate and provides a frequency difference, Δf, signal proportional to the challenge vapor concen-tration. The Δf can be easily measured with a frequency counter. Even though a circuit card of this size is conveniently small for many applications, it can be further miniaturized by arranging the discrete components in a more compact manner. Obviously a very large reduction in size could be achieved if the sensor and module were re-designed using hybrid microelectronic fabrication technology. By this approach the size of the entire module could be reduced to that of a digital watch. However, this would require an expensive engineering effort. Thus the size of a current SAW Vapor Sensor system is convenient for many industrial and field applications, yet there is considerable potential for system miniaturization to meet new and more demanding size and conformation requirements.

Role of Selective Coatings for SAW Devices. The operating frequency of a SAW device is very sensitive to changes in physical properties of the surrounding medium, e.g., density, elastic modulus, and viscosity. However, by themselves they are not inherently sensitive to the chemical properties of the surrounding medium. Thus a chemically selective thin film must be applied to a device surface (Figure 3) in order for it to exhibit selectivity as well as sensitivity to chemical vapors. The coatings my either adsorb or absorb specific vapors, and they may interact reversibly or irreversibly, depending upon their

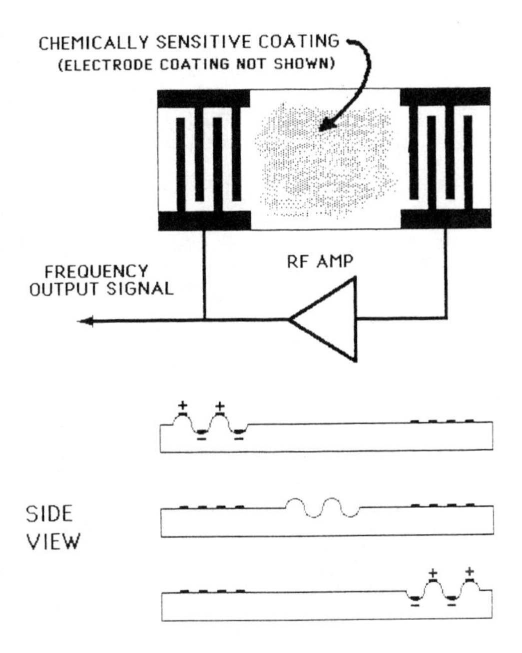

Figure 1. Operation of SAW chemical vapor sensor.

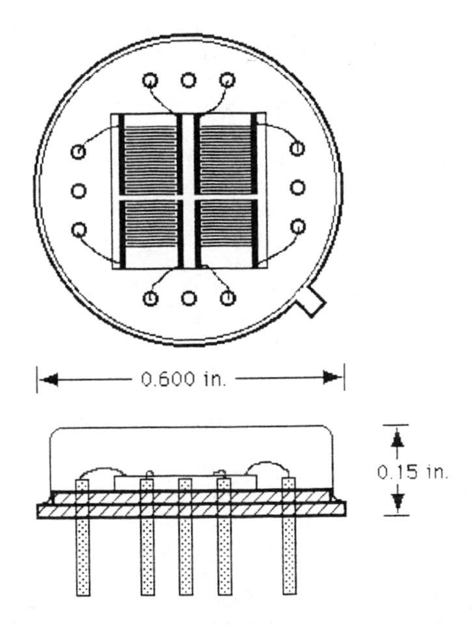

Figure 2. 158 MHz dual SAW device in circular package. T0–8 Header/0.600" diameter, 12 pin package (Part No. SD–158–A).

chemistry. There are two approaches to the development of chemically selective SAW coatings for industrially important vapors: (1) coatings that will selectively and reversibly absorb a selected vapor or gas by matching "solubility" characteristics (and thus be able to continuously monitor the concentration of a selected vapor); and (2) coatings that react chemically and irreversibly with a selected vapor or gas. The irreversible coatings are most suitable for dosimeter or cumulative exposure applications. SAW selectivities in excess of 1,000 to 1 for certain toxic chemical agents have been demonstrated using the "solubility" approach. Greater selectivities should be possible using chemically reactive coating/vapor (gas) combinations.

Mass Sensitivity of SAW Devices. A 158 MHz SAW oscillator having an active area of 8 mm^2 will give a frequency shift of about 365 Hz when perturbed by a surface mass change of 1 nanogram. This sensitivity is predicted theoretically and has been confirmed experimentally using successive depositions of Langmuir-Blodgett monolayers of known dimensions and mass. The same device exhibits a typical frequency "noise" of less than 15 Hz RMS over a 1 second measurement interval (i.e. 1 part in 10^7). Thus, the 1 nanogram mass change gives a signal to noise ratio of about 23 to 1.

It has recently been shown that SAW resonator devices have a distinct advantage over SAW oscillators as vapor sensors (15, 16). At a given frequency (e.g., 158 MHz) both will have the same sensitivity; however, a SAW oscillator will typically exhibit a frequency noise of about 15 Hz RMS over a one second measurement interval, whereas a SAW resonator will exhibit no more than 2 Hz RMS noise. Thus a SAW oscillator will provide a signal to noise of about 23 to 1 for a nanogram mass change, but a resonator will provide a signal to noise ratio closer to 180 to 1.

The SAW frequency change, Δf, produced by a change in mass will increase with the square of the unperturbed resonant frequency of the SAW device, f_o (1). Thus increases in sensitivity can be achieved by using SAW devices of higher frequency. For example, increasing the operating frequency from 158 MHz to approximately 300 MHz will increase the sensitivity by about a factor of 2. By using 300 MHz SAW resonator devices and new, high specificity coatings, it should be straightforward to detect sub-nanogram quantities of many vapors as they react with suitably coated SAW devices.

Performance of SAW Devices as Vapor Sensors

Effect of Temperature. SAW devices are inherently very sensitive to temperature, changing many Hertz per degree, depending on the resonance frequency. In many applications the temperature dependence of the device (as well as pressure dependence) can be sufficiently compensated by using a reference SAW device. Temperature drift of SAW devices can also be controlled by holding them at a constant temperature, for example, with a simple solid-state thermoelectric device. However, there is an effect of temperature than cannot be compensated and that is the effect of temperature on the partition of a vapor between the gas phase and the absorbed phase in a chemically selective SAW coating. As the temperature increases, proportionately less vapor will be absorbed in the coating and the observed mass loading will be less, for a given concentration. It was observed for dimethylmethylphosphonate vapor absorbing in a fluorinated polyol coating on a 158 MHz SAW device that as the temperature increased from 23°C to 42°C the observed frequency shift decreased by a factor of four.

This effect of temperature can have an impact on the potential use of the present state-of-the-art SAW devices in industrial processes. With the coating materials investigated to date SAW devices will be limited to applications where the ambient temperatures are below about 50-80°C, depending upon the specific vapor/coating combination and the concentration to be monitored. In practice, however, this limitation may often be overcome by drawing vapor from the high temperature environment and cooling it appropriately before measurement. At these temperatures it should be possible in many applications to place the sensors directly on-line. Future research may provide coatings that will allow the SAW sensors to operate at the higher temperatures of many industrial applications.

One other factor to consider in the use of SAW sensors at higher temperatures is the performance of the other elec-tronic components. The current SAW electronics module was designed to operate at relatively modest temperatures. Ideally it would be desirable to mount and operate the SAW sensors on extended leads, so that the electronics support package could be held at a lower temperature. This will require a careful re-design of the SAW electronic system.

Response Time of Coated SAW Devices. SAW devices respond rapidly to changes in vapor concentration, as shown in Figure 4 for the absorption dimethylmethylphosphonate by a FPOL coated 158 MHz device. It is seen that at a concentration of 1.0 ppm the SAW sensors respond immediately to the DMMP vapor . It should be noted that the equilibration (response) time in this case includes the time required for the vapor inlet lines and dead volume within the sensor package to equilibrate, as well as the SAW coating. The actual response time of the sensor itself is therefore very rapid, on the order of a few seconds. The time required for desorption of the vapor from the coating (when the vapor challenge is removed) is only slightly longer. The rapid response time of SAW sensors is an obvious advantage in the real time monitoring of many industrial processes.

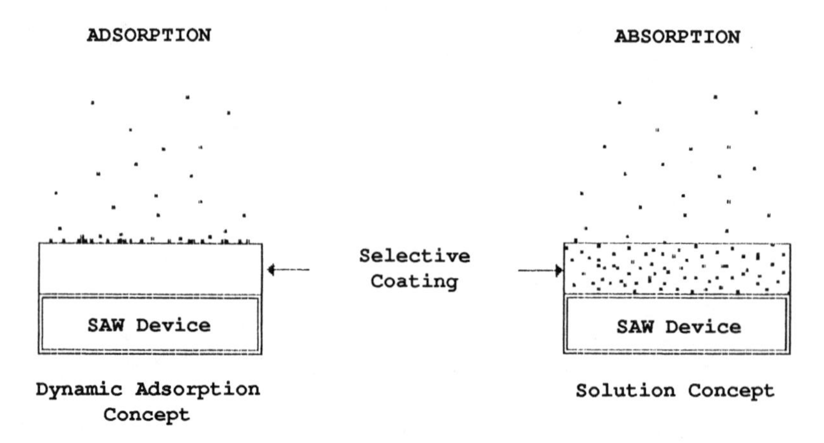

Figure 3. Vapor/coating interactions.

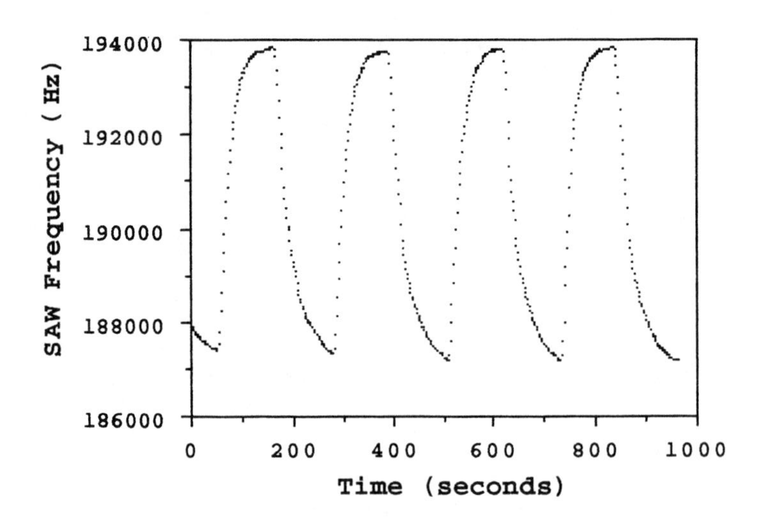

Figure 4. Response of FPOL coated SAW sensor to 1 ppm DMMP.

Selectivity of SAW Coatings. As indicated above, SAW coatings are generally selected for their preferential interaction with a specific chemical vapor or class of chemical vapors. An example of the selectivity of one coating for different chemical vapors is given in Figure 5. Figure 5 shows the response of a FPOL coated sensor to a variety of vapors in addition to a specific vapor it was designed to detect, dimethylacetamide (DMAC) (10). The sensor exhibits an order of magnitude greater sensitivity to DMAC than to any other vapor tested. For example, the selectivity ratio for DMAC vs tributylphosphate (the second largest response) is (358/11.9), or 30, while the selectivity ratio for DMAC vs water vapor is 4,475.

An example of a SAW coating carefully designed to respond irreversibly to a single vapor is poly(ethylene maleate) (PEM) which reacts specifically with cyclopentadiene to form a Diels-Alder adduct. The PEM coated SAW device was also exposed to a variety of other vapors to further demonstrate the selectivity of this film (6). Aliquots measuring 2,000 ppm of each vapor were prepared by addition of the volatile liquid into a sample flask containing the SAW sensor. The SAW frequency shift was recorded as a function of time. The flask was then flushed with clean air to remove the vapor sample. All samples investigated except cyclopentadiene gave relatively small initial responses, which leveled off rapidly after injection and returned to baseline as the vapor was replaced with clean air. Of the vapors studied, only cyclopentadiene gave a large, rapid, irreversible response.

The response of the PEM coated SAW sensor was reversible with all vapors except the cyclopentadiene, which gave a large non-reversible response due to a chemical reaction with the film, forming the Diels-Alder adduct. The behavior of the two coatings, FPOL and PEM, demonstrate the trade-off between vapor selectivity and SAW sensor reversibility. In general SAW sensors that respond reversibly will have coatings that interact physically with the vapors (e.g., the physical forces of attraction associated with solubility). SAW sensors that are highly selective, on the other hand, rely on coatings that react chemically with the vapor, and thus are not reversible.

This shows the potential for developing coatings for a range of applications, e.g., coatings that are reversible and have selectivity ratios up to 1,000 to 1 or so for specific vapors, as well as coatings that are irreversible and have selectivities in excess of 10,000 to 1. A level of selectivity of 1,000 to 1 will be useful for many applications, especially in processes where there are a limited number of known vapors and the vapor to be monitored is present in equal or higher concentration than potential interferents. However, there will be many other applications in which much higher selectivity is required as well as reversibility. There are several techniques that can be used to enhance selectivity of SAW sensors, besides changing the coating chemistry.

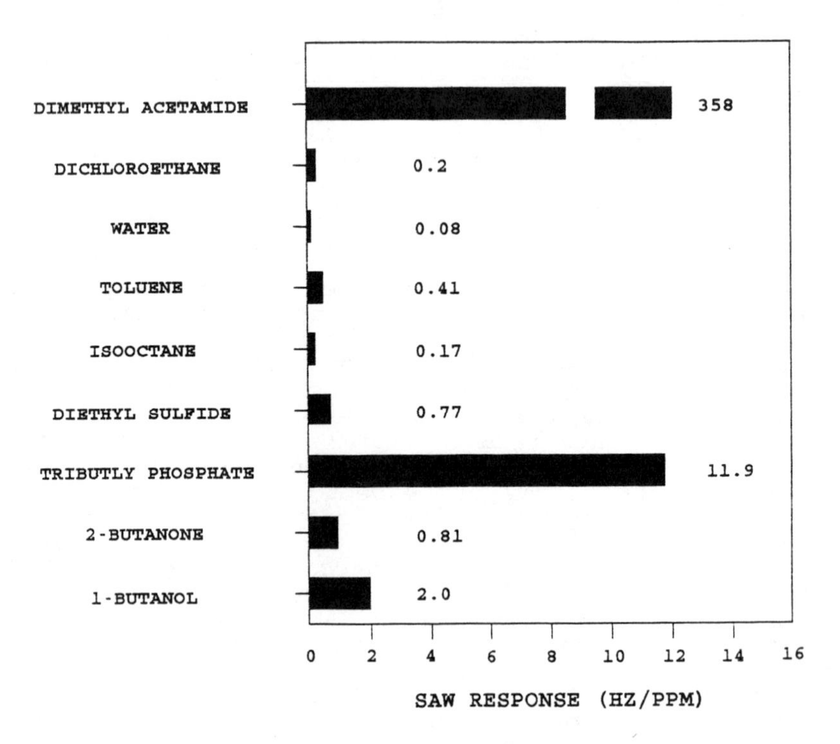

Figure 5. Vapor response of 290 MHz SAW device.

Techniques for Enhancing Sensitivity of SAW Sensors

SAW Sensor Arrays. It can be inferred from the above discussion that high selectivity for many analytes is difficult to achieve without the costly and time consuming development of coatings specific for those analytes. In addition, many applications demand that a number of vapors be detected or monitored simultaneously. SAW sensor arrays have therefore been developed to meet the more complex vapor monitoring applications. In the sensor array approach a number of sensors, each with a different chemically selective coating, are exposed simultaneously to the analyte vapors. Because of their different coating chemistries, each sensor will give a different response to the vapors present. The resulting pattern of responses from the array of sensors can then be analyzed using a variety of pattern recognition algorithms. Patterns of known vapor mixtures can normalized to eliminate the effect of vapor concentration and can be stored in a computer library for use in analyzing vapors associated with an industrial process. A key advantage of a sensor array is that it can usually detect and monitor a number of vapors far greater than the number of sensors in the array. In addition, since the array responds to a large number of vapors, the identification of new vapors can often be accomplished with the pattern recognition software.

A SAW array sensor system is by necessity more complex than a single sensor system. It will not only require additional sensors and electronics, but in order to expose all sensors simultaneously to the same vapor composition, and to periodically establish a clean air baseline, a number of additional system components will be required. These will include one or two onboard pumps, a charcoal or mole sieve trap, valves, and a microcomputer for data acquisition and system operation.

An example of the response patterns obtained from a 4 SAW sensor array are shown in Figure 6. Four 158 MHz SAW sensors were coated with films of poly(ethyleneimine), fluoropolyol, ethyl cellulose, and TENAX GC. The sensor array was then exposed to a variety of vapors. Each sensor responded to the several vapors to a different extent. The magnitude of the signal obtained from each sensor (i.e., channel A, B, C, and D) is plotted in histogram form. The resulting response patterns are significantly different from each other, thereby permitting easy "fingerprinting" of the vapor analytes. As expected, chemically similar compounds produce patterns that are similar but most often different enough to permit identification by the computer (e.g. methanol and propanol). The array will respond to mixtures of vapors and can "fingerprint" them, but determining the individual components of complex mixtures quantitatively is beyond the scope of the present state-of-the-art SAW microsensor technology.

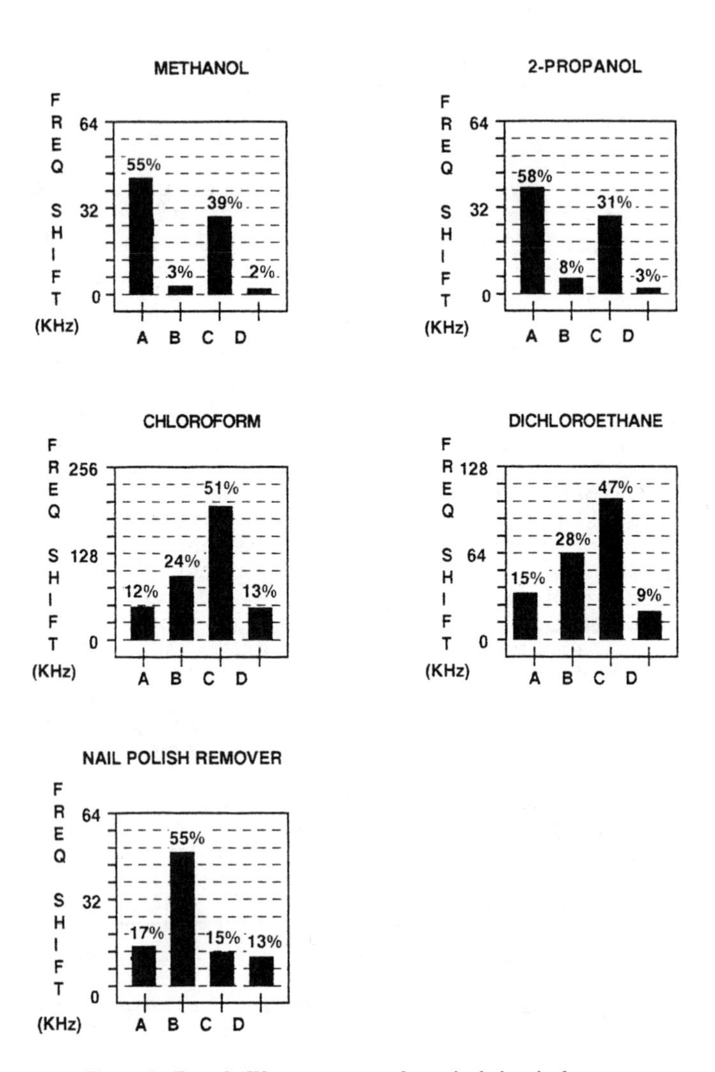

Figure 6. Four SAW array patterns for typical chemical vapors.

In another experiment, 4 SAW sensors with different surface coatings were exposed sequentially to room air and a series of smokes, including smoke from cotton, cigarettes, PVC, and motor oil. The sensor array gave distinctly different response patterns for each smoke. Thus SAW sensor arrays have the potential to identify the source of a smoke as well as detect it, provided it is from a single source, which is very often the case in the initial stage of a fire. The development of sophisticated pattern recognition algorithms to analyze simple mixtures is currently an area of active research (12,13).

SAW/GC Sensor Systems. Another approach to achieving enhanced selectivity for many analytes is to interface the SAW sensor with a simple GC column and select the specific components to be monitored based on column retention time. A commercially available instrument, the MSI-301 Organic Vapor Monitor, has been developed based on this principle. Air samples to be analyzed can be drawn into a sample loop with a small pump, or by syringe injection. The trapped vapor sample is then injected into the GC column. Sample injection, operation of the GC column, and data analysis are controlled by an on-board microcomputer. Once the chromatogram is completed, the microcomputer determines retention times and baseline corrected heights for all peaks. Peak heights can be converted to concentrations by refernce to calibration tables stored in memory.

Instruments such as the MSI-301 are ideally suited for continuous, unattended monitoring of the workplace, industrial and environmental sites for the presence of organic vapors. They are also well suited for the continuous monitoring of industrial chemical processes for quality control and for reducing and/or detecting and monitoring the accidental release of hazardous organic vapors into the environment.

SAW Sensors for Dosimeter Applications

There are applications in industry where it is necessary to determine the cumulative concentration of a given chemical occuring (or accidently released) at a specific site. As indicated above, the primary challenge for for such an application would be the development of coatings for the SAW devices that are sensitive and irreversible to the specific toxic or hazardous vapors and will give an accurate, cumulative, real-time measurement of the vapor being monitored. It is also a challenge to develop SAW coatings that will respond to inorganic or fixed toxic gases, similar to the response of PEM coatings to cylcopentadiene. In a recent study irreversible SAW coatings were proposed for monitoring a number of such gases including NH_3 and HCl (14). The coatings evaluated for NH_3 and HCl were $CoCl_2$, and polyvinylpyridine, respectively. Results of the exposure of coated SAW sensors to these gases are given in the Fig. 7 (a) and (b).

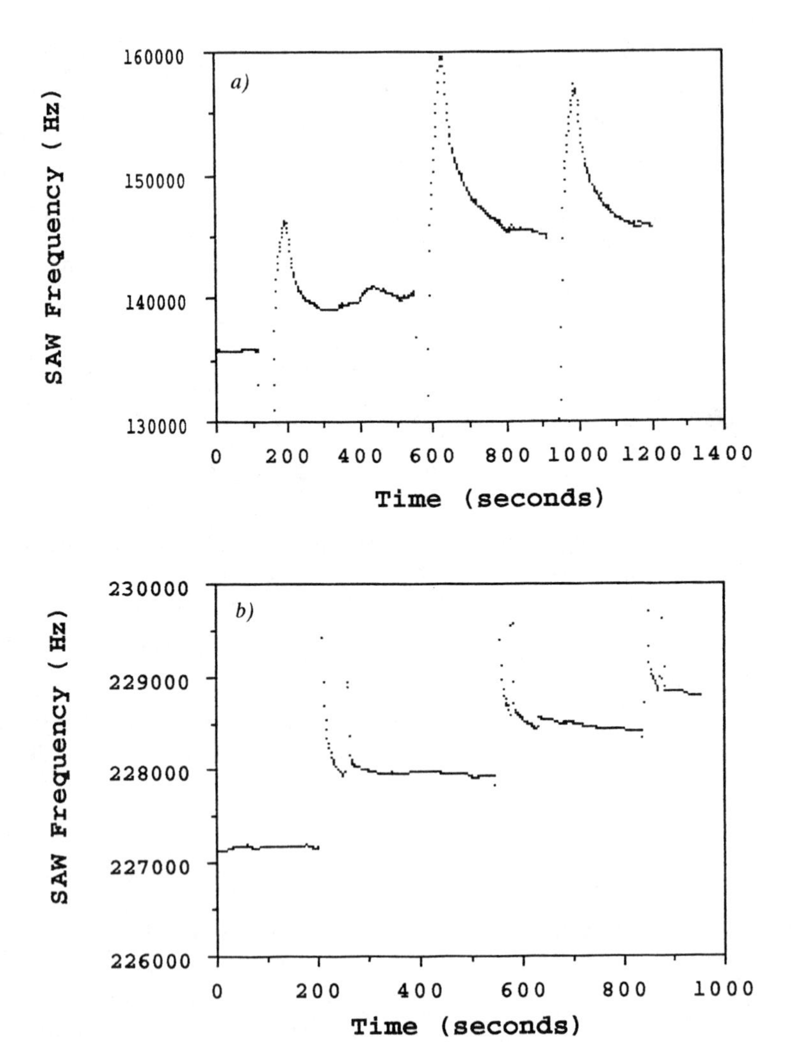

Figure 7. Frequency shift vs. time for exposure of $CoCl_2$ coated SAW devices to 20 ppm NH_3 for 20 sec (a) and PVP coated SAW devices to 20 ppm HCl for 20 sec (b).

The plots of SAW frequency, Δf, with time in each case show a stepwise increase in Δf with each exposure of the coating to 20 ppm of the specific vapor for 20 seconds. The sensitivity of each coating was better than 1 Hz/ppm/sec, sufficient in every case to detect and monitor these gases at concentrations well below their OSHA toxic exposure limits. Thus SAW sensor do indeed have potential as dosimeters for a variety of toxic gases in the environs of an industrial operation. The SAW sensors and much of the support electronics and hardware are readily available to develop this technology.

SAW Sensors for Vapor Monitoring of Industrial Processes

Solid-state SAW chemical microsensors have many physical and performance characteristics that make them ideal as in-situ sensors for the monitoring and control of many industrial processes. Such sensors could be used to optimize chemical processes as well as control the release of hazardous chemicals to the environment. They are very small, rugged, and inexpensive devices, and can be sensitive to ppm (or lower) levels of many organic vapors of industrial importance, such as benzene, ethylene oxide, halogenated hydrocarbons, etc. In addition, individual SAW sensors can be made highly selective for many compounds or classes of compounds by the selection or synthesis of appropriate coating materials for the SAW sensors. The present SAW sensors can operate at temperatures to 50°C or so, and they provide an electronic signal that can be readily integrated into computer networks for real time monitoring at multiple process sites and data for process control.

Single SAW sensors can be very useful for monitoring the concentration of a single chemical vapor, or class of vapors, in an environment where there are few interferents, however in industrial processes there are likely to be considerably more complex chemical environments. The use of multiple sensor arrays would be a truly novel approach to the real time monitoring of complex industrial chemical processes, including feed stocks and potential pollution sources. An array of sensors, each with a different vapor sensitive coating, would provide chemical "fingerprint" patterns that could be used as the basis for identifying the chemical vapors in a mixture as well as their relative concentrations. Thus for a given application a SAW sensor array could be "trained" using a simple pattern recognition algorithm and a computer, to identify and monitor specific vapor combinations. A SAW array could determine when the combination of chemicals in an industrial process are within pre-set limits and provide a signal whenever those limits are exceeded. Signals could also be provided to a process control computer network so that necessary concentration adjustments could be made to keep the process chemicals within the desired concentration limits.

A significant advantage of an array system is that it can in fact easily identify a number of chemical vapors in excess of the number of sensors in the array. Thus a SAW array could detect the appearance of an unknown or undesirable chemical in an industrial process, even though it may not be able to specifically identify it.

From Figure 8 it is apparent that SAW sensors and SAW Sensor Arrays can be used to monitor chemical vapors in various steps and locations in an industrial process. They can monitor chemical vapors associated with feed stocks, or reactants, being introduced into an industrial process, as well as monitor the process itself. SAW sensors could also monitor the products of a process for quality control and for the presence of potential environmental pollutants. SAW sensors systems could readily be engineered to monitor stack gases as well, provided gas samples can be sufficiently reduced in temperature. SAW instrumentation is currently available for monitoring the efficiency (and/or breakthrough) of air purification (filter) systems used to remove hazardous materials from air surrounding industrial processes. Finally, there are now commercially available SAW based chemical vapor monitors designed especially for the continuous, long term perimeter monitoring of industrial sites.

The output signals from SAW sensors and sensor arrays used to monitor any or all aspects of an industrial process, can be readily integrated into a computer network to provide real time information on the performance or status of the operation. Such information can provide information to plant operators for the effective automation of industiral process and at the same time assure that pollution is being prevented at the source.

SAW sensors and SAW based vapor monitors, as described above, are now commercially available and are being evaluated for a number of industrial applications; however, the details of their application and performance tend to be closely held proprietary information by the companies involved. Because the technology is so new, there are still some technical limitations that must be addressed. First, the use of SAW sensors as chemical vapor monitors is dependent upon the availability of chemically specific coatings for the intended application. As new and more effective coatings become available through research and development, SAW devices will find a wider range of applications. Second, SAW sensors are presently limited to monitoring applications at temperatures below about 50°C. At higher temperatures the present organic based coatings lose sensitivity and, depending upon their composition, can become chemically unstable. In addition, new surface acoustic wave sensor and support electronic designs may be required for higher temperature use.

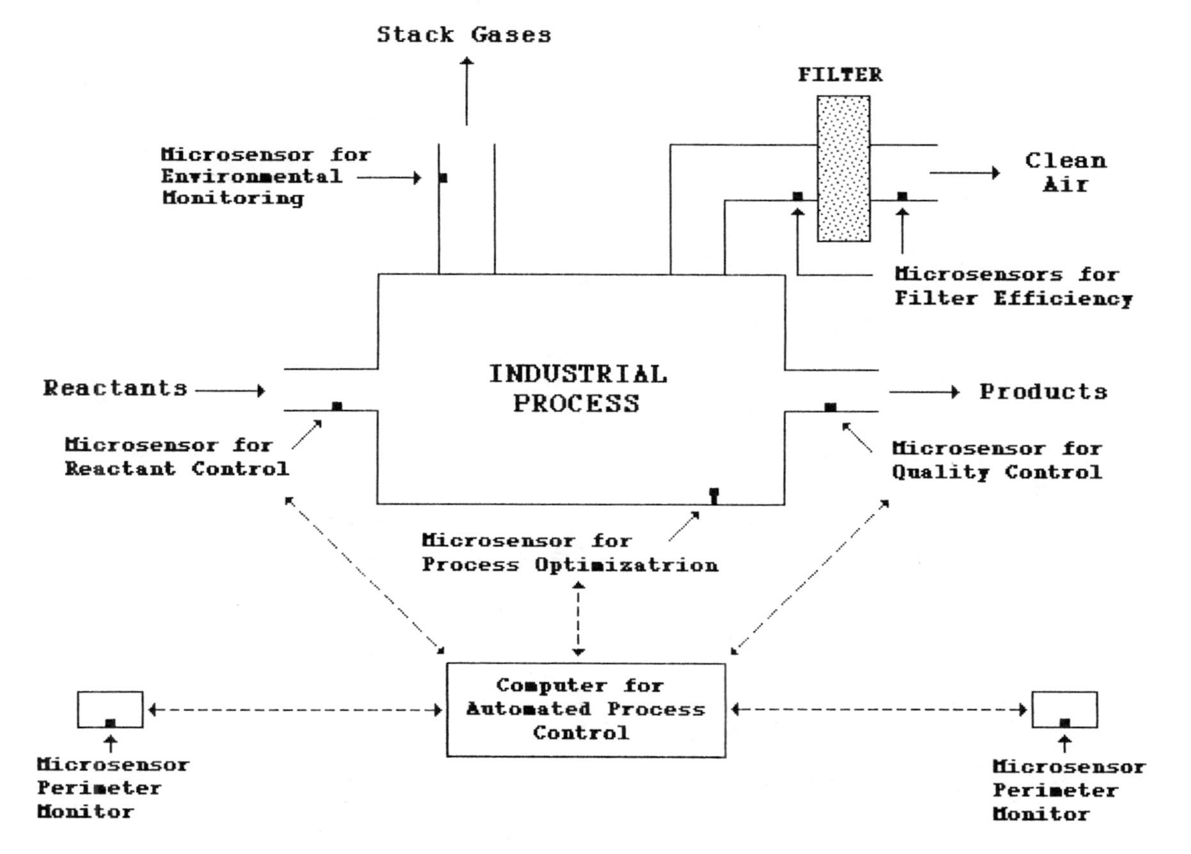

Figure 8. Vapor monitoring of industrial processes.

Literature Cited

1. Wohltjen, H.; Dessy, R. E. Anal. Chem. 1979, 5 (9), pp.1458-1475
2. Bryant, A.; Lee, D. L.; Vetelino, J. F. Proc. IEEE Ultrason. Symp. 1981, pp. 171-174
3. Chung, C. T.; White, R. M.; Bernstein, J. J. 1982, Electron. Dev. Lett., EDL-3 (6), pp. 145-148
4. D'Amico, A.; Palma, A.; Verona, E. Sensors and Actuators 1984, pp. 307-325
5. Wohltjen, H.; Sensors and Actuators 1984, pp. 307-324
6. Snow A. ; Wohltjen H. Anal. Chem., 1984, 56 (8), pp. 1411-1416
7. Martin, S. J.; Schweizer, K. S.; Schwartz, S. S.; Gunshor, R. L. Proc. IEEE Ultrason. Symp., 1984
8. Barendsz, A. W.; et al.Proc IEEE Ultrason. Symp., 1984.
9. Wohltjen, H.; Snow, A. W.; Barger, W. R.; Ballantine, D. S.IEEE Trans. on Ultrasonics, Ferroelectrics, and Frequency Control 1987, UFFC-34 (2), pp. 172-178
10. Wohltjen, H.; Ballantine, D. S. Jr.; Jarvis, N. L. ACS Symposium Series No. 403 1989, pp. 157-175
11. Ballantine, D. L. Jr.; Wohltjen, H. Anal. Chem. 1989, 61 (11) pp.704-713.
12. Ballantine, D. S. Jr.; Rose, S. L.; Grate, J. W.; Wohltjen, H. Anal. Chem. 1986, 58, pp. 3058-3066
13. Rose-Pehrsson, S. L.; Grate, J. W.; Ballantine, D. S. Jr.; Jurs, P. C. Anal. Chem. 1988, 60, pp. 2801-2811.
14. Wohltjen, H.; Jarvis, N. L.; Lint, J. R. Proceedings of the Second International Symposium on "Field Screening Methods for Hazardous Wastes and Toxic Chemicals", Las Vegas, Nevada, Feb 12 -14, 1991, pp. 73-84
15. Bowers, W. D.; Chuan, R. L.; Duong, T. M. Rev. Sci. Instrum.1991, 62, pp.1624-1629
16. Grate, J. W.; Klusty, M.; Naval Research Laboratory Memorandum Report 6829, 1991

RECEIVED May 27, 1992

Chapter 10

The Electrochemical Sensor

One Solution for Pollution

J. R. Stetter[1], W. R. Penrose[1], G. J. Maclay[1], W. J. Buttner[1], M. W. Findlay[1], Z. Cao[2], L. J. Luskus[3], and J. D. Mulik[4]

[1]Transducer Research, Inc., 999 Chicago Avenue, Naperville, IL 60540
[2]Department of Chemistry, Illinois Institute of Technology, Chicago, IL 60616
[3]Armstrong Laboratory, Sustained Operations Branch, Brooks AFB, TX 78235–5301
[4]U.S. Environmental Protection Agency, Atmospheric Research and Exposure Assessment Laboratory, Research Triangle Park, NC 27711

Electrochemical gas sensors represent an expanding field of analytical chemistry. They are employed in instruments for chemical detection, and a wide variety of both portable and fixed-site toxic gas monitors exist today. These sensors have also found applications in monitoring industrial atmospheres, medical gases, process streams, and environmental pollution. Gas chromatographic instrumentation, ventilation controllers, energy management systems, automobile exhaust analyzers, safety alarms, personal exposure monitors, and analytical laboratory gas analyzers also utilize electrochemical sensors.

This is a chapter about electrochemical sensors and their contribution to Pollution Prevention, the new Environmental Protection Agency (EPA) theme. Recent research has shown that human health is not being protected to a large extent by the present strategies employed by the EPA. Several years ago the Science Advisory Board of the EPA recommended a focus on development of real-time instruments and methods for measurement of human exposure. This strategy has yet to be implemented although limited progress has been made (*1,2*).

Examples of Electroanalytical Solutions to Pollution

Amperometric gas sensors for Ozone, NO_2, and CO. As part of an EPA project, a real-time instrument for measuring human exposure to toxic gas was designed and developed. The instrument weighed less than three pounds. This instrument was capable of monitoring for NO_2, O_3, or CO at parts per billion (ppb) levels and contained a fully functional data logger, event marker,

0097–6156/92/0508–0103$06.00/0

temperature sensor and relative humidity sensor, and had room for additional functions and sensors.

The instrument employed an amperometric sensor and some very novel design ideas that extended the capability of the typical amperometric systems from the parts per million (ppm) range into the ppb range.

Amperometric gas sensors are lightweight, rugged, low power, and contain no consumable reagents. This makes them ideal for field measurements. Even though limitations of selectivity and sensitivity exist, the amperometric sensor performance is often sufficient for many human exposure applications. The attractive logistical properties and the many successful applications in medical and industrial environments make this approach worthy of further development.

In an amperometric sensor, the analytical signal is an electric current produced from the oxidation or reduction of the analyte. For example, a schematic diagram of electrochemical sensor for CO is shown in Figure 1. When an analyte, CO, is introduced into the sensor, the CO is oxidized to carbon dioxide ($CO + H_2O \dashrightarrow CO_2 + 2H^+ + 2e$) at the sensing electrode. Electrons are transferred from the analyte to the electrode in the electrochemical reaction. A reduction reaction of oxygen to water ($1/2O_2 + 2H^+ + 2e \dashrightarrow H_2O$) must occur at another electrode (i.e, transfer of electrons from the electrode to solution) to maintain the electrical change balance in the cell. Charge is transported electronically in the electrodes and ionically in the electrolyte. The sensing electrodes are typically made of a Teflon-bonded noble metal composite material. The porous Teflon membranes allow for ready diffusion of analyte vapor but still provide an efficient barrier to prevent electrolyte leakage. The entire sensor is thereby a simple integral system of parts requiring no maintenance and no consumed reagents.

We have previously described methods for measuring nitrogen dioxide (*1*) and ozone (*2*)with a sensitivity at the 10-20 ppb level. These sensors are three-electrode electrochemical cells with working electrodes of gold, vapor-deposited on porous Teflon and operated at a potential of -100 mV versus a platinum-air reference electrode. These amperometric sensors were installed in the Transducer Research Model 2001 Gas Analysis System. In this system, a miniature pump was used to draw the sample gas into the analyzer. The gas concentration data, along with temperature and timing data, were acquired and stored by a powerful on-board microcomputer, which also managed instrument functions. Acquired data could be transferred to a personal computer for analysis. Signal outputs were expressed as microamperes per part-per-million of the relevant gas.

The instrument was engineered to not respond to CO and other common atmospheric gases occurred. Selectivity to nitrogen dioxide (NO_2) and ozone (O_3) was controlled with chemical filters connected to the sample and reference inputs of the 2001 instrument. For ozone analysis, the sample gas was connected directly to the sample inlet. The ambient air was passed through a proprietary ozone-removal filter before being admitted to the reference inlet. The difference between reference inlet and sample inlet is the signal due to the toxic gas, O_3 in this case. This filter removed less than 5% of the NO_2.

For NO_2 analysis, the zero filter contained a combination of activated carbon and Purafil (potassium permanganate on alumina; Purafil, Inc., Doraville,

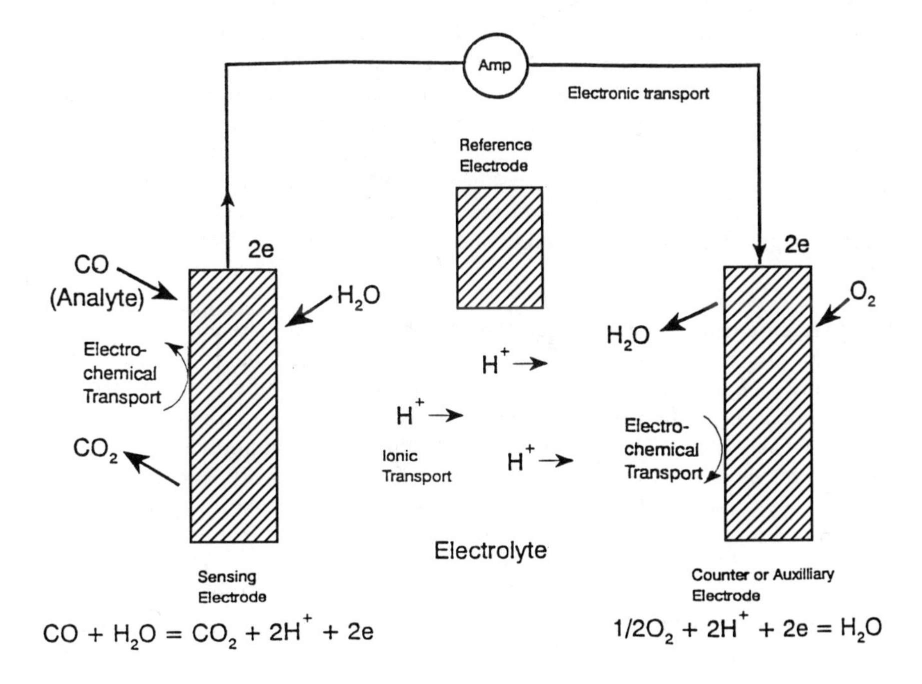

Figure 1. Schematic functional diagram of an amperometric CO sensor with catalytic electrodes illustrating the three mechanisms by which charge is transported within the sensor.

GA). A recent report (*3*) has indicated that this sensor can also be used for very sensitive analysis of nitrous acid vapor (HONO).

Ozone was generated with a Thermo Environmental Corporation Model 565 Ozone Calibrator. The output was passed directly into the analytical system without storage. Nitrogen dioxide samples were generated by diluting a standard mixture of 49 ppm NO_2 with air using a pump and flowmeter. Diluted samples could be stored for a few hours in Tedlar bags.

A series of ozone and nitrogen dioxide mixtures (in room air) with differing concentrations were analyzed, 90 ppb ozone, 90 ppb ozone with 125 ppb NO_2, 190 ppb ozone with 125 ppb NO_2, 190 ppb ozone, and 90 ppb ozone. All measurements were made at 25°C at a gas flow of 400 cm^3/min. The length of the sampling cycle is 30 minutes. The instrument will sample for 25 minutes, then draw the air through the filter for 5 minutes. The results shown in Figure 2 illustrate that the sensor signal is proportional to the ozone concentration and quite repeatable even in the presence of NO_2. This instrument is now commercially available and has been used to measure the O_3 emission from copy machines, personal exposure to NO_2, and NO_2/HONO concentrations in indoor air.

Solid-state Sensor for Chlorinated Hydrocarbons. Solid-state chemical sensing techniques for continuous real-time analysis have attracted intense interest in the last decade (*4,5*). Chlorinated hydrocarbons are potentially toxic vapors that commonly occur in the workplace, environment, and process streams. Recently, the EPA has added 25 organic chemicals to the list of compounds regulated as toxic in wastes under the Resource Conservation & Recovery Act (*6*). Sixteen of these are chlorinated hydrocarbons. Meanwhile, atmospheric scientists have determined that chlorine from chlorofluorocarbons (CFCs) is the primary factor in the seasonal loss of stratospheric ozone (*7*). EPA has already issued rules under the Montreal protocol reducing CFCs production and importation by at least 50% by 1998 (*8*). The increased controls on the use and production of CFCs will require high performance, low cost monitors for regulation, compliance, and safety applications.

On-line accurate process monitoring as well as field screening methods for chlorinated hydrocarbons can provide a low-cost alternative to the available higher cost laboratory methods for these compounds. Membrane technology has developed significantly. Combining membranes with chemical sensors allows the development of new in situ detection and quantification methods for pollutants and trace chemicals in waste water. Recent reports (9,10,11) illustrate the feasibility of a rugged permeable membrane in combination with the low cost, small, selective, solid-state sensor for on-line and real-time analysis of chlorinated hydrocarbons (R-Cl) in an aqueous phase or in a vapor phase.

This solid-state sensor was prepared by sintering a mixture of rare earth compounds. Two electrodes and a heater are embedded in the material. A dc potential of about 3 V is maintained across the electrodes and the output signal is the conductance change of the rare earth material. In the absence of a chlorinated hydrocarbon vapor, the conductance (G_0) between the two electrodes is very small. But in the presence of a chlorinated hydrocarbon vapor, the conductance (G) increases significantly. The relative conductance change (G/G_0) in the presence of chlorinated hydrocarbon vapors is the analytical "signal" from this sensor, i.e., the conductance ratio is a function of the concentration of the chlorinated compound vapor present.

Since this solid-state sensor is a gas sensor, it is necessary to convert an aqueous sample containing the analyte into a vapor sample. A permeation apparatus was assembled to evaluate the sensor. Figure 3 presents a block diagram of the experimental apparatus for the measurement of the concentration of chlorinated hydrocarbons in water. The carrier gas (typically air) is pumped through the inside of the semipermeable tubing to the sensor. When the permeable tubing is submerged in the water, organic compounds permeate through the silicone wall and are picked up in the carrier gas. The gas sensor can detect the chlorinated organic vapors and the sample flow is maintained at above 170 cm^3/min during analysis. A sample of 10 ppm trichloroethylene in water was measured using this system. The result shown in Figure 4 illustrates that the chlorinated hydrocarbon sensor and permeator system has a high sensitivity and selectivity and can be used for the determination of chlorinated hydrocarbons in aqueous samples. A detection limit for the trichloroethylene well below 1 ppm is observed even without preconcentration (4,11). Using a pre-concentrator would allow the sensitivity below 1 ppb for this system.

This sensor has a high sensitivity and good selectivity to chlorinated organic vapors such as chlorobenzene, trichloroethylene, dichloromethane, chloroform, and even the warfare agent simulant, chloroethyl ethyl sulfide. Yet it is not sensitive to commonly occurring gases CO, O_2, NO_2, hydrocarbon vapors such as hexane, benzene, and phenol (*4*). Since this sensor has virtually no interference from water vapor and O_2, it may provide a more robust portable and lower cost alternative to EPA methods that now use the Hall or electron capture detectors.

Sensor Arrays. Electrochemical sensors respond to a limited number of chemicals, and each one responds with limited selectivity. One way to overcome these limitations, or at least to improve the sensor's capabilities, is to construct sensor arrays (*12*). The device, called the Chemical Parameter Spectrometer (CPS-100), can provide a "fingerprint" of a sample gas. In this array, catalytic, hot-wire filaments are used in conjunction with four electrochemical sensors. The filament promotes oxidation, the use of heat to oxidize the gas in air, so that the electrochemical sensors can sense the filament products. The sensor's working electrodes, some gold and some platinum, are set at different electrical potentials to give different selective responses to various chemical vapors. The different responses provided by the sensors are used to generate a fingerprint that is representative of the sample composition and concentration.

Four sets of readings from each of the four sensors produce sixteen values that provide a "fingerprint" of the sample gas. This fingerprint is a reflection of the type and amount of a substance passing through the sensors. About 100 compounds have been tested, and each has yielded a unique combination of 16 readings. These compounds include sulfur-, nitrogen-, oxygen and carbon-containing organic and inorganic vapors. They also include room-temperature gases like carbon monoxide (CO), nitric oxide (NO), nitrogen dioxide (NO_2), hydrogen sulfide (H_2S), sulfur dioxide (SO_2) and hydrogen cyanide (HCN). The normal constituents of clean air (oxygen, nitrogen, carbon dioxide and argon) produce a negligible response.

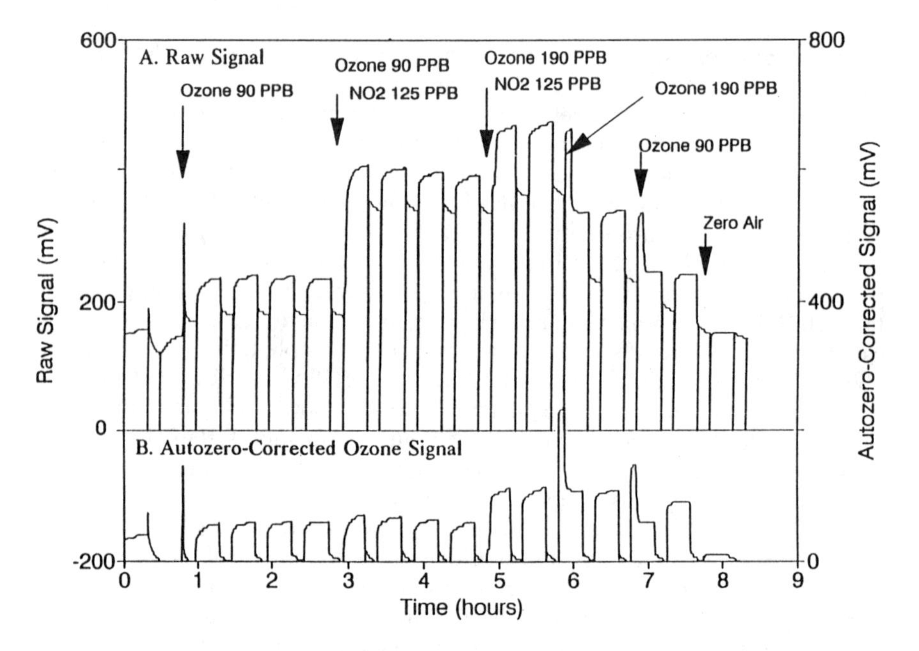

Figure 2. A. Signal vs time for a series of NO_2 + O_3 + Air mixtures. B. Signal from difference between sample and reference inputs with ozone filter on reference input. The reference input is sampled 5 minutes and the sample input 25 minutes in each cycle.

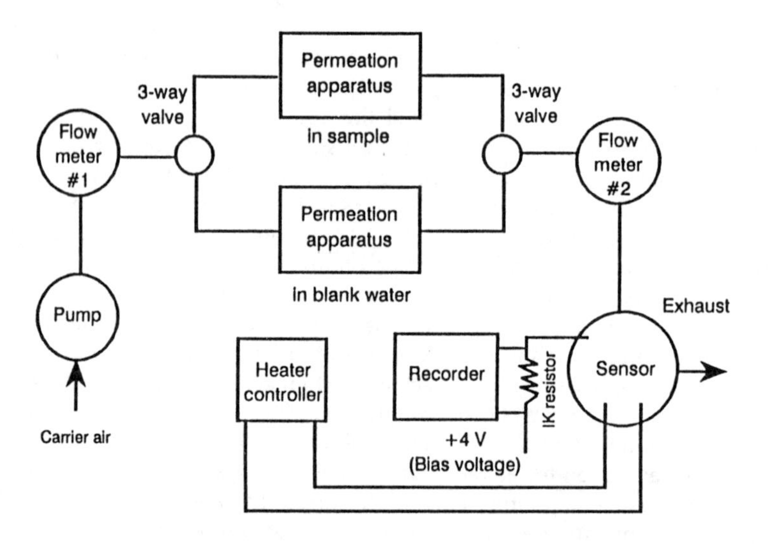

Figure 3. Experimental apparatus for selective analysis of aqueous chlorinated hydrocarbons using a gas sensor.

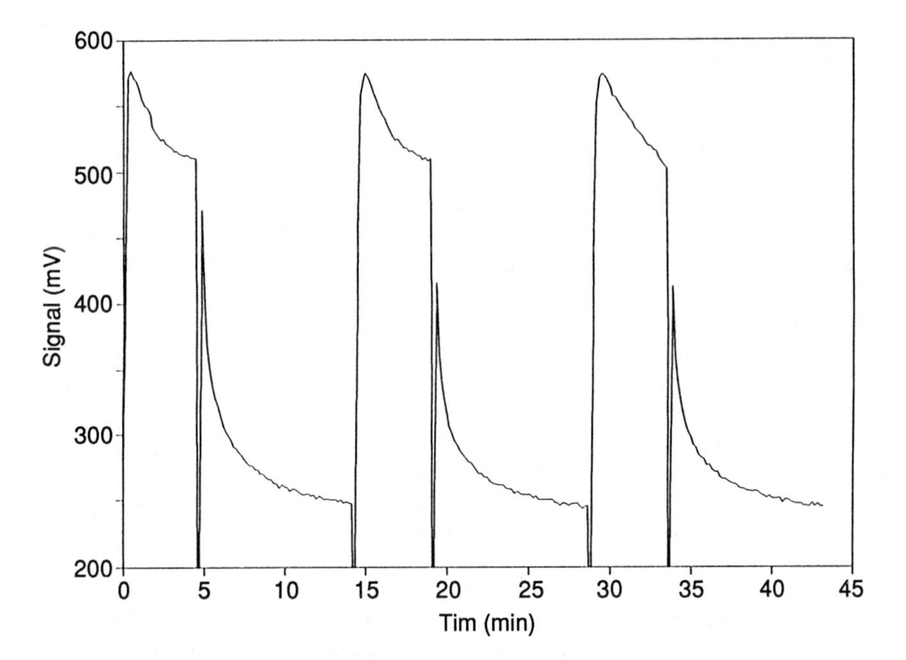

Figure 4. Responses of the chlorinated hydrocarbon sensor and permeator system to 10 ppm trichloroethylene in water.

Four sets of sensor readings are taken. The first is taken by exposing the test gas to a platinum filament heated to 900°C (about 1,650 F) before the gas is drawn past the sensors. The second set of readings is taken after exposing the gas to the filament at 750°C. A third set is taken after exposure to the filament at about 600°C and a final set of data is acquired with the filaments off. Sensor readings are normalized (divided by the strongest response to eliminate the concentration dependence) and compared against a preprogrammed library of known responses. The closet match is determined and the result is shown on a liquid crystal display, along with the concentration of the identified gas. The entire operation is controlled by a microprocessor and takes about four minutes.

One applications of this sensor-array instrument is for detecting and classifying grain odors (13,14). The United States Department of Agriculture, specifically the Federal Grain Inspection Service, has the task of inspecting grain quality for several parameters, including odor. Currently, inspectors grade the grain odor by placing their face directly in a sample and inhaling lightly. The perceived odor is classified as Good, Sour, Musty, and COFO (Commercially Objectionable Foreign Odors), with additional notice if insect or pesticide odors are detected. While olfaction is extremely sensitive, it is a very subjective process, depending upon the inspector and the conditions as well as the odor and its strength. It is often the case that several inspectors may classify the same sample differently, or even disagree whether a foreign odor is present. Complications arise in part because an "odor" is not a simple chemical compound but a very complex mixture of odorous chemicals.

A sensor-array-based instrument can detect and identify gases using patterns of responses rather than analysis of the individual compounds. In principle, a sample elicits a response from a multiple number of sensors. Each sensor response provides a "channel" of information. The channels of data are analyzed as a group, or "pattern". If the grain odor contains detectable compounds, and the concentrations of these are related to the odors, and the patterns are different for the different odors, then the sensors will produce patterns that can be correlated to both the identity and the intensity of the odor.

Figure 5 shows the CPS-100 data, in histogram form, for wheat samples (one "sour" (S3) and one "insect" (I3) odor). Although not so dramatic as the extremes in the different sorghum odor samples (Figure 6), there are visible differences between the patterns for different odors. Thus, different odors produce different patterns (a requirement of the pattern recognition scheme) and there is some evidence that the patterns are related to the different odors specifically (i.e., a pattern for "good" could not be the same as a "sour", but could be different from another "good" because of differences in the vapor matrix). These results indicate that the difference in patterns created by the sensor array are visibly obvious. Indeed, the array can "smell" the difference in grain odors. The results of this work clearly indicated that the concept of a portable low-cost sensor array based instrument with an adaptive learning algorithm (neural network) to detect and identify grain odors is possible (*13*).

A second example is the identification or classification of unknown organic waste. A CPS-100 was demonstrated at an EPA site in 1989. The waste site contained primarily paint thinners and solvents, so it was expected that hydrocarbons and C-O compounds would be encountered. A two sensor array

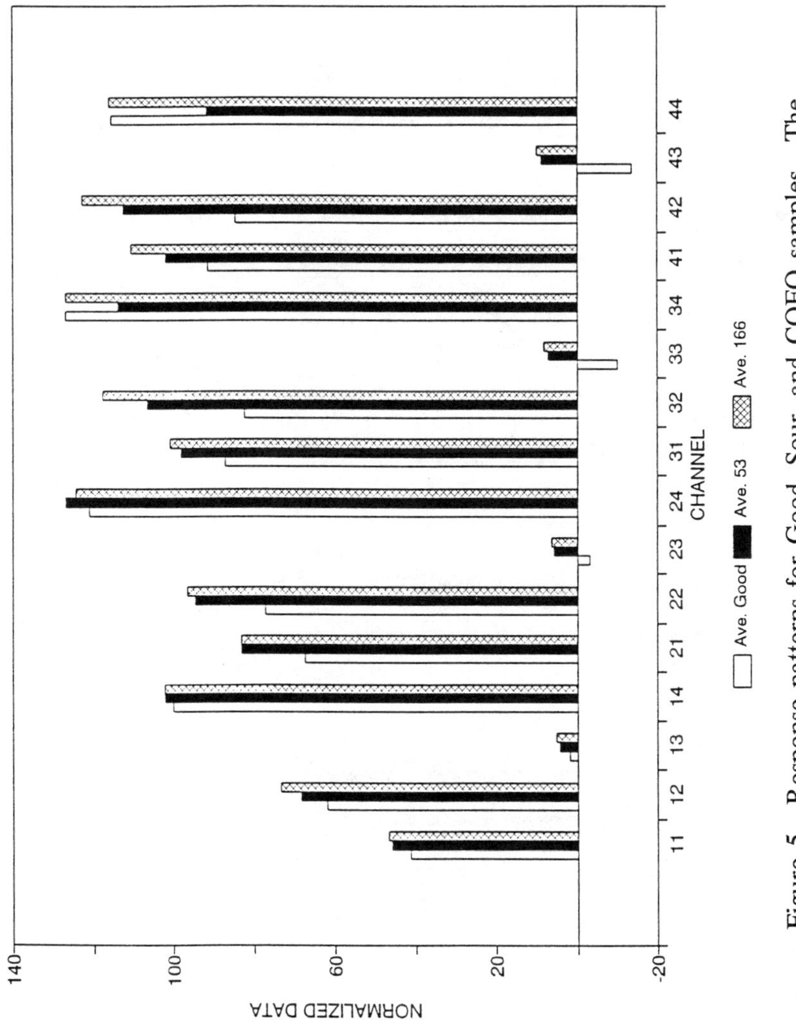

Figure 5. Response patterns for Good, Sour, and COFO samples. The pattern shown for "Good" is the average of the two sound sorghum samples analyzed.

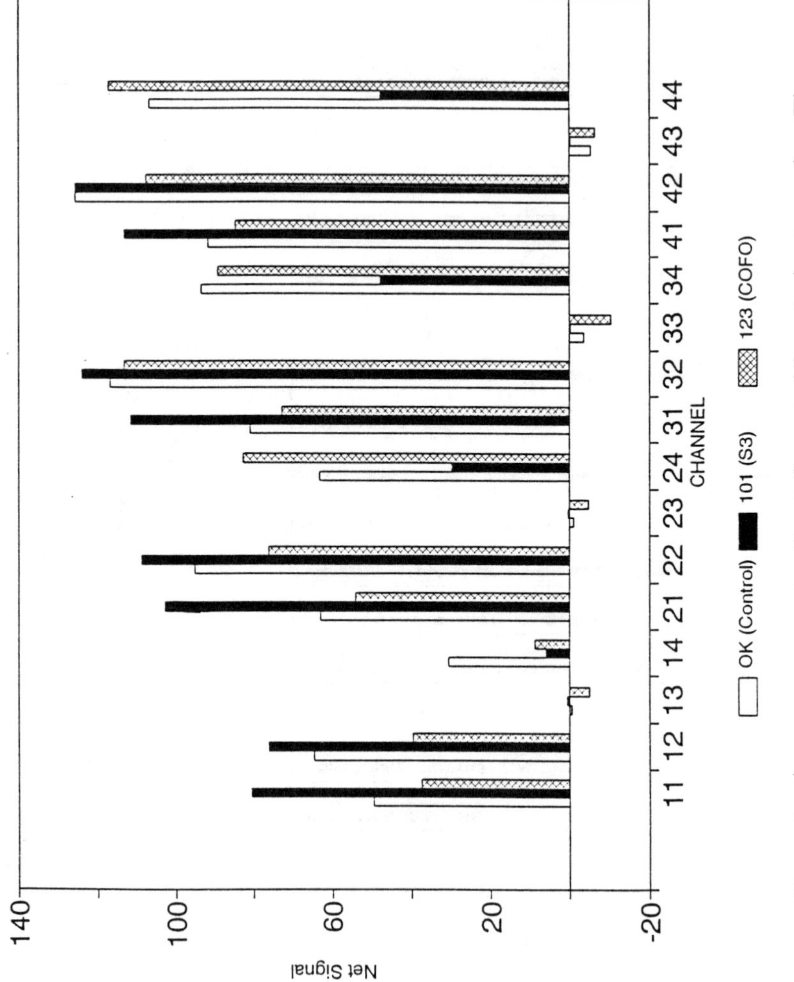

Figure 6. Average response for Good, Sour, and Insect wheat samples. The graph is the normalized response patterns.

with four filament temperatures was used for this application. Figure 7 shows a comparison of two such unknowns to methyl isobutyl ketone (MIBK). As can be seen, the pattern match is nearly exact. Since other solvents like methanol are very different, these unknowns are probably ketone like MIBK. This array provides some class distinction between hydrocarbons, ketone, and alcohols. However, patterns from chlorinated compounds were similar to those from the aliphatic carbon analog. In order to screen for chlorinated compounds, a sensor having a different response to the chlorinated functional group would be required.

Microsensors and Future Possibilities. An amperometric microsensor was fabricated on a thermally oxidized layer of silicon dioxide on a silicon wafer. A gold layer of 3000 Å thick was deposited by thermal evaporation and then patterned by photolithography and etching. In order to obtain good adhesion of the gold to the silicon dioxide it was necessary to first deposit a thin layer of chromium. The working electrode (W.E.) was square, 5 mm along each edge. A thin film of Nafion (a Du Pont trade mark) was spin-coated over the electrodes and served as the electrolyte (*15*). A Au-air reference electrode (R.E.) was used in this sensor design.

Microsensors were mounted in a housing that provided a gas exposure chamber over the working electrodes and contacts for the electronic signal (see Figure 8). For this microsensor, the Nafion film was wetted by flowing humidified air over the W.E. prior to and during electrochemical measurements. Humidification of the air was achieved by flowing dry cylinder air through a 14 cm length of Gortex tubing immersed in distilled water. At a flow rate of 100 cm^3/min this procedure results in nearly 100% relative humidity in the gas sample stream (*16*). The sensors were maintained under potentiostat control using a PAR 273 potentiostat, and were biased to 1.3 V versus the standard hydrogen electrode in acid electrolyte (0.3 V versus a Pt-air R.E. or 0.5 V versus a Au-air R.E.).

Upon exposure to 100 ppm of nitric oxide, the sensor exhibited a rapid, large, and reversible response. As the humidification period increased, the sensitivity increased even more. Presumably, this was due to the improved conductivity of the Nafion. The maximum response of this sensor occurred after approximately 45 min of humidification. The maximum response to 100 ppm of NO at 100 cm^3/min was 490 *u*A. Unfortunately, this microsensor has a short lifetime.

Micrographic analysis revealed the probable cause of sensor failure was corrosion can be attributed to the use of chromium in the structure. While chromium improves adhesion of the gold thin film to the silicon substrate, etching of the gold and chromium film exposes the chromium to the Nafion. Under the applied potential in the aqueous environment, the chromium corrodes. This degrades the gold/silicon interface and causes the ultimate failure of the sensor. Morita et al. (*17*), who also used chromium as an adhesion layer, did not observe corrosion with a non-aqueous electrolyte system consisting of poly(ethylene oxide) with $LiCF_3SO_3$ and acetonitrile. Similarly, when we used a water free electrolyte ($LiClO_4$ in propylene carbonate) instead of humidified

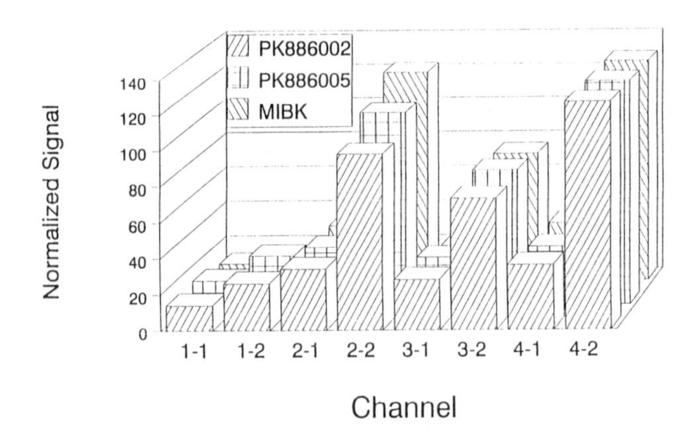

Figure 7. A comparison of two waste unknowns (primarily paint thinners and solvents) to methyl isobutyl ketone (MIBK).

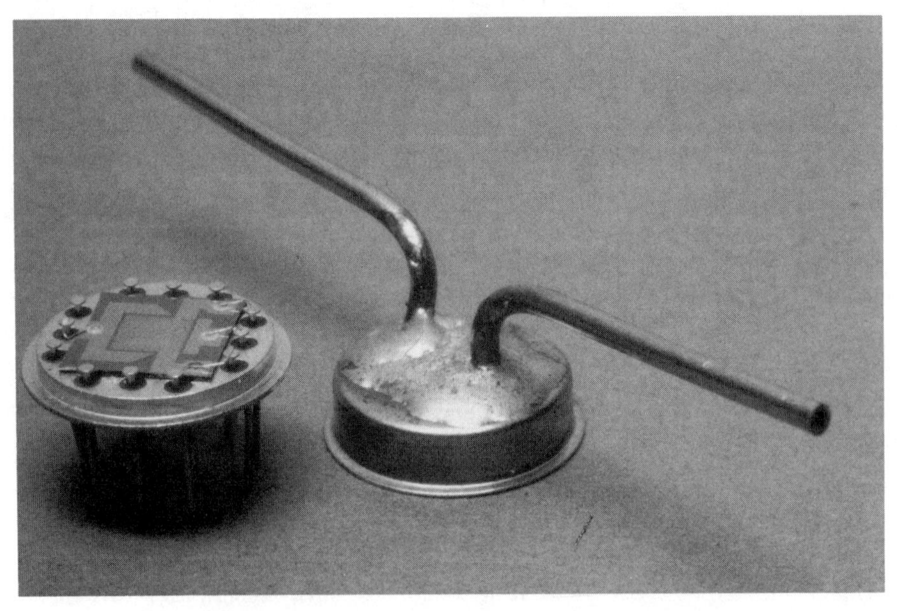

Figure 8. Planar three-electrode amperometric sensor fabricated on a TO-8 header with gas exposure cap.

Nafion, the chromium corrosion did not occur and the sensor exhibited a much longer lifetime.

Conclusions

Electrochemical sensors are playing an increasingly larger role in the pollution analysis. They provide real-time, on-site information about the presence and concentration of chemicals in a given environment or process stream. Their application and role is central to many operations and is illustrated in Figure 9. In views of above specific examples, several conclusions can be made. Reliable sensors are key to the issues of pollution prevention, pollution reduction, and pollution control. Sensors can impact the cost, safety, effectiveness, versatility and speed of jobs we must do to prevent pollution. Sensors are our key to obtaining high quality epidemiological data to define many of the pollutant threats to human health. Since a variety of sensors now exist, applications engineering of existing sensors into systems can make significant contributions to resolve existing problems with only minimum investment. New and advanced sensor technology is waiting to be applied to EPA problems in pollution and EPA need only seize this opportunity.

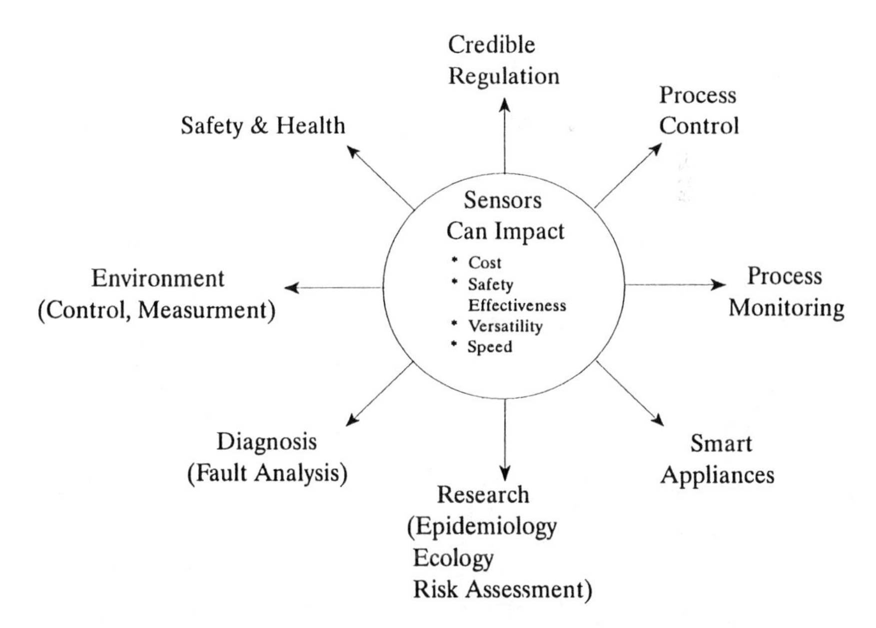

Figure 9. A block diagram of electrochemical sensor application.

Literature Cited

1. Penrose, W.R.; Pan, L.; Maclay, G.J.; Stetter, J.R.; Mulik, J.D.; Kronmiller, K.G.; Personal exposure monitoring of nitrogen dioxide at part-per-billion levels: Autozero and other corrections to electrochemical measurements. Proceedings, EPA/AWMA Symposium on "Total Exposure Assessment Methodology: A New Horizon", Las Vegas, NV, Nov. 27-30, 1989.
2. Penrose, W.R.; Pan, L.; Maclay, G.J.; Stetter, J.R.; Mulik, J.D.; Kronmiller, K.G.; Portable ozone detection in the ppb range using electrochemical sensors. Proceedings of "Sensors Expo'90", September 11-13, Chicago, IL, Helmers Publishing, Peterborough, NH 03458, pp. 305B1-4, 1990.
3. Spicer, C.W.; Ward, G.F.; Kenny, D.V.; Leslie, N.P.; Billick, I.H.; Measurement of Oxidized Nitrogen Compounds in Indoor Air. Proceedings, EPA/AWMA Symposium on "Measurement of Toxic and Related Air Pollutants", May 6-10, 1991, Durham, North Carolina.
4. Cao, Z.; A Solid-state Sensor for Halogenated Compounds, Ph.D. dissertation, Illinois Institute of Technology, Chicago, Illinois, 1990.
5. Callis, J.B.; Illman, D.L.; Kowalski, B.R.; Anal. Chem. 1987, 59, 624A-637A.
6. Hanson, D.; C&E News, March 12, 1990, 4.
7. Zurer, P.; C&E News, October 9, 1989, 4.
8. Zurer, P.; C&E News, May 7, 1990, 44.
9. Stetter, J.R.; Cao, Z.; Anal. Chem. 1990, 62, 182.
10. Stetter, J.R.; Cao, Z.; "An Analytical System for On-line Analysis of Chlorinated Hydrocarbons in Waste Water", Presented at On-line Analysis Symposium and Exhibition, April 10-12, 1989, Chicago, Illinois.
11. Cao, Z.; Stetter, J.R.; "A Real-Time Monitor for Chlorinated Organics in Water", Proceedings of the 1990 EPA/A&WMA International Symposium on "Measurement of Toxic and Related Air Pollutants", U.S. EPA Air & Waste Management Association VIP-17, 1990, 836-848.
12. Stetter, J.R.; "Electrochemical Sensors, Sensors Arrays, and Computer Algorithms," in FUNDAMENTALS AND APPLICATIONS OF CHEMICAL SENSORS, Schuetzle, D.; Hammerle, R.; Butler, J.; eds., ACS Symposium Series No. 309, 1986, 299-308.
13. Findlay, M.W.; Stetter, J.R.; Schroeder, K.; Yue, C.; "Detection and Identification of Foreign Odors in Grain," Final Report, USDA SBIR Agreement No.: 90-33610-5088, February, 1991.
14. Penrose, W.R.; Stetter, J.R.; Findlay, M.W.; Buttner, W.J.; Cao, Z.; "Arrays of Sensors and Microsensors for Field Screening of Unknown Chemical Wastes," Presented at and Published in Proceedings of the Second International Symposium on "Field Screening Methods for Hazardous Wastes and Toxic Chemicals", February 12-14, 1991, Las Vegas, NV.
15. Buttner, W.J.; Maclay, G.J.; Stetter, J.R.; Sensors and Actuators, 1990, B1, 303-307.

16. Zaromb, S.; Woo, C.S.; Quandt, K.; Rice, L.M.; Fermaint, M; Mitnaul, L.; Simple permeation absorber for sampling and preconcentrating hazardous air contaminants, J. Chromatogr., 1988, 439, 283.
17. Morita, M.; Longmire, M.L.; Murray, R.W.; Solid-State Voltammetry in a Three Electrode Electrochemical Cell-on-a-Chip with Microlithography Defined Microelectrode, Anal. Chem., 1988, 60, 2770-2775.

RECEIVED June 25, 1992

Chapter 11

Advanced Testing Line for Actinide Separations

An Integrated Approach to Waste Minimization via On-Line–At-Line Process Analytical Chemistry

L. R. Austin

Los Alamos National Laboratory, Nuclear Materials Processing Technology, P.O. Box 1663, MS E–501, Los Alamos, NM 87545

A full scale Actinide Recovery Line has been designed and installed in the Plutonium Facility at Los Alamos National Laboratory. This system, called ATLAS (Advanced Testing Line for Actinide Separations) comprises operations ranging from dissolution, purification, and calcination, to waste stream polishing. The system is modular, such that new recovery technologies can be introduced into ATLAS and evaluated in an integrated manner with up-stream and down-stream recovery operations. A full spectrum of on-line and at-line analytical capabilities exist. The objective of this system is to demonstrate advanced technologies that can minimize or eliminate nuclear waste generation at the recovery site. The capabilities of the facility, as well as some of the development activities aimed at waste reduction and minimization, will be discussed.

The principal Department of Energy (DOE) site for the production of nuclear weapons is the Rocky Flats Plant, near Golden, Colorado. Residues from the production operations containing recoverable quantities of plutonium are processed as part of the Rocky Flats operations. Discharges from these secondary recovery operations comprise a significant part of the nuclear waste from Rocky Flats.

Los Alamos National Laboratory, which operates a full-scale plutonium recovery facility, has been assigned an important technical support function to the recovery operations at Rocky Flats. The Los Alamos facility has been designated as the Lead Laboratory for developing and demonstrating improved processing technologies for the entire DOE complex. Therefore, aqueous recovery operations at Los Alamos have two objectives: (1) Improved recovery technologies are tested and demonstrated at full scale before they are transferred to other DOE production

0097–6156/92/0508–0118$06.00/0

facilities, and (2) residues are processed to recover plutonium and to minimize the introduction of hazardous wastes to the environment.

Previous Approach To Operations

Although plutonium recovery operations have been in use for over forty years, the specific chemical and technical details required to recover plutonium with high efficiency are not well known. Plutonium operations generally are far from optimal, and involve the addition of excessive chemical reagents that produce excessive liquid and solid waste volume.

One of the major reasons for not operating processes in a manner that minimizes waste generation is the complexity of the basic chemistry. A given impure feed material may contain a wide range of unknown metal contaminants and anionic impurities that can interfere with the plutonium separation process. If the plutonium product is not sufficiently pure, the entire batch must be recycled and reprocessed, with a corresponding increase in processing time, personnel exposure, and waste generated. Thus, to assure a product of acceptable purity, we have focused on plutonium recovery at the expense of waste minimization.

The lack of timely analytical data also contributes to process inefficiency. At present, almost all analytical samples must be removed from the process and sent to another building, several miles distant, for analyses. The normal delay before analytical results are available is much too long to benefit our process operations. Moreover, the specific analysis needed for on-line process control often has either not been identified, or the analytical technique required for rapid assay does not exist.

For example, the presence of excess fluoride in the ion exchange feed solution directly affects process performance, yet, we are just beginning to appreciate the complex and competitive interactions of fluoride, aluminum, nitrate, and phosphate on the formation of the anionic plutonium nitrate complex. Until recently, the nitric acid concentration of feed solutions was determined by a free-acid titration. This lengthy procedure yielded a value for hydrogen ion concentration, whereas what we need is nitrate ion concentration. Likewise, because fluoride is masked by elements such as silicon and aluminum, a total fluoride assay does not provide what we need, which is the amount of fluoride that will interfere with the formation of the sought plutonium nitrate complex.

The concept of ATLAS provides a tool useful in solving these complex problems. Process analytical instruments have been integrated into the unit operations required for purification. Full scale recovery can then be performed in an experimental mode to develop and optimize the processes. In addition, ATLAS can be used as the final full-scale testing and demonstrating of new recovery technologies as they are developed. ATLAS is a modular system, and new unit operations can be inserted and evaluated in an integrated mode to optimize the entire flowsheet.

Special Constraints

Nuclear materials processing is subject to a unique set of constraints that set it apart from other manufacturing and chemical processing operations. Among these constraints are: (1) nuclear materials control and accountability, (2) radioactive contamination control, and (3) exposure of personnel to radiation.

Technologies such as continuous processing, automatic process control, robotics, and other forms of automation are well developed and routinely integrated into the design of most modern processing facilities. Although a few good examples of automatic process control have existed for many years (PUREX at Hanford and Savannah River) in general, advanced control and automation technologies have not found wide spread use within the DOE complex because of these cited constraints. Although these technologies could contribute significantly to nuclear materials process efficiency and waste minimization, the ability to integrate them into facilities within the complex today represents a very difficult and challenging technical goal. For example, reduced exposure for operators at the cost of increased exposure for maintenance personnel is not necessarily a good trade-off.

Nuclear materials accountability poses a particularly demanding set of problems and constraints. Almost by definition, all process operations must be batch, or at best, semi-continuous/batch in order to keep an accurate account of the nuclear material inventory. Quite often, accountability requirements, process efficiency considerations, and safety regulations conflict with each other. Good nuclear materials accountability practice requires that two people oversee each nuclear material transfer. Such required duplication of effort prevents efficient use of personnel. Likewise, this is in direct conflict with our policy of keeping radiation exposure of personnel to levels as low as reasonably achievable (ALARA). A large fraction of our total effort is devoted to accountability activities.

The general accountability approach is simple and straight forward. Each nuclear material item is to be measured at every step in the process, while maintaining accurate and up-to-date accountability records. Although nuclear materials accountability is not a direct waste minimization issue, obtaining accurate accountability measurements on a wide variety of waste presents a formidable technical challenge. Sound technical solutions to these difficult accountability problems could produce significant gains in worker productivity and a simultaneous reduction in radiation exposure. Improved accountability measurement techniques therefore could contribute significantly to process efficiency.

At the Los Alamos Plutonium Facility we have just begun an integrated program to apply our best technical effort toward these accountability problems. Reliable on-line and at-line process analytical chemistry is expected to play an important role in the solution to this problem.

Our challenge is to address the important technical needs, while simultaneously meeting the cited constraints, in an integrated development program. The goal is to provide the DOE complex with a combination of technologies that will significantly improve recovery efficiencies, reduce waste, and comply with all of the accountability, safety, and security regulations.

The Approach

The approach used at the Plutonium Facility at Los Alamos National Laboratory is to develop processes that produce the least amount of waste when evaluated on a total integrated facility. Our comprehensive waste minimization program has three elements:

1. Optimize existing processes to minimize the waste generation, including recycling when possible.

2. Develop additional treatment or polishing operations that will convert a significant fraction of the total waste volume to effluents that can meet discharge limits for the plutonium facility, or can be discharged directly to the environment without harm.

3. Develop revolutionary new technologies that could result in significantly lower total waste generation.

The ATLAS concept embodies all three aspects of waste minimization described above. The goal is to provide process optimization and waste minimization for the facility as a whole. Often what may appear best for one operation may have a net over-all negative effect on the entire facility. Process modeling is an important tool that can provide this over-all assessment.

What is ATLAS?

Approximately four years ago, we recognized a need within the DOE Complex for a facility capable of reprocessing many of the unusual test items that were collecting within nuclear materials storage vaults. These items continue to be stored because of the uncertainty of recovery success, as insufficient processing knowledge exists. Also, insufficient material exists to warrant development of a recovery flow sheet, and then scale up for the recovery campaign. In addition, we recognized the need to evaluate new full-scale recovery flow sheets in a fully integrated manner. Combining these two needs the idea of an Advanced Testing Line for Actinide Separation (ATLAS) was born.

ATLAS is an aqueous recovery facility that includes all operations ranging from dissolution to waste stream polishing. Two of the glove boxes are Kynar lined so both nitrate and chloride based flow sheets can be evaluated. The facility is full scale, that is kilogram quantities of plutonium can be processed. If additional capacity is required, the line would be duplicated rather than scaled up. ATLAS is designed in a modular way such that new technological developments can be easily incorporated. All unit operations are integrated so that the effects of a single unit on other up and down stream operations can be measured. Each unit is linked together with an advanced on line data gathering and control computer.

ATLAS has a full spectrum of on-line/at-line process analytical instrumentation. Instruments include a Quantum 1200 UV-VIS spectrophotometer, gamma monitor, alpha monitor, ion chromatograph, X-ray fluorescence unit, and acid sensor. Analytical capabilities included plutonium valence, actinide concentration, trace metal and anion concentrations, and acidity determination.

A chemical sensor program augments the process analytical program. After the fundamental technical understanding of the process is obtained, the next step is being able to obtain real-time data from the appropriate control point. Often one finds that at certain critical control points, a sensor is not available to extract the required information. Thus, to augment the process analytical effort, a program on sensor development exists at the Laboratory. We have recently demonstrated an on-line acid sensor that has the capability to measure acidity up to 11 M nitric acid concentrations.

ATLAS provides one of the keys to linking most of our aqueous development programs. Aqueous development activities include electrochemical dissolution, improved ion exchange, solvent extraction, and precipitation. In addition, we have an extensive program in the area of organic ligand development. This program is directed at understanding the mechanism of selective metal ion capture with organic ligands, and custom building ligands for specialized selectivity. There is a strong tie between the ligand extraction and the sensor programs. If a ligand can be synthesized that is very selective for a given species, then a theoretical basis exist for sensing that species.

Technical Developments

The concept of ATLAS embodies several new technical developments. These cover such activities as dissolution, precipitation, purification using advanced separation technologies to include membrane development and synthesis of extraction ligands for heavy metals separation. The two development activities that will be discussed here are focused on process analytical chemistry.

The On-line Acid Sensor. One recent development involves a new sensor to measure acidity below a pH of 1, a crucial measurement for many high acid chemical processing operations. We have developed an optical sensor that can make high acidity measurements on-line automatically.

The sensing material utilizes a polymer (polybenzimidazole) and an indicator (chromazurol-S). The polymer is chemically bound to the surface of a fused silica lens. The indicator is physically entrapped within the polymer. The absorbance band of chromazurol-S at 554 nm increases in intensity with increasing acidity starting at 4M hydrogen ion concentration. The peak sensitivity of this material is between 4 and 12M hydrogen ion concentration. The sensor is very selective for hydrogen ions because it uses an acid/base indicator. Furthermore, the polymer's small pore size prevents most other cations from reaching the indicator.

This sensor has proven to be quite stable in high acid solutions. We have observed a signal degradation of less than 10% over a two month period. Acid concentrations obtained with this sensor were within 3% of values obtained by titration.

The sensor uses fiber optic spectrophotometry. Light is focused through a lens, from a fiber optic cable. The light passes through the acidic solution, and is refocused with a second lens into another fiber optic cable. One or both lens are coated with the sensing material. The attenuated light proceeds to the spectrometer's fiber optic detector. Acidic solutions are pumped perpendicular to the light path. The acidic solution is in direct contact with the sensing material. The sensor absorbance equilibrates rapidly, usually within one minute. Alternatively, a probe configuration can be used where the sensing material is directly deposited onto an optical fiber.

Phenol red is an indicator that responds well between .1 and 4M hydrogen ion concentration. Initial indications show that by properly combining the phenol red and chromazurol-S indicators into the same coating, the range of sensing can be extended from .1 to 12 M hydrogen ion concentration.

Advanced Neural Net Pattern Recognition. Another development involves the use of "intelligent" process control using neural networks. A neural net pattern recognition system is being used to obtain process control parameters from abstract knowledge and complex spectral data. Spectral data from concentrated nitric acid solutions containing a wide array of anions and cations can readily be obtained. The highly varying concentrations of these other anions can complex dissolved species and produce dramatic shifts in the spectral peaks, thus making the interpretation of the data extremely difficult.

The thrust of this work is to use the entire spectral range rather than focus of specific peaks. This presents a complex data set, which will be feed into the pattern recognition algorithm. Once the system is trained using a known data set with predictable recovery results, parameters will be set that can be used to extract information from spectra obtained on each new item to be purified. The data extracted will then provide key parameters that will be used to optimize feed treatment and ion exchange operations to optimize plutonium recovery and minimize waste generation.

Summary

The DOE faces an enormous task of rebuilding the weapons complex with cost effective and efficient operations. Some of the base technologies required to accomplish this task also apply to the environmental remediation of the DOE site, which is even a larger and more costly project. The technologies being developed here are focused on aqueous recovery of nuclear materials for the DOE Complex, however, the broader application to other industrial problems is also readily apparent.

At Los Alamos National Laboratory a systematic approach has been developed to integrating unit operations and providing total process optimization for waste minimization. On-line and at-line process analytical chemistry is one of the major keys to realizing this goal. Many other developments under investigation at Los Alamos, such as membrane and organic ligand synthesis, sensor development, process control, modeling, and artificial intelligence/neural net data analyses and process control are part of the total integrated approach.

RECEIVED December 9, 1991

MEMBRANE INTERFACES AND PROCESS ANALYTICAL CHEMISTRY

Chapter 12

Membrane Introduction Mass Spectrometry in Environmental Analysis

R. G. Cooks and T. Kotiaho[1]

Department of Chemistry, Purdue University, West Lafayette, IN 47907

A semi-permeable membrane, mounted in a direct insertion probe, is used to introduce aqueous samples into a triple quadrupole mass spectrometer while a simpler version of this device is used to introduce samples into an ion trap detector. Both instruments are equipped with flow injection analysis fluid handling systems and this allows on-line monitoring of aqueous solutions at low levels. Low ppb level detection limits are achieved for some typical compounds of environmental interest, and response to changes in analyte concentration is very rapid, especially when the membrane is heated. This means that reacting systems can be monitored on-line, a capability demonstrated for the interconversion of the chloramines NH_2Cl, $NHCl_2$ and NCl_3. This methodology is also well-suited to determination of organochloramines at sub-ppm levels in water and the successive chlorinations can be followed as the solution pH is adjusted. Typical aliphatic amines are chlorinated at nitrogen but aniline is ring chlorinated. The reactions which accompany ozonation of contaminated water are also studied and the pentafluorobenzyl hydroxylamine derivatives of aldehydes are studied by negative chemical ionization using membrane introduction mass spectrometry. Membrane introduction mass spectrometry is also demonstrated to be applicable in air monitoring.

The increased awareness of the general public of environmental issues and growing knowledge of the toxicity of some compounds exerts increased demands for knowledge regarding environmental risks. An essential requirement for

[1]On leave from Technical Research Center of Finland, Chemical Laboratory, Biologinkuja 7, 02150 Espoo, Finland

0097–6156/92/0508–0126$08.25/0

logical risk reduction is on-line analysis of environmentally significant compounds present in waste water streams and in air emissions from factories, power plants and other sources. The knowledge gained from the analysis of environmentally significant compounds can also be used to minimize production of environmental pollutants and to control the manufacturing processes so as to optimize conditions with respect to this as well as other factors. Only by knowing the constituents of waste streams, air emissions and process streams in a molecularly specific fashion can the above aims be fully met.

This paper reviews some of the experiments done in our laboratory aimed at developing and characterizing instrumentation for trace level, on-line molecular analysis of aqueous and gaseous waste and process streams. The methodology used is a unique combination of a membrane probe interface and a mass spectrometer together with flow injection analysis (FIA) methods for efficient sample handling. Results of its applicability to some chemical compounds of environmental significance are discussed, and refinements to the technique of membrane introduction mass spectrometry (MIMS) which should further enhance its capabilities are presented.

Membranes for Sample Introduction into Mass Spectrometers

Background Information. Pervaporation (*1*), the process whereby an analyte is transferred from the solution phase on one side of a semi-permeable membrane, into the vapor phase on the other, occurs at widely varying rates for different compounds and hence can be used as a selective means of sample transfer. Membrane separators were once widely used for coupling gas chromatographs to mass spectrometers (*2*) and this led to the continuing interest in membrane interfaces for analysis of organic compounds in air by mass spectrometry (*3,4*). Several investigators have employed such devices for breath analysis by mass spectrometry and even for *in vivo* analysis of blood gases (*5,6*). Following work by Tou and coworkers (*7*), a number of groups became interested in pervaporation of dissolved analytes through the walls of capillary membrane tubes immersed in solution. When the tubes are evacuated by the mass spectrometer vacuum system, compounds which permeate the membrane are transported in the vapor phase into the source of the mass spectrometer and there ionized (*7,8*). Silicone rubber membranes have been used almost exclusively in this and other membrane introduction devices (Figure 1); they have very low permeabilities for water and for polar molecules and this facilitates sampling from aqueous solutions. Although a combination of physical and chemical properties, including molecular weight, size, and polarity, affect permeability of organic compounds, an explicit expression for permeability can be written at steady state (*9*). For a capillary membrane, the permeability Q can defined by the equation 1 (*9*). In this expression,

$$Q = \frac{Fy \ln(R_2/R_1)}{2\pi l(P_0 x - fPy/\beta)} \tag{1}$$

(a)

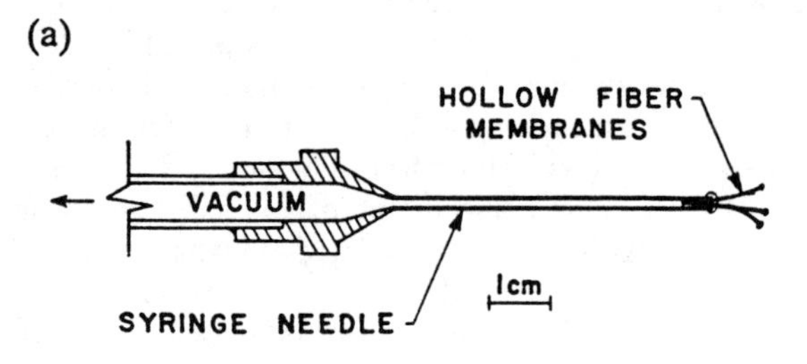

(b)

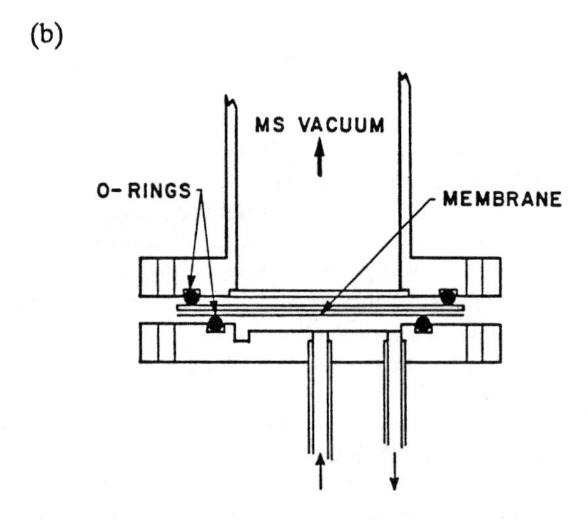

Figure 1. Early versions of a) hollow fiber and b) sheet membrane interfaces as reviewed by Bier for examining aqueous solutions by mass spectrometry. (Reprinted with permission from M. E. Bier Ph.D. thesis, Purdue University, 1988.)

F is permeation rate (mole/s), y is the mole fraction of the analyte in the gas phase, x is the mole fraction in the liquid, f is the fugacity, P_0 is the vapor pressure of the pure analyte, P is the pressure of the gas phase, β is the activity coefficient, R_2 and R_1 are the outer and inner radii and l is the exposed length of the capillary membrane. Note, that the object of membrane separation is not simply high permeability to the analyte(s), it is equally important to discriminate against the matrix. It is principally the very low permeability to water which continues to recommend silicone rubber (dimethyl vinyl silicone polymer) membranes for the analysis of organic compounds in aqueous solutions.

All the early designs for membrane introduction systems positioned the membrane separator external to the mass spectrometer ion source. In work particularly pertinent to this study, thin film membranes have been mounted inside reactors (*10,11,12*) although the long transfer lines cause some problems including, memory effects. Some of these disadvantages are overcome by more recent designs in which transport is assisted using inert gases such as helium (*13*).

Flow-through Membrane Probes. An alternative to the sampling of aqueous solutions by immersing a hollow fiber probe into the solution (flow-past configuration) is to allow the solution to flow through the capillary tube (flow-through configuration, see Figure 2). This concept was introduced by our group at Purdue in collaboration with the Dow group (*14-16*). It has been embodied in a direct insertion probe (Figure 3) which allows ready interfacing of a reactor or sample stream to the mass spectrometer (*16*).

The use of the flow-through principle, in which the analyte solution passes across the inner surface of the membrane while the outer surface is exposed to the vacuum of the mass spectrometer, improves response times while detection limits are dramatically lowered (*17-20*). The flow-through technique also facilitates the use of flow injection analysis procedures which allow precise repetitive sampling of flow streams or external standard solutions. This means that on-line monitoring and quantitation can be performed by injecting aliquots of the sample mixture or standard solution into the continuous water stream supplied by a peristaltic pump. In these experiments, quantitation is performed using an external standard. We have described elsewhere some aspects of the performance of a triple quadrupole mass spectrometer (*21*) and an ion trap (*22*) fitted with a membrane interface and flow injection sampling.

Figure 4 compares a simple membrane probe used in conjuction with an ion trap with an early version of membrane probe used with the triple quadrupole. In both cases the membrane used was a dimethyl vinyl silicone polymer (Dow Corning Silastic Medical Grade). In both the triple quadrupole and ion trap instruments, experience showed that it was desirable to maintain the probe at a temperature of approximately 50 to 65°C for optimal performance. The performance characteristics of both probe designs are very similar.

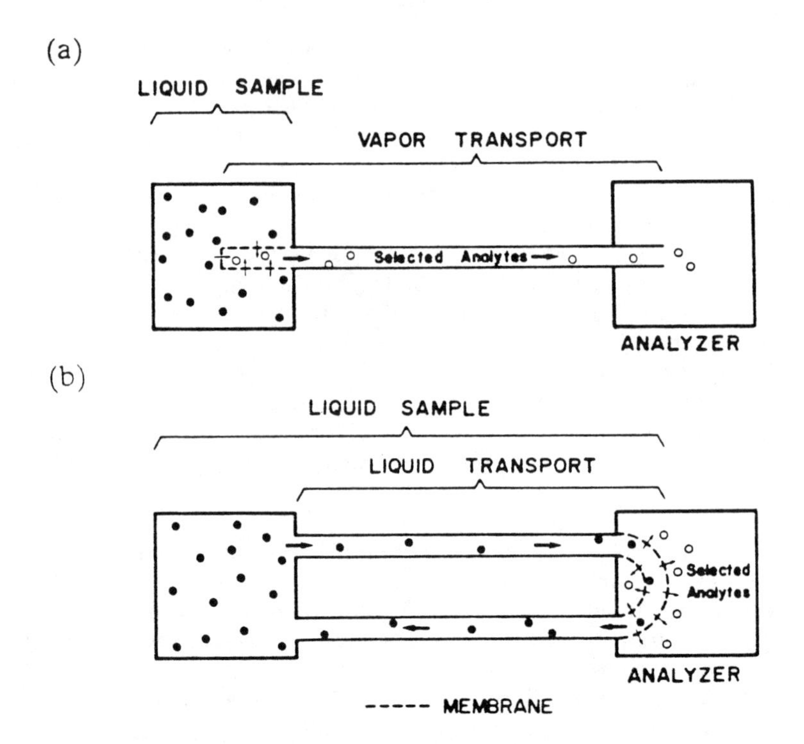

Figure 2. Difference between (a) the flow-by and (b) the flow-through configurations. In the first case the capillary membrane is evacuated, in the second it carries the analyte solution through the ion source.
(Reprinted from reference 16. Copyright 1987 American Chemical Society.)

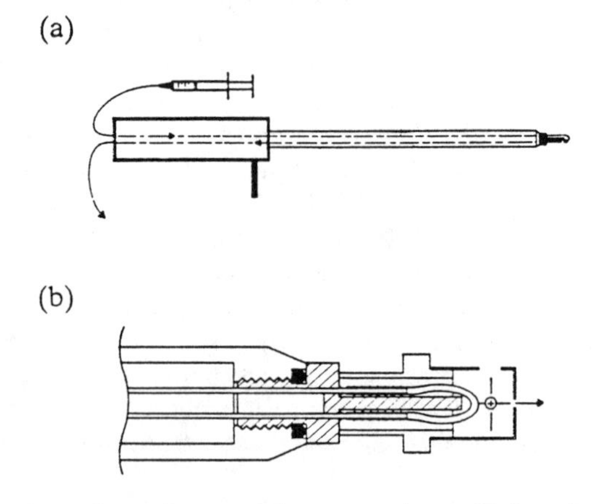

Figure 3. Direct insertion membrane probe utilizing the flow-through concept; (a) shows whole device, solution being introduced via a syringe, (b) shows detail of probe tip illustrating how it enters the ion source.

(a)

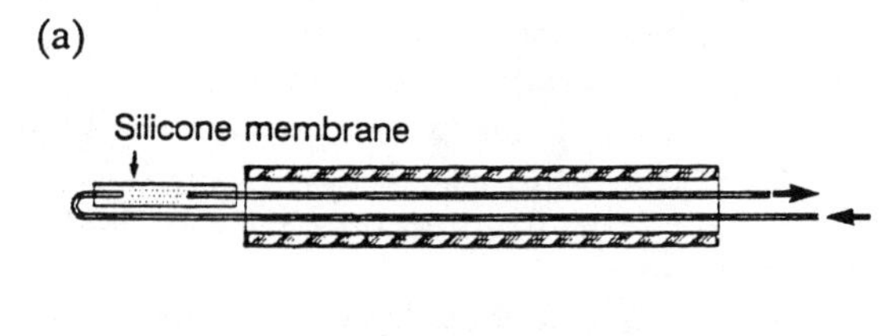

(b)

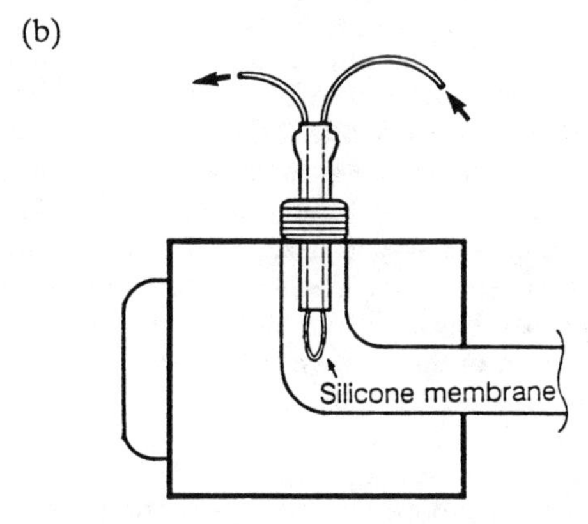

Figure 4. Comparison between (a) the simple membrane probe used in an ion trap and (b) an early version of membrane probe used with a triple quadrupole instrument.

A third design employed a modified form of interface using the membrane in the form of a sheet, mounted in a direct insertion membrane probe *(23)* instead of the earlier capillary tube interface. As in previous work, the membrane was a dimethyl vinyl silicone. The temperature of the membrane probe, which was controlled independently of that of the ion source, was optimally 70°C.

Mass Spectrometry. The triple quadrupole mass spectrometer employed was a Finnigan 4500 mass spectrometer equipped with an INCOS data system and operated under electron impact (70 eV) ionization conditions or isobutane chemical ionization conditions (ion source pressure 0.4 torr). Both full mass spectra as well as selected ion monitoring experiments were typically performed. The identities of the reaction products were confirmed where necessary by measuring the spectra of fragment ions generated by collisions at an energy of 20 eV and a collision gas (argon) pressure of 1 mtorr, which corresponds to multiple collision conditions.

Control of the Finnigan ion trap detector and data storage was handled through an IBM PC computer with a modified version of Finnigan's standard software as described in a previous publication *(24)*. The ion trap detector (ITD) although inexpensive, is extraordinarily sensitive and its mass/charge (m/z) range of 650 daltons/charge is well-suited for membrane introduction mass spectrometry. The ion trap detector was operated under chemical ionization conditions, which were accomplished through methods introduced by Brodbelt et al. *(24)*. The specialized inlet system designed for the ion trap detector was built in such a way that calibration gas, chemical ionization reagent gas, helium buffer gas as well an analyte vapor from the membrane probe could all be introduced *(22)*. The design of the inlet system allowed also the heating of the membrane independently of the ion trap.

Detection limits. Table I summarizes detection limits measured for a set of compounds using a silicone rubber membrane probe in conjunction with an ion trap mass spectrometer and a triple quadrupole mass spectrometer *(19,23,25,26)*. Detection limits vary widely from compound to compound, although small (less than 300 dalton molecular weight) compounds which are not polar (less polar than for example 2,3-butanediol) can usually be detected at the 100 ppb level or below. Detection limits could not be measured for hexachlorobenzene and pentachlorophenol, probably because of the low diffusivities of these compounds in the polymer. The highest molecular weight compound we have detected, 1-phenyltridecane (MW 260) is monitored easily using the sheet direct insertion membrane probe with the silicone membrane *(23)*.

In the case of the triple quadrupole mass spectrometer, detection limits are often better, by up to an order of magnitude, for MS/MS than for single stage mass spectrometry, due to the decreased chemical background. The lowest detection limits are obtained by operating in the single reaction monitoring mode. Note the low detection limits measured with the triple quadrupole

Table I. Detection limits (S/N = 3) using membrane introduction into an ion trap or a triple quadrupole mass spectrometer[a]

Compound	*Detection Limit (ppm)*
Ion trap[b]	
Benzene	0.001[c]
Carbon tetrachloride	0.1
Chloroform	0.11
o-cresol	0.40
1,4-dichlorobenzene	0.10
1,2-dichloroethane	0.0048
Hexachloro-1,3-butadiene	0.23
Methyl Ethyl Ketone	0.091
Nitrobenzene	0.024
Pyridine	0.53
Tetrachloroethene	0.26
2,4,5-trichlorophenol	0.95
Triple quadrupole[d]	
Benzene	0.01
Chlorobenzene	0.01
PFBOA deriv. of formaldehyde	0.001[c]
PFBOA deriv. of acetaldehyde	0.010[c]
Tetrahydrofuran	0.08
Dichloromethane	0.10
Acrolein	0.05[c]
Acrylonitrile	0.01[c]

[a]Data from refs. 17, 19, 23, 25 and 26.

[b]Selected ion monitoring mode.

[c]S/N ratio 2.

[d]Multiple reaction monitoring mode.

instrument for acrolein and acrylonitrile, which are very difficult to analyze using conventional methods. As an example of the performance of the triple quadrupole instrument, the EI mass spectrum of acrylonitrile and reaction monitoring data measured during detection limit measurements are presented in Figure 5 (*25*).

Quantitative accuracy. The linear dynamic range accessible by membrane introduction mass spectrometry covers three or more orders of magnitude in concentration, beginning at the low ppb levels. As an example, a calibration curve for benzene is presented in Figure 6 (*17*). To some extent the matrix effects encountered with MIMS have also been studied. At ppb levels, responses of analytes have been observed to be independent of matrix composition (*17*), but at the part-per-thousand levels the presence of other compounds in the sample solution can have an effect on the permeation rates of the analytes (*23,27*). Also at these high concentration levels, when compounds with widely varying proton affinities are examined, selective protonation of higher proton affinity compounds may lead to discrimination effects (*23,27*).

Permeation times. Figure 7 illustrates the effect of membrane temperature on the pervaporation rise and fall times (*16*). Higher membrane temperatures result in increased diffusivities of the analytes and therefore faster response times are achieved. Short rise and fall times are necessary to resolve consecutive injections, especially when working at low analyte concentrations in the presence of high matrix concentrations. Typical rise times obtained with direct insertion membrane probes vary from a few seconds up to about ten seconds, depending on the analyte and typical fall times are somewhere between 10 and 40 seconds. These short response times facilitate the use of membrane introduction mass spectrometry together with flow injection analysis sampling methods for on-line monitoring applications.

On-line Monitoring of Environmental Pollutants Using an Ion Trap Detector

The potential value of membrane introduction mass spectrometry as a new approach to pollutant monitoring through the detection and on-line monitoring of compounds in aqueous solutions was a motivation of interfacing a membrane probe to an ion trap detector (*19*). The results of detection limit measurements (S/N ratio 3) for some environmentally significant compounds (*28*), measured using a simple ion trap membrane introduction apparatus, are given in Table I.

The applicability of the FIA membrane introduction mass spectrometry apparatus as an on-line monitoring device was tested by an experiment, in which the concentration of benzene, chloroform and 1,2-dichloroethane in water was changed gradually and continuously (*19*). The analyte concentrations in the reservoir were zero initially. The flow rate of the analyte feed stream into the reservoir was held at 1 mL/min for the first 73 minutes and then increased to 10

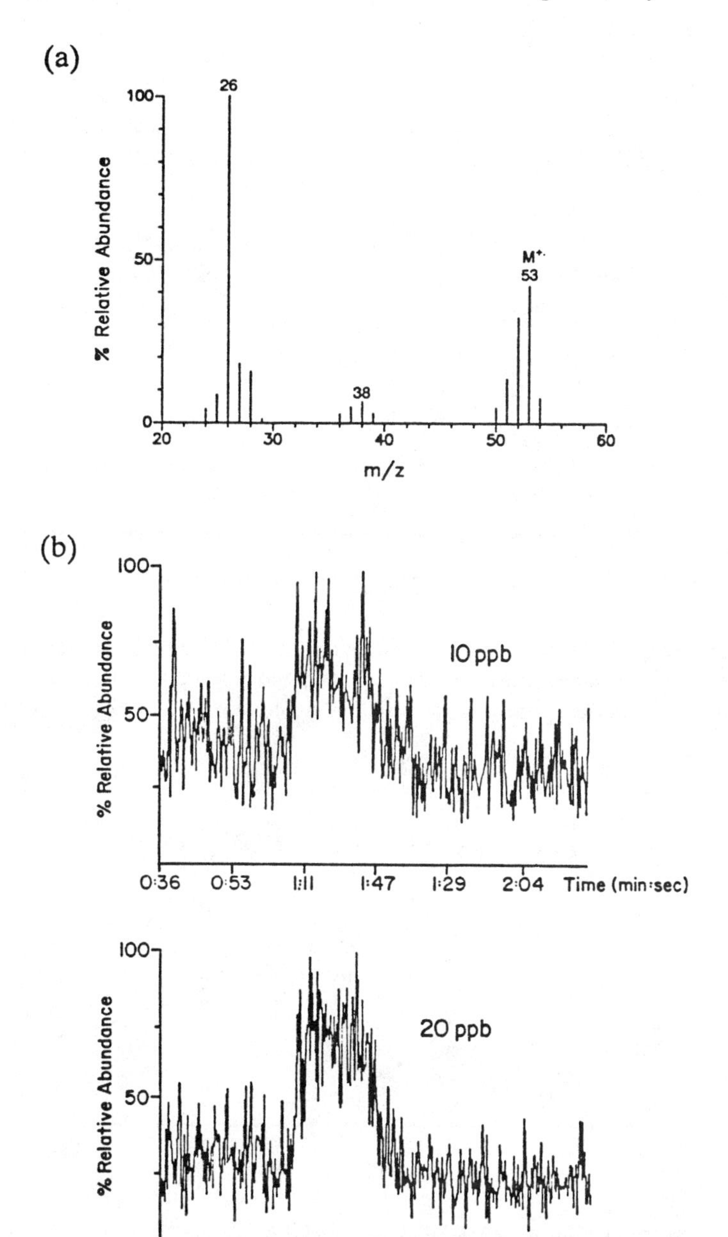

Figure 5. (a) Electron ionization mass spectrum and (b) reaction monitoring (m/z 53 ---> m/z 52,26,25) data measured for acrylonitrile using a triple quadrupole instrument with the sheet direct insertion membrane probe for sample introduction. Electron ionization (70 eV), collision energy 16 eV and collision gas argon.

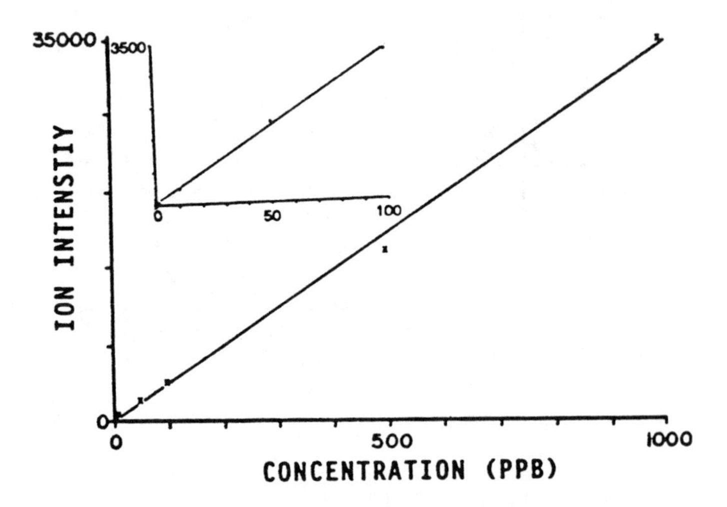

Figure 6. Calibration curve for benzene (m/z 79, the protonated molecule) when introduced into an ITD mass spectrometer via a simple capillary membrane. (Reprinted with permission from reference 17. Copyright 1989 John Wiley & Sons.)

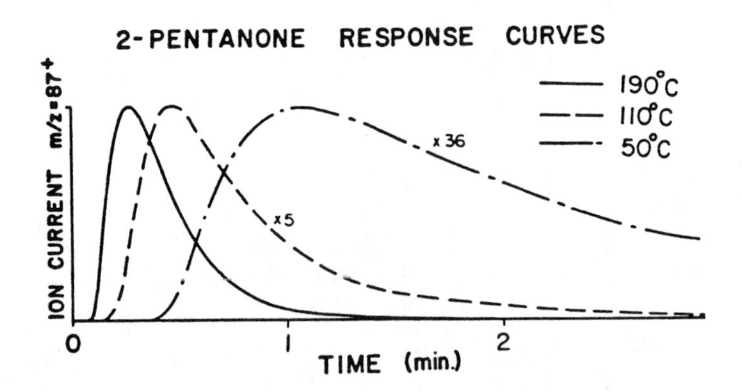

Figure 7. Effect of ion source temperature on signal due to 2-pentanone (10^{-2} M; m/z 87) introduced via a flow through capillary membrane. Data are normalized to peak response which is 36 times greater at 190°C than at 50°C. (Reprinted from reference 16. Copyright 1987 American Chemical Society.)

mL/min. A constant volume was maintained by continuously removing liquid from the reservoir at a rate equal to the analyte feed rate.

The expected analyte concentration present in the reservoir at any time is presented by the solid line and the measured concentrations are presented by the open circles in Figure 8 (*19*). As it can be seen from Figure 8 the measured concentration values and the expected concentration values correspond well to each other.

The on-line monitoring capability was also tested by examination of the response signal to instantaneous changes in concentration. The concentration of the analyte stream was changed dramatically over a wide range as shown in Figure 9. The measured concentrations agree very well with expected values and typical response times were found to be less than the sampling time which was one sample per three minutes.

Although the procedures used here are still being improved, the ability to make rapid, quantitative, on-line measurements of regulated compounds at relevant levels in an aqueous matrix is clearly demonstrated. The method is applicable to a wide range of organic compounds and the accuracy is excellent. The role of this method as an early warning device, its low detection limits, and its high sample throughput all recommend continued investigation.

On-line Monitoring of Formation and Reactions of Inorganic and Organic Chloramines

Inorganic Chloramines. One of the most significant applications of membrane introduction mass spectrometry (MIMS) is to follow in real time changes in reacting systems or sample streams. The applicability of MIMS to monitor pollutants at the necessary low levels in a direct analysis procedure has already been demonstrated in static experiments e.g. in monitoring for benzene in water at ppb levels (*17*). The flow-injection analysis (FIA) capability has also been applied, in conjunction with triple quadrupole mass spectrometry, to monitor chloramines in aqueous solution under dynamic conditions (*29*), that is under conditions in which interconversion of the chloramines is occurring on the time-scale of the analysis. Clearly such a capability should be helpful in elucidating the complex chemistry of the chloramine system or in the general case, in following rapid changes in concentration of analytes in aqueous solution. The individual chloramines are characterized with high chemical specificity at low (sub-ppm) levels by this procedure, something not possible by alternative methods.

The main reason to measure inorganic chloramines in aqueous samples is that monochloramine is often used as a final disinfectant in drinking water (typical levels 1-2 ppm) and there is concern over the possible toxicity of monochloramine, dichloramine and trichloramine (*30-35*). The standard methods used for measuring these compounds are colorimetric, spectrophotometric and titrimetric ones (*30-34,36*). All the above methods are quite susceptible to interferences, and differentiation between chloramines can be difficult.

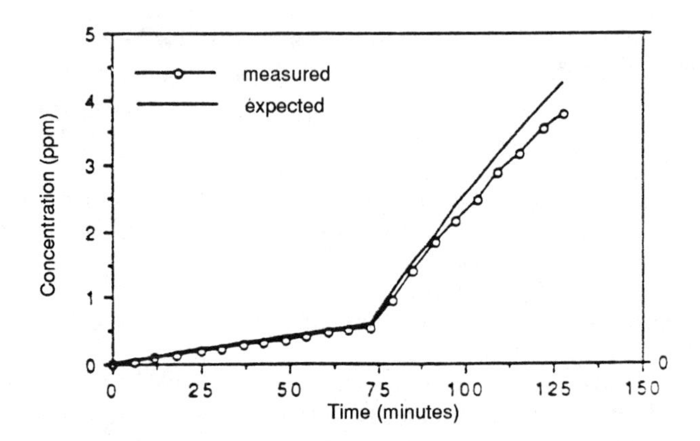

Figure 8. Measured and expected concentrations of 1,2-dichloroethane in a soil/water matrix. The break at seventy five minutes is due to increase in flow rate of the mixture of organics into the reservoir (see text). The concentration of 1,2-dichloroethane increased from 10-times below to 10-times above the environmentally significant level over a two hour period.

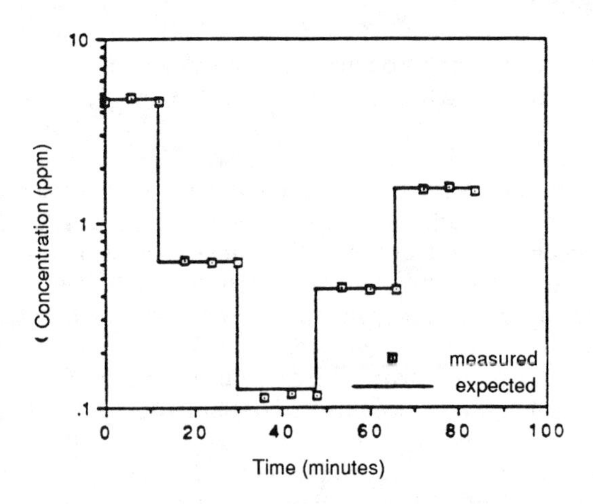

Figure 9. Measured and expected step changes in concentration of 1,2-dichloroethane in a soil/water matrix.

The measured EI mass spectrum for an aqueous solution (540 ppm) of monochloramine is presented in Table II (*29*). The daughter spectrum recorded for both isotopic forms of the molecular ion of monochloramine confirms the remarkably stability of the molecular ion, $NH_2Cl^{+\cdot}$, and the agreement between the daughter spectra of the two isotopic forms of the molecular ion confirms the assignments made in Table II (*29*). The detection limit (S/N = 3) for monochloramine by membrane introduction mass spectrometry is 0.7 ppm (*29*).The detection limit was measured using flow injection analysis (sample size 250 μl), EI and selected ion monitoring.

The reactions of monochloramine with hydrogen chloride were studied on-line (*29*) and they clearly show formation of dichloramine and trichloramine (Figure 10) as Margerum et al. demonstrated earlier using spectrophotometric methods (*31-33*). Data presented in Figure 10 were collected by scanning full EI mass spectra in real time as the reaction, induced by addition of aliquots of HCl, proceeded. The initial rise of monochloramine signal represents the time during which water was replaced by a 540 ppm monochloramine solution. After the first addition of HCl (added at 3.1 min) formation of dichloramine was observed and a short time later small amounts of trichloramine were also detected. The full EI mass spectrum (Figure 11a) at this point (4.4 min) shows ions characteristic of dichloramine and traces of ions characteristic of trichloramine. The measured daughter ion mass spectra of m/z 85 and 87 confirm the production of dichloramine. Note that almost all monochloramine has been reacted, as demonstrated by the negligible abundance of m/z 53, which is the heavy isotopic form of the molecular ion, m/z 51, of monochloramine.

The production of trichloramine occurred in good yield after the second addition of HCl at 9.1 minutes (Figure 10) (*29*). It shows up clearly in the mass spectrum taken at this time (Figure 11b) and is confirmed by its daughter spectra. The daughter spectra of the molecular ions of dichloramine and trichloramine both show preferential formation of the ionized divalent species $NHCl^+$ and NCl_2^+. Note that Figure 11b demonstrates the generation of molecular chlorine (ions m/z 70, $^{35}Cl_2^{+\cdot}$; m/z 72, $^{35}Cl^{37}Cl^{+\cdot}$ and m/z 74, $^{37}Cl_2^{+\cdot}$) simultaneously with the trichloramine. Addition of further aliquots of HCl increased the concentration of Cl_2 which is generated in a reversible reaction.

Organic Chloramines. Chlorine is widely used in water treatment and, during chlorination the inorganic chloramines discussed above are formed. In the presence of organic amines, formation of their chlorinated derivatives is possible, and the N-Cl bond is expected to make these molecules both labile and difficult to characterize. Not only has this proved to be the case, but there is evidence that many of these compounds are also toxic (*37*), although difficulties in preparing pure samples limit the information available. A further complexity expected to be encountered in the analysis of organochloramines is their ready interconversion, expected if they mimic the parent inorganic chloramines. Each of these considerations recommends study of these systems by membrane introduction mass spectrometry, especially since current chromatographic methods (*38-40*) often require derivatization, are time-consuming and

Table II. EI mass spectrum of monochloramine (mass range 33 to 100) [a]

m/z	*% Rel. Abund.*	*Ion Species*[b]
35	<0.5	$^{35}Cl^{+}$
36	0.6	$H^{35}Cl^{+\cdot}$
37	<0.5	$^{37}Cl^{+\cdot}$
38	<0.5	$N^{37}Cl^{+\cdot}$
49	5	$N^{35}Cl^{+\cdot}$
50	1	$NH^{35}Cl^{+}$
51	100	$NH_2{}^{35}Cl^{+\cdot}$; $N^{37}Cl^{+\cdot}$ [c]
52	2	$NH_3{}^{35}Cl^{+}$; $NH^{37}Cl^{+}$ [d]
53	31	$NH_2{}^{37}Cl^{+\cdot}$
54	0.5	$NH_3{}^{37}Cl^{+}$

[a]Data from ref. 29.

[b]Inferred from isotopic abundance distribution

[c]Ca. 1.5%

[d]Ca. 0.3%

decomposition of target chloramines is possible during the chemical manipulations which are necessary.

The FIA/MIMS system used to study the chlorination of 2-aminobutane gave the results shown in Figure 12 (*41*). The 2-aminobutane solution was introduced into the FIA system at the point indicated by the asterisk, calcium hypochlorite was added at the small arrow and HCl at the heavy arrows. The ions monitored, m/z 74, 108, 110, 142 and 144 represent the protonated amine, the two isotopic versions of the protonated monochloro derivative of 2-aminobutane and the two protonated versions of the dichloro derivative of 2-aminobutane. Note that the amine is observed after the addition of hypochlorite because the associated change in pH liberates the free base which crosses the membrane and can therefore be ionized by isobutane chemical ionization. Note also that the monochloro derivative is generated immediately upon addition of hypochlorite but that the dichloro derivative is only generated when the pH is dropped further.

The data shown in Figure 12 were measured at the starting amine concentration of 0.2 ppm (*41*). This concentration level represents the limit of concentration at which the entire reaction sequence can be followed when continuous membrane introduction mass spectrometry is used. When individual organic chloramines are measured by MIMS the detection limits are significantly lower (*41*). The structures of the chlorinated amines were determined by recording daughter spectra and that for the monochloro derivative of 2-

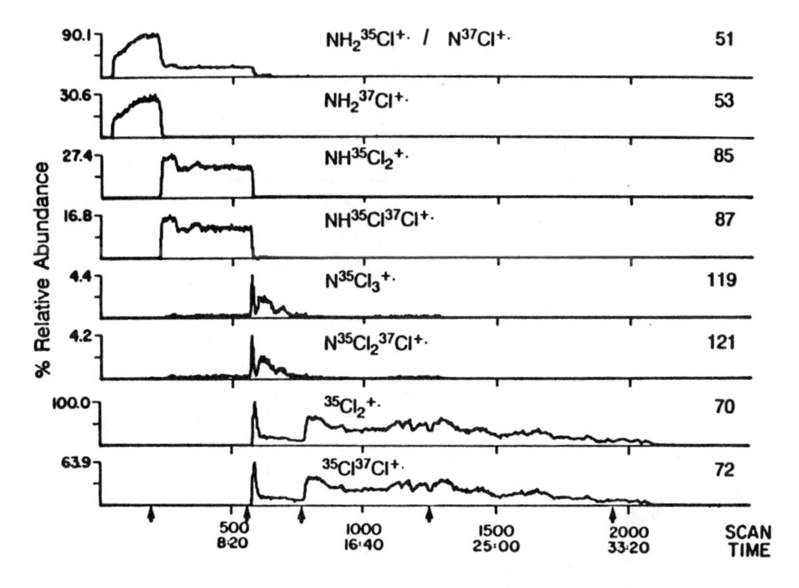

Figure 10. On-line monitoring of the reactions of monochloramine with HCl. Mass chromatograms presented are m/z 51 ($NH_2{}^{35}Cl^{+\cdot}$) and m/z 53 ($NH_2{}^{37}Cl^{+\cdot}$) for NH_2Cl, m/z 70 ($^{35}Cl_2{}^{+\cdot}$) and m/z 72 ($^{35}Cl^{37}Cl^{+\cdot}$) for chlorine, m/z 85 ($NH^{35}Cl_2{}^{+\cdot}$) and m/z 87 ($NH^{35}Cl^{37}Cl^{+\cdot}$) for $NHCl_2$, and m/z 119 ($NH^{35}Cl_3{}^{+\cdot}$) and m/z 121 ($NH^{35}Cl_2{}^{37}Cl^{+\cdot}$) for NCl_3. HCl aliquot additions occurred at the times indicated by the arrows. Data were collected by scanning the full EI mass spectra (m/z 33 to 300). Note that the ion m/z 51 is both due to $NH_2{}^{35}Cl^{+\cdot}$ and $N^{37}Cl^{+\cdot}$ in the proportion 4:1.

(Reprinted with permission from reference 29. Copyright 1991 Pergamon Press.)

a)

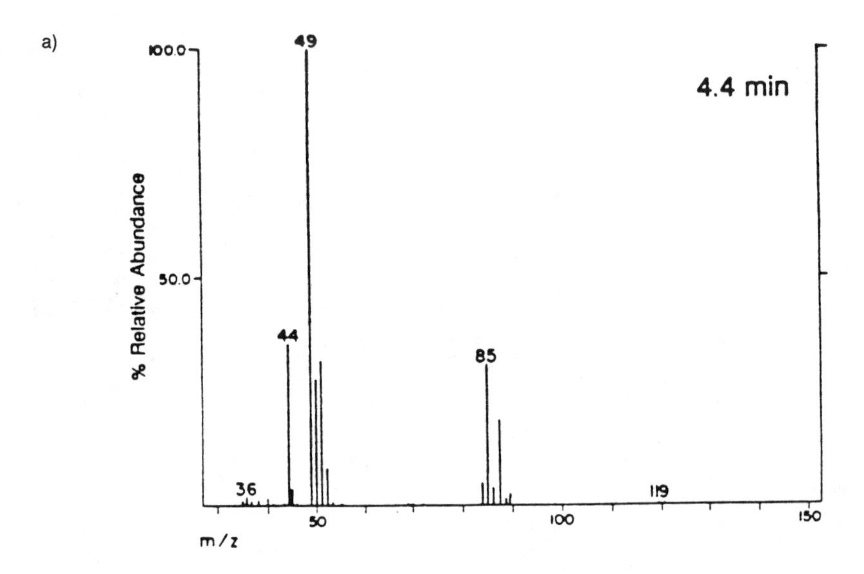

b)

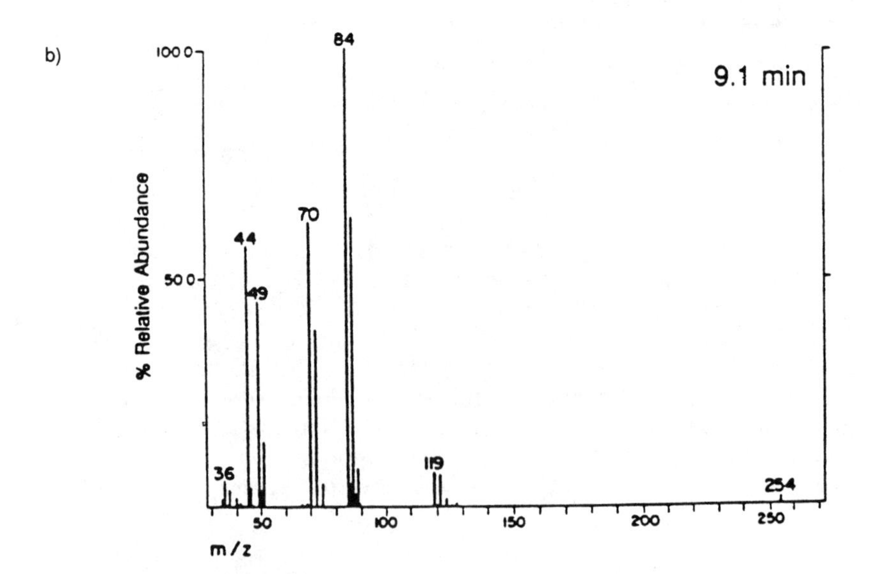

Figure 11. Representative EI mass spectra for the on-line monitoring experiment (see Figure 10). Figure 11 (a) shows the mass spectrum acquired during scans 266 to 270 (4.4 min) and Figure 11 (b) shows the mass spectrum acquired during scans 580 to 585 (9.1 min).
(Reprinted with permission from reference 29. Copyright 1991 Pergamon Press.)

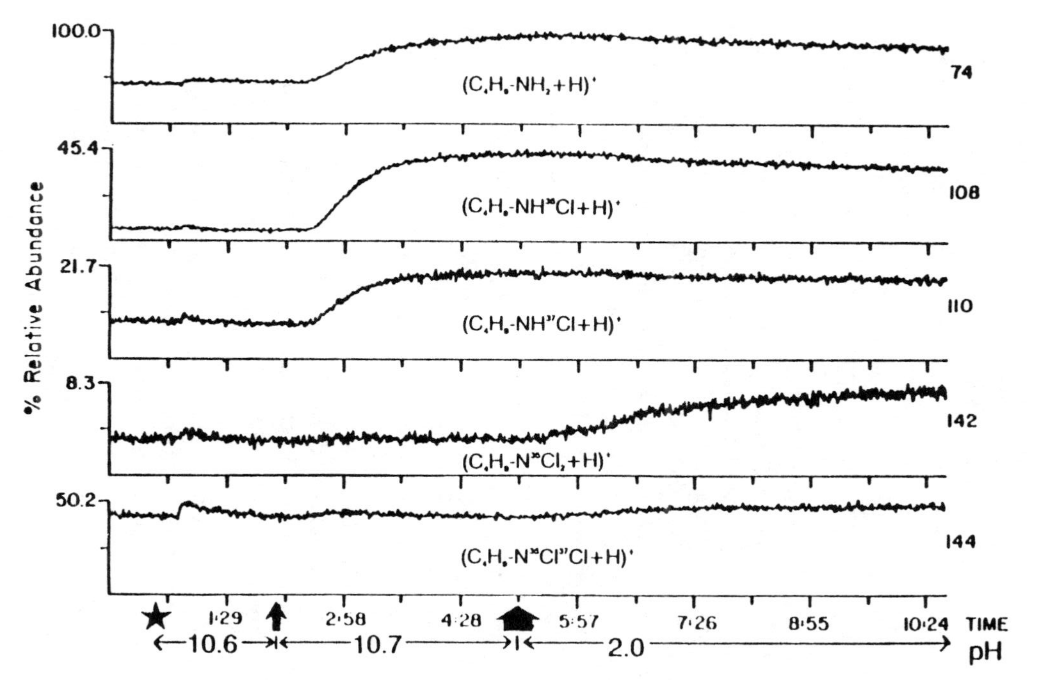

Figure 12. On-line monitoring of the chlorination reactions of 0.2 ppm 2-aminobutane. The mass chromatograms are for the following ions for 2-aminobutane: m/z 74 $(C_4H_9\text{-}NH_2+H)^+$, m/z 108 $(C_4H_9\text{-}NH^{35}Cl+H)^+$, m/z 110 $(C_4H_9\text{-}NH^{37}Cl+H)^+$, m/z 142 $(C_4H_9\text{-}N^{35}Cl_2+H)^+$, and m/z 144 $(C_4H_9\text{-}N^{35}Cl^{37}Cl+H)^+$. Replacement of water by the corresponding amine solution occurred at the time represented by the asterisk. Addition of calcium hypochlorite solution (small arrow) and HCl solution (heavy arrows) addition occurred at the times indicated. Note also that the pH of the reaction mixtures is presented at the bottom of the figure. Data were collected using CI(isobutane) while monitoring only the ions indicated. (Reprinted from reference 41. Copyright 1991 American Chemical Society.)

aminobutane is shown as Figure 13. The ion at m/z 52 shifts cleanly to m/z 54 when the ^{37}Cl-parent is chosen instead of the ^{35}Cl-parent. This demands that it contain Cl and it is assigned as NH_3Cl^+. This fact, and the fact that the normal butyl-derived ions at m/z 57, 55, 41 and 29 are unshifted, requires N-chlorination. Similar MS/MS data established that the second product is N,N-dichloro-2-aminobutane. In a separate experiment 1,3-diaminopropane was chlorinated and found also to yield the N-chlorinated products, in this case accepting up to 4 chlorine atoms as the pH is lowered (*41*).

An aromatic amine, aniline, behaves quite differently (*41*). The first indication of a mechanistic difference is provided by the fact that chlorination of aniline requires acidification. A second difference is that, depending on the conditions used, one, two or three chlorine atoms can be incorporated. The experiment is very sensitive as the 0.2 ppm data shown in Figure 14 demonstrate.

It is relatively simple, using MS/MS, to demonstrate that ring and not N-chlorination occurs in aniline (*41*). This assignment was facilitated by the fact that the authentic compounds are available. As an example the MS/MS spectra of the monochlorinated reaction product and authentic compounds are reproduced in Table III. The positions of isomeric substitution are also discernable from the MS/MS data because strong ortho effects operate to favor HCl elimination when the chlorine is ortho to the amino group.

Although product analysis could be based on mass spectrometry alone, it was judged desirable to confirm the aniline results chromatographically (*41*). The ITS-40 ion trap detector was used in the GC/MS mode for these experiments, and confirmed the MS/MS data. Typically, monochlorination yielded 80-95% of the 2-chloro isomer and 5-20% of the 4-chloro isomer and dichlorination yielded the 2,4-dichloro isomer in greater amount (ca. 2:1) than the 2,6-isomer. The GC/MS data also confirmed that the trichlorinated product was 2,4,6-trichloroaniline.

In these studies (*29,41*), the direct analysis of inorganic and organic chloramines is achieved without any derivatization or isolation. Even more significantly, the formation and interconversions of these labile and toxic compounds could be monitored in situ at sub-part-per-million levels. The main disadvantage of MIMS in chloramine analysis is that ionic compounds do not pass through the membrane and therefore analysis of some target compounds might not be possible. The main advantages of membrane introduction mass spectrometry for inorganic and organic chloramine analysis are that target compounds can be analyzed intact, different chloramines can be easily distinguished using tandem mass spectrometry and a high rate of analysis can be maintained when flow injection methods are used.

By-products of Ozonation of Drinking Water

Ozonation of drinking water is believed to increase concentrations of aldehydes, especially formaldehyde and acetaldehyde (*42,43*). This, together with the rapid increase in the use of ozonation in the U.S., means that it is desirable that

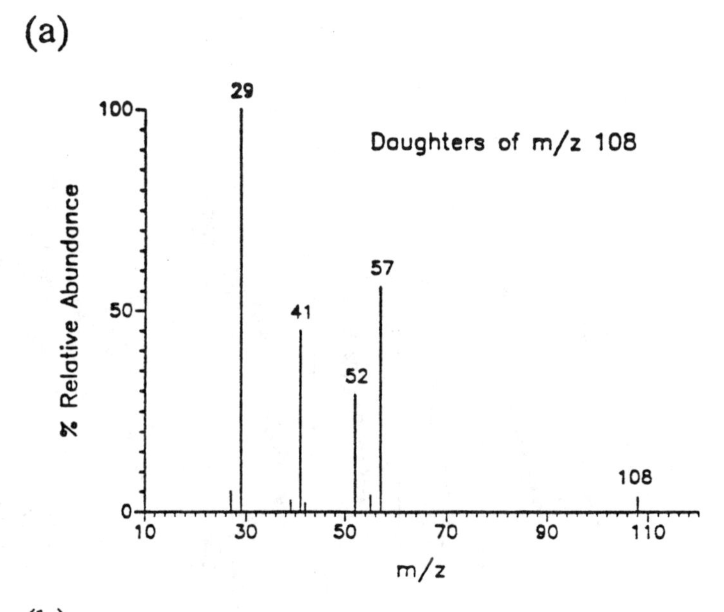

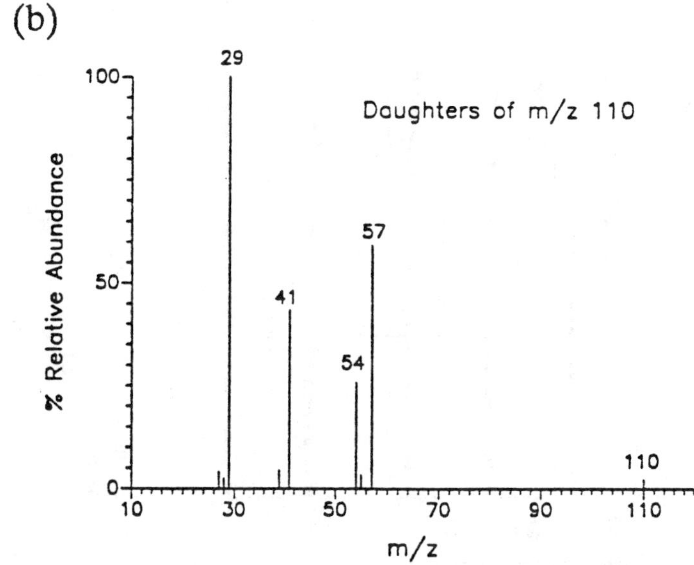

Figure 13. Product ion spectra of different isotopic forms of the protonated molecular ion (MH^+) of the monochloro derivative of 2-aminobutane. Figure (a) presents the spectrum measured for the isotope $(C_4H_9\text{-}NH^{35}Cl+H)^+$ m/z 108 and Figure (b) presents the spectrum measured for the isotope $(C_4H_9\text{-}NH^{37}Cl+H)^+$ m/z 110. Collision energy 20 eV and collision gas (argon) pressure 1 mtorr.

(Reprinted from reference 41.

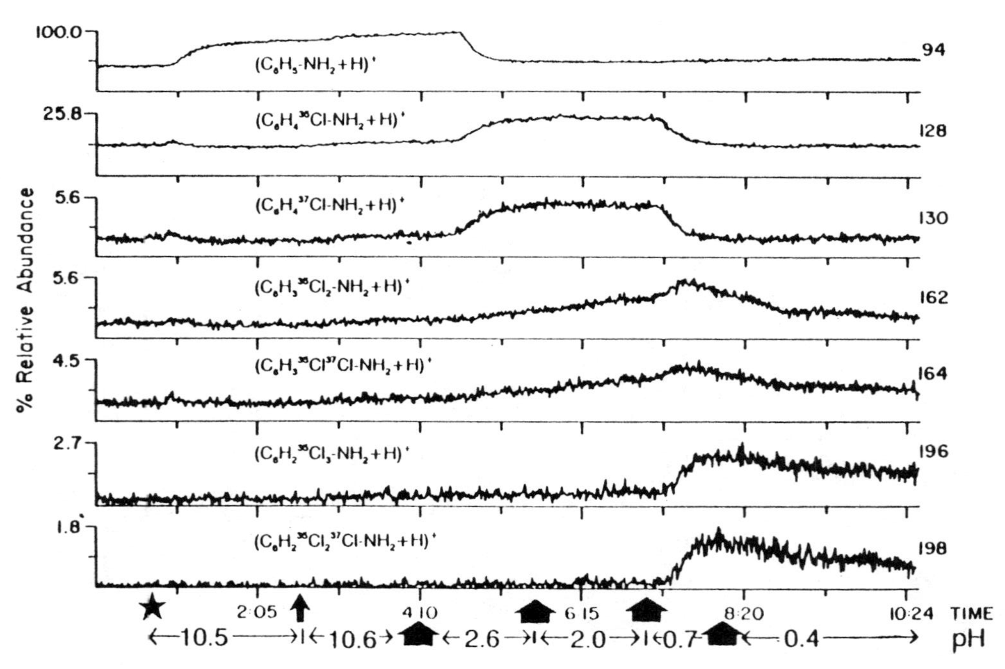

Figure 14. On-line monitoring of the chlorination reactions of 0.2 ppm aniline. The mass chromatograms are for the following ions of aniline and its derivatives: m/z 94 $(C_6H_5\text{-}NH_2+H)^+$, m/z 128 $(C_6H_4{}^{35}Cl\text{-}NH_2+H)^+$, m/z 130 $(C_6H_4{}^{37}Cl\text{-}NH_2+H)^+$, m/z 162 $(C_6H_3{}^{35}Cl_2\text{-}NH_2+H)^+$, m/z 164 $(C_6H_3{}^{35}Cl^{37}Cl\text{-}NH_2+H)^+$, m/z 196 $(C_6H_2{}^{35}Cl_3\text{-}NH_2+H)^+$, and m/z 198 $(C_6H_2{}^{35}Cl_2{}^{37}Cl\text{-}NH_2+H)^+$ are displayed. Replacement of water by the corresponding amine solution occurred at the time represented by the asterisk. Addition of calcium hypochlorite solution (small arrow) and HCl solution (heavy arrows) additions occurred at the times indicated. Note also that the pH of the reaction mixtures is presented at the bottom of the figures. Data were collected using CI(isobutane) while monitoring only the ions indicated. (Reprinted from reference 41. Copyright 1991 American Chemical Society.)

Table III. Product ion spectrum of different isotopic forms of the protonated molecule of 2-chloroaniline, 4-chloroaniline, and monochlorinated reaction product formed during chlorination of aniline [a,b]

Parent ion		Ions characteristic of chlorination position. m/z (%RA)			
m/z (%RA)	Ion Species	$(MH-NH_3)^+$	$(MH-Cl)^{+\cdot}$	$(MH-HCl)^+$	m/z (%RA)
2-chloroaniline					
128(67)	$(C_6H_4{}^{35}Cl-NH_2+H)^+$	111(0.4)	93(16)	92(100)	75(0.6), 66(1.3), 65(83), 39(3.7)
130(64)	$(C_6H_4{}^{37}Cl-NH_2+H)^+$	113(0.2)	93(12)	92(100)	75(0.5), 66(1.4), 65(76), 39(3.4)
4-chloroaniline					
128(56)	$(C_6H_4{}^{35}Cl-NH_2+H)^+$	111(20)	93(100)	92(1.3)	75(4.3), 66(1.1), 65(1.4)
130(56)	$(C_6H_4{}^{37}Cl-NH_2+H)^+$	113(24)	93(100)	92(0.9)	75(4.5), 66(0.9), 65(1.0)
Monochloro product					
128(66)	$(C_6H_4{}^{35}Cl-NH_2+H)^+$	111(1.1)	93(20)	92(100)	75(0.8), 66(1.8), 65(87), 39(4.1)
130(59)	$(C_6H_4{}^{37}Cl-NH_2+H)^+$	113(1.0)	93(17)	92(100)	75(0.8), 66(1.3), 65(81), 39(4.2)

[a]Collision energy 20 eV and collision gas argon.

[b]Data from ref. 41.

analytical methods which allow molecularly specific, low-level detection of ozonation by-products, preferably in an on-line fashion, be developed. Although this could be attempted using direct analysis of the water samples via the FIA/MIMS methodology utilized in the experiments described above, the need for still lower levels of detection suggested that a rapid on-line derivatization reaction be incorporated into the protocol. Further, because O-(pentafluorobenzyl)hydroxylamine (PFBOA) is a useful derivatizing agent for determination of low levels of ketones and aldehydes by electron capture GC (*43,44*), it was chosen for these experiments. The derivative has already been shown to allow low ppb detection in off-line GC/ECD and GC/MS experiments (*42*).

As a first step in the on-line FIA/MIMS experiment with on-line derivatization, we have studied the ozonation of aqueous solutions of organic compounds using on-line monitoring by membrane introduction mass spectrometry (*45*). The direct insertion membrane probe (*23*) was used on a triple quadrupole mass spectrometer. Typical of the data obtained are results for 10 ppm styrene which was treated with O_3 at room temperature for four minutes. The styrene ions (EI, 70 eV) at m/z 104 (100%) m/z 103 (30%) and m/z 78 (25%) were quantitatively replaced by ions due to benzaldehyde (m/z 106, 62%; m/z 105, 92%; m/z 77, 100%; m/z 51, 42%). Confirmation of the identity of the product was obtained by recording the MS/MS product ion spectrum of the product m/z 106 by collision with argon at 20 eV in the triple quadrupole (Figure 15).

Experiments have also been done on ozonation of mixtures, e.g. a mixture of CH_2Cl_2, $C_2H_5OC_2H_5$, $C_6H_5CH_3$, C_6H_5Cl and C_6H_6 (*45*). Among these compounds, toluene is most readily oxidized. On-line monitoring of such competitive processes is now possible using the flow-injection analysis system together with membrane probe sampling.

The study of the formation of PFBOA derivatives and their mass spectrometric properties is also underway (*26*). Negative chemical ionization is used and for example a detection limit of 1 ppb for formaldehyde and 10 ppb for acetaldehyde PFBOA derivatives has been measured. The fragmentation behavior under collisional activation has also been established (*26*). Aldehydes do not give highly abundant molecular anions ($M^{-\cdot}$) however, the HF elimination product is relatively abundant and serves as to indicate the nature of the aldehyde. Fragmentations which are characteristic of the aldehyde are also observed, e.g. CH_3CN loss occurs from the $(M\text{-}HF)^{-\cdot}$ ion of the acetaldehyde derivative (Figure 16). On the basis of this result an MS/MS single reaction monitoring experiment is being developed for low level monitoring. Another MS/MS experiment of great potential value is a constant neutral loss 20 dalton scan, which should serve to reveal the molecular weights of all derivatized aldehydes and ketones in the sample.

The next step in these experiments is the study, on-line, of the oxidation processes accompanying ozonation. This work needs to be done at very low levels and the ability to perform on-line reactions of aldehydes and ketones, using the apparatus already available (*21*), to yield PFBOA derivatives and to detect these by NCI will be pursued. The quantitation of the reaction products

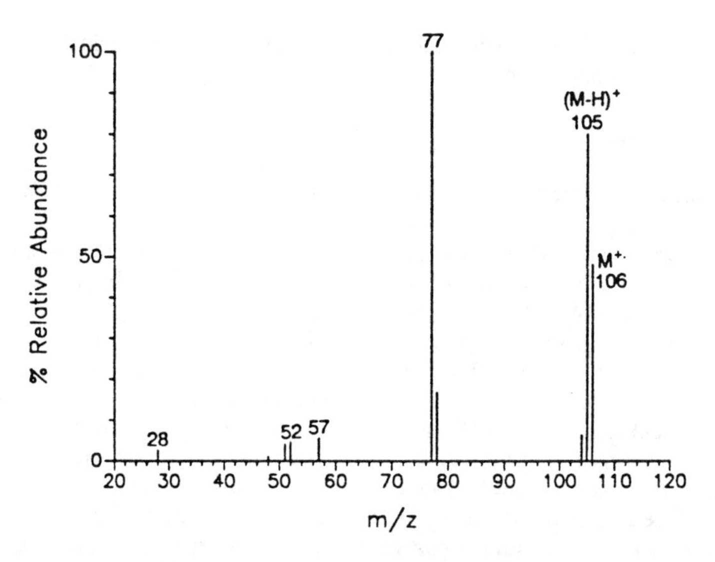

Figure 15. Product spectrum of ion m/z 106 present in a 10 ppm styrene solution after treatment with O_3 for four minutes. Electron ionization (70eV), collision energy 20 eV and collision gas argon.

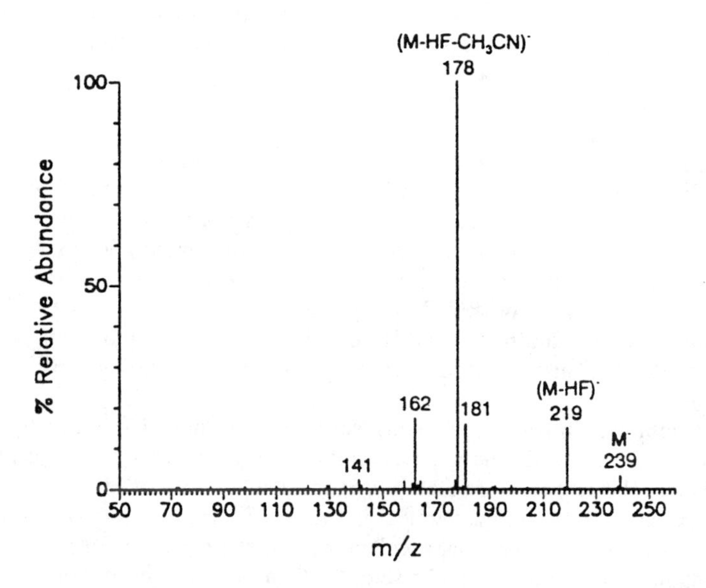

Figure 16. Product spectrum of $M^{-\cdot}$ ion pf PFBOA derivative of acetaldehyde. Negative ion chemical ionization (isobutane), collision energy 20 eV and collision gas argon.

will be attempted using external standards, an experiment which has already been used in other cases.

Air Monitoring

Membrane introduction mass spectrometry has also been used to monitor environmentally significant compounds directly from air (*3,4,18*), but the experiment is not widely used. In our laboratory some initial air monitoring experiments have also been done (Figure 17) (*46*). In Figure 17 ion chromatograms of protonated molecules, MH^+, of diethyl ether (m/z 75), benzene (m/z 79), toluene (m/z 93) and chlorobenzene (m/z 115, ^{37}Cl isotopic form) are presented (*46*). This experiment was done using a triple quadrupole instrument, the sheet direct insertion membrane probe for sample introduction and chemical ionization (methane). During the experiment laboratory air (light arrows) and an air sample (heavy arrows) containing diethylether (110 ppm), benzene (0.13 ppm), methylene chloride (9 ppm), toluene (11 ppm), and chlorobenzene (1.1 ppm) was introduced through the membrane probe. All the compounds except for methylene chloride were detected. Methylene chloride was not detected, because laboratory air has very high background signal level at m/z 85 i.e. the protonated molecule of methylene chloride. For this same reason the ion m/z 115 (^{37}Cl isotope) for chlorobenzene gave a better signal to noise ratio than the ion m/z 113 (^{35}Cl isotope). This experiment and other work clearly demonstrate that membrane introduction mass spectrometry can be used for real time monitoring of environmentally significant compounds from air and that detection limits at ppb levels can achieved without preconcentration.

Conclusions

Membrane introduction mass spectrometry shows great promise as an analytical method for the analysis of environmentally significant compounds particularly from aqueous solutions. The main advantage is that these compounds can be analyzed directly from aqueous samples at part-per-billion levels without any isolation or derivatization steps. Even though the detection limits for most of the compounds are at low ppb levels, lower detection limits still are desirable. One method of addressing this problem is to employ rapid on-line preconcentration and therefore several methods for doing this are being explored.

Membrane introduction mass spectrometry depends critically on the properties of the membrane used as an interface between aqueous solution and gas phase. The temperature stability, ready availability in the form of sheets or as microcapillaries, and the chemical inertness of silicone rubber are advantages of this material which also has excellent rejection properties for water. On the other hand, this polymer is limited to transport of more volatile organic compounds of relatively low molecular weight. For example, compounds greater than 300 dalton molecular weight are seldom successfully transmitted and ionic or highly polar compounds are not passed by the membrane. Development of

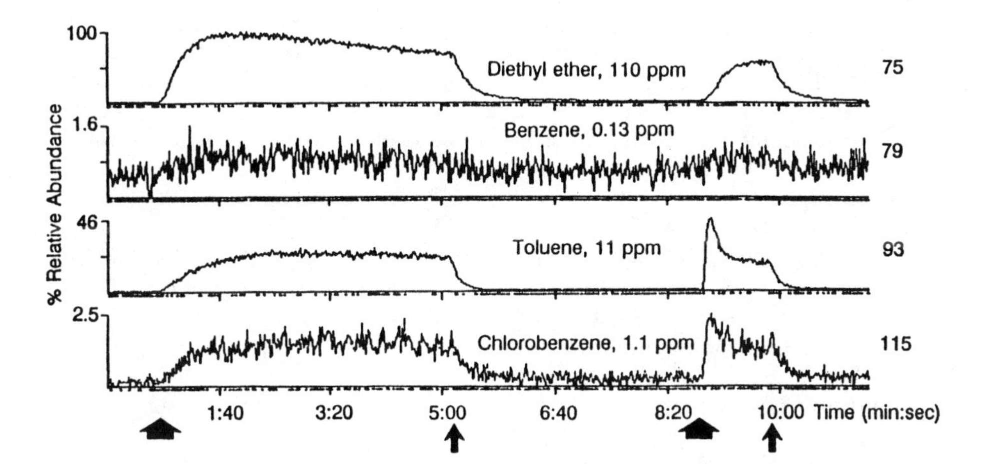

Figure 17. An air monitoring experiment measured using a triple quadrupole instrument, the sheet direct insertion membrane probe for sample introduction and chemical ionization (methane). Ion chromatograms presented are the following: m/z 75, MH^+ of diethyl ether; m/z 79, MH^+ of benzene; m/z 93, MH^+ of toluene and m/z 115, MH^+ of chlorobenzene (^{37}Cl isotopic form). Introduction of the air sample into the membrane probe is indicated by heavy arrows and introduction of laboratory air is indicated by light arrows.

alternative membranes with selectivity toward particular classes of organic compounds is desirable and work on this problem is currently underway.

Summarizing, membrane introduction mass spectrometry is an excellent method for direct analysis of environmentally significant compounds from aqueous samples at ppb concentration levels. Analysis of these compounds from air samples at ppb levels may also be possible. The low detection limits obtained are facilitated by the use of flow-through operation mode of membrane introduction device, which gives also minimal memory effects and fast response times. Together with flow injection analysis methods, membrane introduction mass spectrometry provides the high speed of analysis necessary for high sample throughput. It is also demonstrated that the direct analysis capability of membrane introduction mass spectrometry allows the analysis of toxic and labile compounds such as acrolein, acrylonitrile and inorganic and organic chloramines. Taking all these advantages into account one would expect a large increase in the near future in the utilization of membrane introduction mass spectrometry in environmental analysis as well as in on-line reaction and process monitoring.

Acknowledgments

The support of the U.S. Environmental Protection Agency (EPA CR-815749-01-0) is acknowledged. The contributions to these developments of the coworkers cited in the references are gratefully acknowledged. The support provided for T.K. by the Emil Aaltonen Foundation is greatly appreciated.

Literature Cited

1. Hwang, S. T.; Kammermeyer, K. *Membrances in Separation*; Techniques in Chemistry Series, Vol. VII; John Wiley & Sons:New York, NY, 1975.
2. Llewellyn, P. M.; Littlejohn, D. P. U.S. patent 3,429,105, February 1969.
3. Tou, J. C.; Kallos, G. J. *Anal. Chem.* **1976**, *46*, 1866.
4. Evans, J. E.; Arnold, J. T. *Environ. Sci. Technol.* **1975**, *9*, 1134.
5. Woldring, S.; Owens, G.; Woolford, D. C. *Science* **1966**, *153* 885.
6. Brodbelt, J. S.; Cooks, R. G.; Tou, J. C.; Kallos, G. J.; Dryzga, M. D. *Anal. Chem.* **1987**, *59*, 454.
7. Westover, L. B.; Tou, J. C.; Mark, J. H. *Anal. Chem.* **1974**, *46*, 568.
8. Calvo, K. C.; Weisenberger, C. R.; Anderson, L. B.; Klapper, M. H. *Anal. Chem.* **1981**, *53*, 981.
9. Hoover, K. C.; Hwang, S.-T. *J. Membrane Sci.* **1982**, *10*, 253.
10. Pungor, E.; Pecs, M.; Szigeti, L.; Nyeste, L. *Anal. Chim. Acta* **1984**, *163*, 185.
11. Heinzle, E.; Moes, J.; Griot, M.; Kramer, H.; Dunn, I.J.; Bourne, J. R. *Anal. Chim. Acta* **1984**, *163*, 219.
12. Reuss, M.; Piehl, H.; Wagner, F. *Eur. J. Appl. Microbiol.* **1975**, *1*, 323.
13. Slivon, L.E.; Bauer, M.R.; Ho, J.S.; Budde, W.L. *Anal. Chem.* **1991**, *63*, 1335.
14. Brodbelt, J. S.; Cooks, R. G. *Anal. Chem.* **1985**, *57*, 1153.

15. Bier, M. E.; Cooks, R. G.; Brodbelt, J. S.; Tou, J. C.; Westover, L. G. U.S. Pat., 4791292, 1989.
16. Bier, M. E.; Cooks, R. G. *Anal. Chem.* **1987**, *59*, 597.
17. Lister, A. K., Wood, K. V.; Cooks, R. G.; Noon, K. R. *Biomed. Envir. Mass Spectrom.* **1989**, *18*, 1063.
18. LaPack, M.A.; Tou, J. C.; Enke, C.E. *Anal. Chem.* **1990**, *62*, 1265.
19. Bauer, S.J; Hayward, M.J.; Riederer, D.E; Kotiaho, T.; Cooks, R.G.; Austin, G.D.; Tsao, G.T., unpublished.
20. Lauritsen, F.R. *Int. J. Mass Spectrom. Ion Proc.* **1990**, *95*, 259.
21. Hayward, M. J., Kotiaho, T.; Lister, A. K.; Cooks, R. G.; Austin, G. D.; Narayan, R.; Tsao, G. T. *Anal. Chem.* **1990**, *62*, 1798.
22. Hayward, M. J.; Riederer, D. E.; Kotiaho, T.; Cooks, R. G.; Austin, G. D.; Syu, M.; Tsao, G. T. *Proc. Cont. Qual.* **1991**, *1*, 105.
23. Bier, M. E.; Kotiaho, T.; Cooks, R. G. *Anal. Chim. Acta* **1990**, *231*, 175.
24. Brodbelt, J. S.; Louris, J. N.; Cooks, R. G. *Anal. Chem.* **1987**, *59*, 1278.
25. Choudhury, T.K.; Cooks, R.G unpublished results.
26. Choudhury, T.K.; Kotiaho, T.; Cooks, R.G. unpublished results.
27. Lister, A.K. Ph.D. thesis, Purdue University, 1990.
28. Hanson, D. *C&E News* **1990**, *March 12*, 5.
29. Kotiaho, T.; Lister, A.K.; Hayward, M.J.; Cooks, R.G. *Talanta* **1991**, *38*, 195.
30. White, G. C. *The Handbook of Chlorination 2nd ed*; Van Nostrand Reinhold Company Inc.:New York, NY, 1986.
31. Gray, E. T. Jr., Margerum, D. W.; Huffman, R. P. *Chloramine Equilibrium and Kinetics of Disproportionation in Aqueous Solution. Organometals and Organometalloids, Occurrence and Fate in the Environment.*; Brinckman, F. E.; Bellama, J. M., Eds.; ACS Symp. Ser. No. 82; American Chemical Society:Washington, DC, 1978. p. 264.
32. Margerum, D. W.; Gray, E. T. Jr.; Huffman, R. P. *Chlorination and the Formation of N-Chloro Compounds in Water Treatment. Organometals and Organometalloids, Occurrence and Fate in the Environment.*; Brinckman, F. E.; Bellama, J. M., Eds.; ACS Symp. Ser. No 82, American Chemical Society:Washington, DC, 1978. p. 278.
33. Hand, V. C.; Margerum, D. W. *Inorg. Chem.* **1983**, *22*, 1449.
34. Kumar, K.; Shinnes, R. W.; Margerum, D. W. *Inorg. Chem.* **1987**, *26*, 3430.
35. Helz, G. R.; Kosak-Channing, L. *Environ. Sci. Technol.* **1984**, *18*, 48A.
36. Aoki, T *Environ. Sci. Technol.* **1989**, *23*, 46.
37. Scully, F. E. Jr.; Mazina, K.; Sonenshine, D. E.; Ringhand, H. P. *Toxicological Significance of the Chemical Reactions of Aqueous Chlorine and Chloramines. Biohazards of Drinking Water Treatment*; Larson, R. A., Ed.; Lewis Publishers, Inc.:xxx, Michigan, 1989, p. 141.
38. Kearney, T.J.; Sansone, F.J. *Water Chlorination Chemistry, Environmental Impact and Health Effects*, Jolley, R.L.; Bull, R.J.; Davis, W.P.; Katz, S.; Roberts, Jr., M.H.; Jacobs, V.A., Eds.; Vol. 5, Lewis Publishers Inc., Michigan, 1985, p. 965.

39. Lukasewycz, M.T.; Bieringer, C.M.; Liuokkonen, R.J.; Fitzsimmons, M.E.; Corcoran, H.F.; Lin, S.; Carlson, R.M. *Environ. Sci. Technol.*, **1989**, *23*, 196.
40. Nweke, A.; Scully, Jr., F.E. *Environ. Sci. Technol.*, **1989**, *23*, 989.
41. Kotiaho, T.; Hayward, M.J.; Cooks, R.G. *Anal. Chem.* **1991**, *63*, 1794.
42. Glaze, W. H.; Koga, M.; Cancilla, D. *Envir. Sci. Technol.* **1989**, *23*, 838.
43. Yamada, H.; Somiya, I. *Ozone Science & Engineering* **1989**, *11*, 127.
44. Koshy, K. T.; Kasier, D. G.; Van Der Slik, A. L. *J. Chrom. Sci.* **1975**, *13*, 97.
45. Kotiaho, T.; Cooks, R.G unpublished results.
46. Kotiaho, T.; Hayward, M.J.; Cooks, R.G. unpublished results.

RECEIVED April 23, 1992

Chapter 13

Hollow Fiber Membrane System for the Continuous Extraction and Concentration of Organic Compounds from Water

Caryl L. Fish[1], Ian S. McEachren[2], and John P. Hassett[3]

Chemistry Department, State University of New York, College of Environmental Science and Forestry, Syracuse, NY 13210

A system has been developed and evaluated which extracts and concentrates semi-volatile organic compounds from water using hollow fiber membranes. The extraction and concentration components have been evaluated using naphthalene as a test compound. Both components were designed using polydimethylsiloxane hollow fiber membranes and hexane as the extraction solvent. In the extraction component the stir rate in water was found to have a significant effect on the permeation coefficient. A linear response of permeation to water temperature was also found. In the concentration component naphthalene was found to be retained in the solvent. Concentration factors for an optimized system could reach 2000.

Control of process and wastewater streams to minimize or prevent discharge of contaminants to the environment requires analytical methods which can provide continuous, real-time information. While most instruments for analysis of organic chemicals (e.g. GC, MS, UV, IR) can provide results on time-scales of less than an hour, they usually require that the sample be processed in some manner before it is presented to the instrument. This sample processing can require several hours to days. For example, conventional analysis of organic chemicals in water involves a batch liquid-liquid extraction to remove the chemicals from the water, concentration of the extract by evaporation and clean-up of the concentrate by

[1]Current address: Chemistry Department, St. Vincent College, Latrobe, PA 15650
[2]Current address: Engineering Science, Inc., 290 Elwood Davis Road, Suite 312, Liverpool, NY 13088
[3]Corresponding author

0097-6156/92/0508-0155$06.00/0

adsorption chromatography before chromatographic analysis (*1*). This approach is unsuitable for process control since it is discontinuous, slow, labor-intensive, expensive, and it generates a large amount of waste as a byproduct.

The goal of the research described here has been to develop a method for continuous, on-line extraction and concentration of organic chemicals from water which yields an extract stream compatible with existing analytical instruments. The conventional purge and trap method (*1*) already offers a relatively simple and rapid analysis of volatile compounds in water, and experimental methods such as membrane-separator/mass spectrometry are also very promising for these compounds (*2,3*), but these approaches are not suitable for less volatile compounds. The target analytes for this research are semi-volatile organic compounds that are analyzed by EPA Method 625.

The device which has been developed is shown schematically in Figure 1. It contains an extraction module and a concentration module, both incorporating hollow-fiber membranes. In the extraction module, solvent is continuously pumped through the inside of a hollow fiber, and water is continuously pumped from the sampling point past the outside of the fiber. Extraction takes place by diffusion of analytes from the water through the membrane and into the solvent. The extract flows from the extraction module to the concentration module, where it passes through a hollow fiber, the outside of which is under vacuum. Solvent molecules diffuse through the membrane while the more slowly diffusing analyte molecules are selectively retained and concentrated in the remaining solvent. The concentrated extract flows from the concentration module to an analytical instrument compatible with the solvent and extract. For example, this could be a flow-through absorbance monitor or a chromatograph equipped with an autoinjector. Thus, continuous extraction, concentration and analysis is accomplished without the need for manual manipulation of the sample. Background and the theoretical bases for the extraction and concentration modules are presented below.

Extraction. The extraction component is based on the diffusion of hydrophobic compounds through a non-porous hollow fiber membrane. This is the same process used in the passive sampling of organic compounds from water using flat sheet membranes. (*4-6*) In passive sampling the compounds in water diffuse through the membrane and are collected in a solvent reservoir. This is a batch process which produces an integrated sample over the sampling period. It can not detect concentration changes on a real time basis. A continuous extractor could detect these changes providing feedback for pollution prevention. Continuous systems have been developed by D'Elia et al. (*7*) and Sirkar and Prasad (*8*) for extraction of industrial components using porous hollow fibers. The pores allow direct contact of the water and solvent. This gives high diffusion rates, but the pressure must be carefully controlled to avoid emulsion formation. Non-porous hollow fibers would eliminate the emulsion problem and allow sufficient extraction for analytical purposes. Non-porous, hydrophobic hollow fibers would also exclude the polar natural organic matter eliminating the need for sample clean-up. Melcher (*9*) used non-porous hollow fibers in a flow injection apparatus to

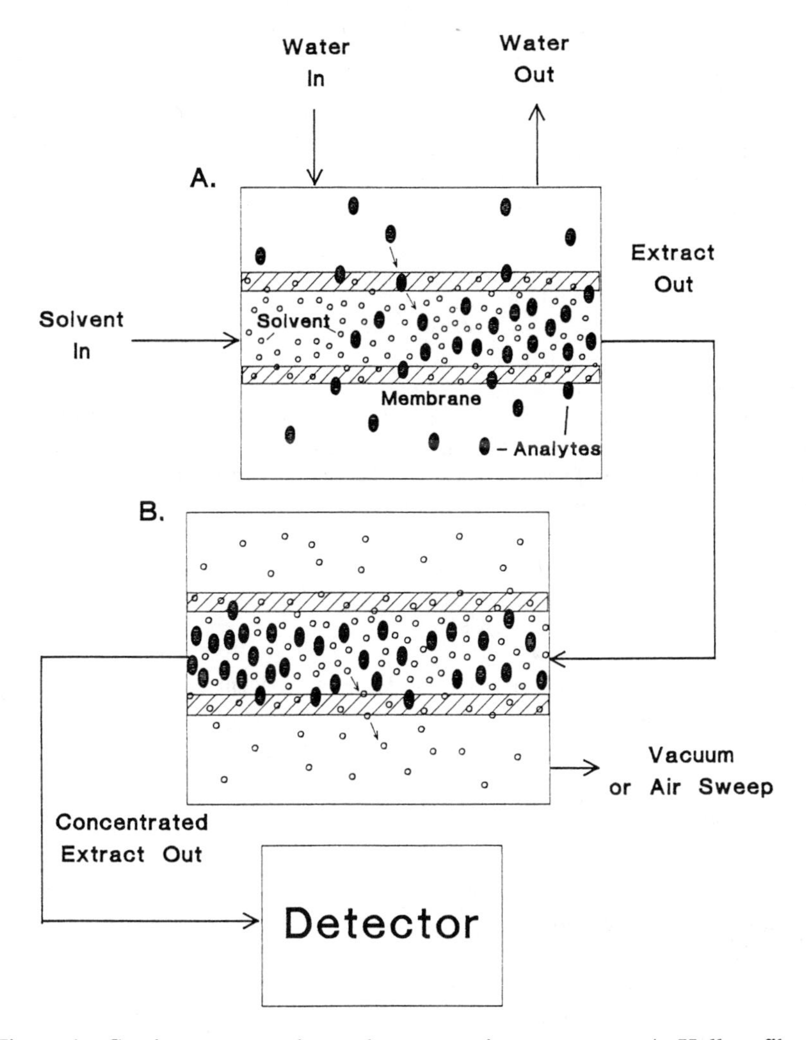

Figure 1. Continuous extraction and concentration apparatus. A. Hollow fiber extraction process. B. Hollow fiber pervaporation process.

measure concentrations of phenol in water. Melcher and Morabito (*10*) used a similar hollow fiber extraction system interfaced to a GC and evaluated it for extraction of chlorobenzenes and pesticides.

The extraction process is shown schematically in Figure 1A. The extraction of these compounds occurs in three steps. First, the hydrophobic compounds from the water partition to the membrane surface. The compounds then diffuse through the membrane and finally partition from the membrane to the solvent inside the hollow fiber. The solvent containing the extracted compounds continues to the concentration hollow fiber.

The diffusion process at steady state is described by Fick's law (Equation 1) where J is the flux through the membrane, D is the diffusion coefficient,

$$J = \frac{D\ A}{L} (Cmw - Cms) \tag{1}$$

A is the surface area of the membrane, L is the thickness of the membrane, and Cmw and Cms are the concentrations in the membrane at the membrane-water and membrane-solvent interface, respectively. The partition coefficients between the membrane and the water (Kw) and solvent (Ks) are defined in Equations 2 and 3.

$$Kw = Cmw/Cw \tag{2}$$

$$Ks = Cms/Cs \tag{3}$$

Cw is the concentration in the water and Cs is the concentration in the solvent. Substitution gives Equation 4.

$$J = \frac{D\ A}{L} (Kw\ Cw - Ks\ Cs) \tag{4}$$

If the system does not approach equilibrium, then $Kw\ Cw >> Ks\ Cs$ and the equation simplifies to Equation 5.

$$J = \frac{D\ A\ Kw}{L}\ Cw \tag{5}$$

The permeation coefficient (P) can be defined as in Equation 6

$$P = D\ Kw \tag{6}$$

giving Equation 7.

$$J = \frac{P\,A}{L}\,Cw \tag{7}$$

Equation 7 can be used to determine P by measuring the flux through the membrane, the concentration of the compound in water, and knowing the dimensions of the hollow fiber. Then in the field the concentration in the water can be determined using P and measuring the flux through the membrane. The effect of turbulence and temperature on P must also be determined.

Pervaporation. The concentration component of the system utilizes the process of pervaporation. This process separates liquid mixtures by selective permeation of one or more components through a non-porous membrane into the vapor phase. Improved separation can be achieved by altering the compostion of the membrane or the mixture, manipulating the temperature and reducing the vapor phase pressure. Pervaporation has been explored to achieve ethanol dehydration (*11,12*), separation of azeotropes (*13-16*) and compounds with similar chemical properties (*16-18*), and trace organic compounds from water for analysis by mass spectroscopy (*2,3*).

Pervaporation is used in this apparatus to concentrate the extract prior to analysis. This process is illustrated in Figure 1B. The mechanism involves the sorption of a portion of the solvent into the membrane, diffusion through the membrane, and desorption into the gas phase. The less volatile analytes are retained in the remaining solvent.

The steady state pervaporation process can be described by Fick's law of diffusion (Equation 8).

$$J = \frac{D\,A}{L}\,(Cml - Cmv) \tag{8}$$

J, D, A, and L are as defined above. Cml and Cmv are the concentrations of the solvent at the membrane-liquid and membrane-vapor interfaces. Cml and Cmv can be expressed in terms of liquid solvent-membrane and vapor solvent-membrane partition coefficients (Kl and Kg) as shown in Equations 9 and 10. The concentrations in the liquid and gas phases are given in terms of the mole fraction (X), and the partial pressure (p) of the solvent.

$$Kl = Cml\,/\,X \tag{9}$$

$$Kg = Cmv\,/\,p \tag{10}$$

Substitution into Equation 8 gives Equation 11.

$$J = \frac{D\,A}{L}\,(Kl\,X - Kg\,p) \tag{11}$$

Since the solvent solution contains only trace solute concentrations, the mole fraction of solvent is one. Since the vapor side of the membrane is maintained at a low pressure, Equation 11 can be simplified to Equation 12.

$$J = \frac{D\,A}{L}\,Kl \qquad (12)$$

Thus the flux of the solvent through the membrane is a function of the diffusion coefficient, the solubility of the solvent in the membrane, and the physical characteristics of the membrane.

Methods and Materials

Chemicals. Solvents used were hexane, (Resi-analyzed, J.T. Baker, Inc.) toluene, (SpectraAnalyzed, Fisher Scientific, Inc.) dichloromethane, (Resi-analyzed, J.T. Baker, Inc.) ethyl acetate, (Certified, ACS, Fisher Scientific, Inc.) iso-octane, (pesticide grade, Fisher Scientific, Inc.) pentane, (Resi-analyzed, J.T. Baker, Inc.) and methanol (Baker analyzed, J.T. Baker, Inc.). Naphthalene was 99+ %, Aldrich "gold labeled", Aldrich Chemical Co., Inc.

Hollow Fibers. The membrane materials were polydimethylsiloxane (PDMS) (Silastic brand medical grade, Dow Corning, Inc., OD = 0.22 cm, ID = 0.10 cm), polyethylene (PE) (in house by I. Cabasso, OD = 0.125 cm, ID = 0.100 cm, density = 0.923 g/cm^3), aminated polyethylene (A-PE) and sulfonated polyethylene (S-PE) (in house by I. Cabasso, OD = 0.114, cm ID = 0.088 cm, 1.1 meq SO_3^- /g and 0.25 meq NH_4^+/g).

Extraction Apparatus. The extraction apparatus is shown in Figure 2. The hollow fiber was contained in a 1 L glass chamber. Water was pumped into the hollow fiber chamber where it was extracted. A continuous source of water with a constant concentration of naphthalene was produced by spiking purified tap water. The tap water flowed through an ion exchange resin followed by an activated carbon cartridge (Barnstead, Inc.). Water samples for analysis were collected directly from the hollow fiber chamber. Solvent was pumped through the inside of the hollow fiber and after extraction flowed through a UV detector (V4 absorbance detector, Isco, Inc.) at 275 nm wavelength to determine if steady state had been reached. Solvent samples were collected at steady state.

All water tubing was 0.64 cm O.D. copper or polypropylene. The solvent tubing was 0.16 cm O.D. teflon or stainless steel tubing. The hollow fiber membrane was 30.0 cm long. The water was temperature controlled by flowing through a thermostated water bath (Fisher Scientific, Inc.). The water was pumped at constant flow of 500 ml/min with a gear pump (Micropump, Cole Parmer, Inc.). A spiking solution of 3.0 mg/ml naphthalene in methanol was spiked into the water with a metering pump (Milton Roy, Inc.) at a constant flow rate of 0.170 ml/min.

The solvent was pumped (model RHSY, Fluid Metering, Inc.) at a flow rate of 1.80 ml/min unless otherwise noted. For the evaluations of solvents and membranes the concentration in the solvent was determined by GC analysis. (Hewlett-Packard model 5890) using a capilliary column and FID detector. The temperature program was as follows: initial temperature of 100 °C for 3 minutes, then ramp to a final temperature of 130 °C at 3 °C/min and hold for 4.5 min. Naphthalene concentrations in the water and in the solvent for the other experiments were determined by UV analysis at 275 nm (Perkin-Elmer, model Lambda 5).

Concentration Apparatus. The pervaporative concentration apparatus is shown in Figure 3. Hollow fiber membranes (length = 15.0 - 75.0 cm) were contained in a sealed glass vacuum chamber immersed in a constant temperature water bath. Vacuum was maintained in the hollow fiber chamber at 0.036 - 0.038 atm with a vacuum pump (teflon diaphragm pump, KNF Neuberger, Inc.). The solvent solution to be concentrated was pumped (Lab Pump Jr., Fluid Metering, Inc.) through the interior of the hollow fiber and the concentrated extract was collected in a fraction collector (Cygnet Model, Isco, Inc.). A liquid nitrogen cold trap collected the pervaporated solvent. Flow rates were determined from the weight of the solvent samples which were collected over a given time period. The flux of the solvent through the membrane was determined from the difference between the inlet solvent flow rate and the outlet solvent flow rate. The naphthalene concentrations before and after pervaporation were determined by UV analysis at 275 nm as above.

Combined System. The extraction and concentration systems were combined by attaching the outlet of the extraction system to the inlet of the concentration system. Samples were collected from the outlets of the extraction and pervaporation systems. Water samples were taken from the extraction chamber. All samples were analyzed by UV absorption at 275 nm. For this experiment the flow rate of the hexane was 0.659 ml/min., the water temperature was 25 °C, and the pervaporation system temperature was 60 °C.

Results and Discussion

Extraction System. Solvents and membranes were evaluated by comparing the permeation coefficient for each combination of membrane and solvent. Permeation coefficients (P) are shown in Table I. All permeation coefficients are based on the swollen length and surface area. The swelling factor (swollen length/unswollen length) for PDMS in hexane is 1.52. Unswollen permeation coefficients could be obtained by multiplying P by this factor. The PDMS membrane had permeation coefficients which were much higher than the other membranes with all the solvents tested. The solvent giving the highest permeation coefficients was dichloromethane. However, problems with this solvent in the concentration component of the system required the use of hexane in the remaining experiments. S-PE membranes, with charged SO_3^- groups, had very low values of

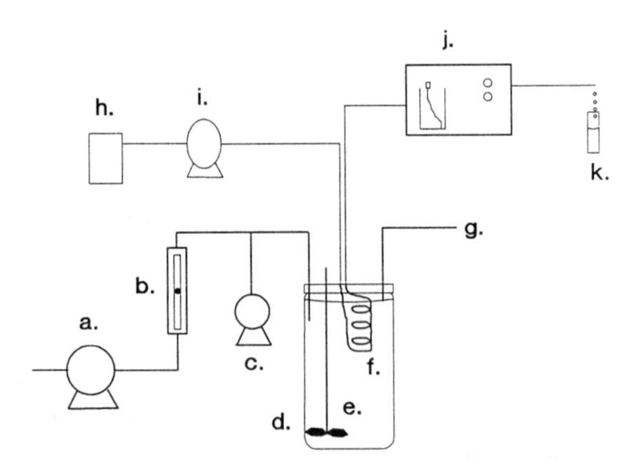

Figure 2. Extraction apparatus: (a) water pump, (b) flow meter, (c) spiking pump, (d) extraction chamber, (e) mechanical stirrer, (f) hollow fiber, (g) waste water, (h) solvent reservoir, (i) solvent pump, (j) flow-through UV detector, (k) fraction collector.

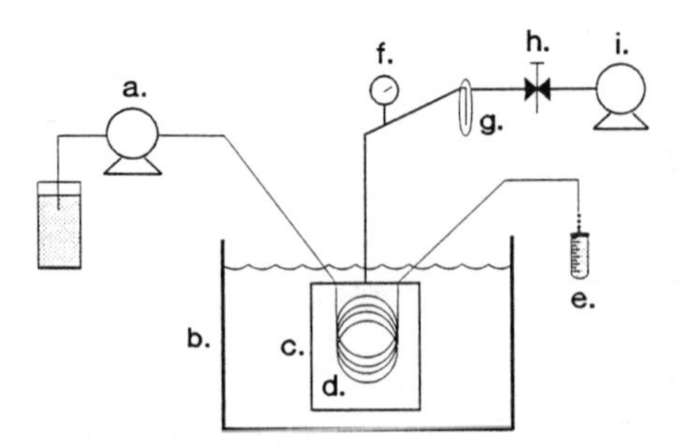

Figure 3. Pervaporation apparatus: (a) solvent pump, (b) thermostated water bath, (c) pervaporation chamber, (d) hollow fiber, (e) fraction collector, (f) vacuum gauge, (g) cold trap, (h) valve, (i) vacuum pump.

P. This is probably due to a low degree of swelling and decreased partitioning of the hydrophobic compounds to the more hydrophilic membrane.

Table I. Comparison of membranes and solvents for extraction system

	Permeation Coefficients (cm^2/min)			
	Hexane	Toluene	DCM[a]	Ethyl Acetate
PDMS	4.54×10^{-3}	4.95×10^{-3}	6.91×10^{-3}	3.99×10^{-3}
PE	1.99×10^{-3}	1.92×10^{-3}		
S-PE	1.00×10^{-9}	1.33×10^{-5}	1.69×10^{-4}	1.45×10^{-4}

[a] dichloromethane

A linear relationship between Cs and Cw was found for water concentrations from 0.10 mg/l to 0.86 mg/l. The regression equation for the line is Cs = 19.1 Cw + 0.16 and the R value is 0.999. The intercept is not significantly different from zero. A linear relationship is expected from Equation 7.

The effect of turbulence on the inside of the hollow fiber was evaluated using solvent flow rates from 0.100 ml/min to 1.70 ml/min. There was no effect of solvent flow rate on P. Turbulence in the water was changed by stirring the water in the hollow fiber chamber with a mechanical stirrer at various stir rates. The effect of stir rate on P is shown in Figure 4 at a temperature of 30 °C. An increase in P is seen from no stirring to about 1200 rpm where the value of P reaches a plateau. The value of P which is measured is actually a combination of the mass transfer through the water and the permeation through the membrane. At low levels of turbulence the mass transfer in the water is a significant component of P. As the stir rate increases the compound is transported more quickly to the membrane surface and the permeation through the membrane becomes limiting. At that point, P is no longer dependent on stirring rate. It is at this stir rate and higher where P, the permeation coefficient through the membrane, is actually measured. Therefore, in the remaining experiments the stir rate will be in the plateau region.

The effect of water temperature on P is shown in Figure 5. The water temperature was varied from 5 °C to 30 °C. There is a linear increase in P with increase in temperature. The percent change in P per degree temperature change was 3.0%. Blanchard and Hardy (*19*) also found a linear increase with temperature for the permeation of volatile organics through silicone polycarbonate membranes. However, the temperature dependence was much higher because they were diffusing into the vapor phase.

Concentration System. Combinations of membranes and solvents were evaluated by determining the solvent flux through the membranes. PE and PDMS had the

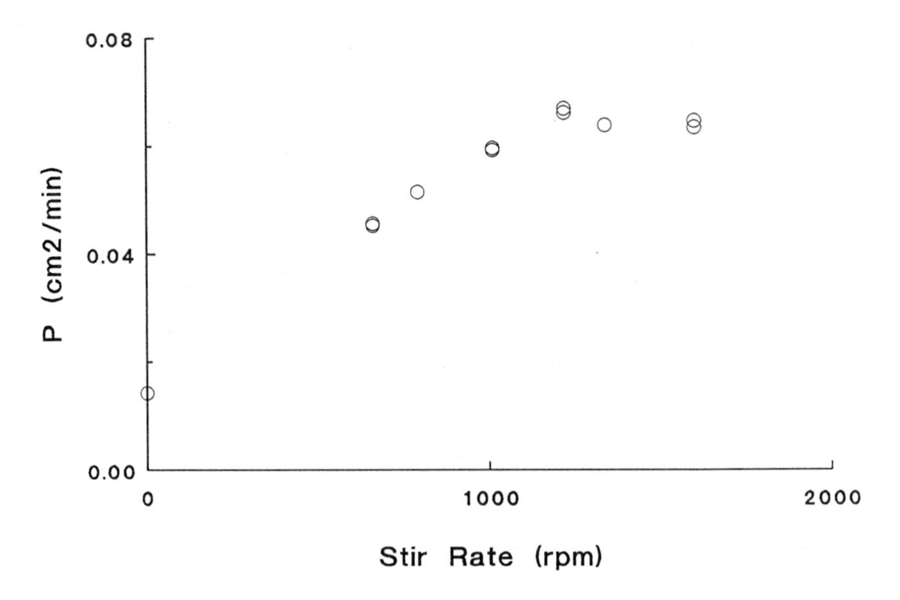

Figure 4. Stir rate in the extraction chamber vs. permeation coefficient.

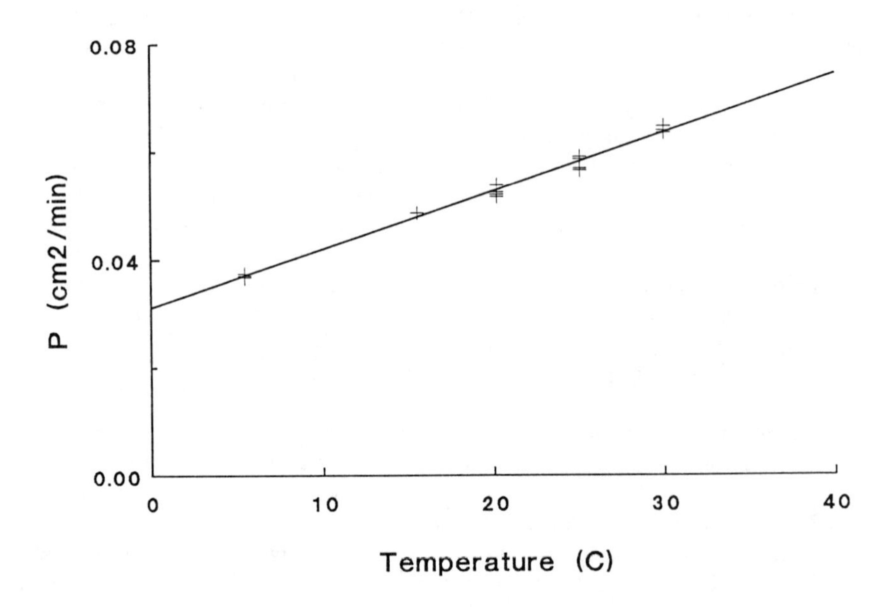

Figure 5. Water temperature vs. permeation coefficient.

largest fluxes and best physical integrity of the membranes evaluated. Table II shows the fluxes (per unit surface area) through these membranes at 60 °C.

Table II. Solvent Flux (ug/min/cm^2) for Pervaporation System

Membrane	Iso-Octane	Benzene	Toluene	Hexane
PE	0.24	0.96	1.23	0.85
PDMS	3.56	3.14	3.21	5.56

Solvent flux through PDMS was much greater than PE for all of the solvents. Hexane had the greatest flux through PDMS. Solvent fluxes for S-PE and A-PE membranes were of the same magnitude as PE, but they were rejected because they were brittle and split easily. Two solvents, pentane and dichloromethane, did not work in the system. The fluxes of these solvents were difficult to control because of vapor formation in the hollow fiber and saturation of the solvent in the vacuum chamber. This was due to the low boiling points of these solvents: 36.0 °C for pentane and 40.5 °C for dichloromethane. From these results PDMS and hexane were used in the pervaporative concentration system.

To test the pervaporative concentration system, a 1.79 mg/l solution of naphthalene in hexane was concentrated at 60 °C. The average inlet solvent flow rate was 540 (+/- 38.4) µl/min. Figure 6 shows the solvent concentration factor and the naphthalene concentration factor over the time period of the test. The solvent concentration factor is the inlet solvent flow divided by the outlet solvent flow. The naphthalene concentration factor is the naphthalene concentration in the outlet samples divided by the inlet solution concentration. If the analyte is being retained, both methods of calculating the concentration factor should give the same results.

For the first two samples the napthalene solution was pumped through the system without heat or vacuum applied, so the concentration factors were close to one. After the second sample was taken, the vacuum chamber was immersed in the water bath and vacuum was applied. The outlet flow rate decreased and the concentration factor rose as the hexane permeated through the membrane. A steady state was then reached, giving an average outlet flow of 152.1 (+/- 11.3) µl/min and an average analyte concentration factor of 3.55 (+/- 0.26). At steady state, the average percent difference in the two concentration factors was only 1.8%. Since the percent difference was low and naphthalene was not detected in the pervaporate, the naphthalene seems to be retained in the solvent.

Combined System. The final analysis was to put the two systems together and evaluate the performance of the combined apparatus. Since the permeation coefficient and pervaporation concentration factors were already determined,

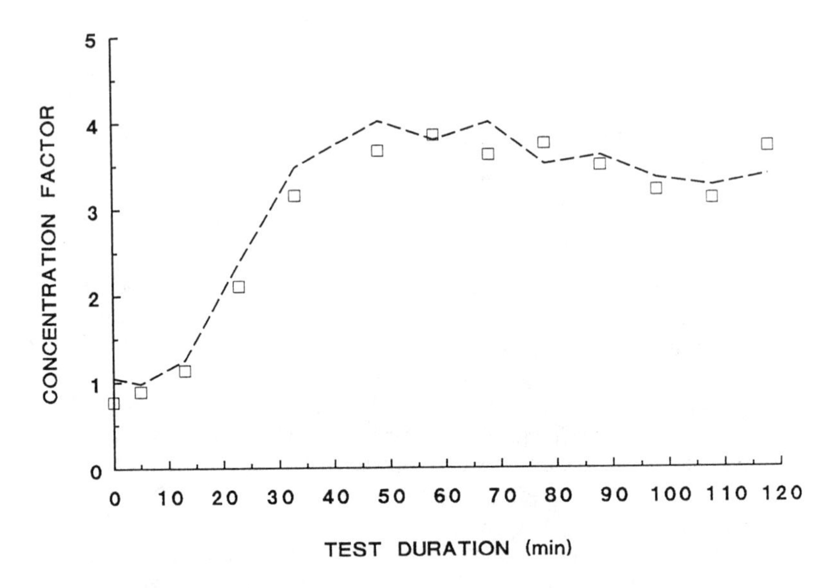

Figure 6. Concentration factors for naphthalene during the pervaporation process. (□) naphthalene concentration factor, (---) solvent concentration factor.

expected concentration factors for the individual systems and the combined system could be calculated. The expected and observed concentration factors are shown in Table III.

Table III. Concentration factors for combined extraction and concentration apparatus

	Expected Concentration Factor	Observed Concentration Factor
Extraction	48.0	64.1
Concentration	3.1	2.6
Combined	149	166

This shows good agreement between the expected and observed values. The deviations are probably due to analytical problems in accurately measuring the flow rates needed to calculate the observed values.

Concentration factors (Cs/Cw) for an optimized extraction system could be optimized to increase the concentration factor greater than 166 by increasing the hollow fiber surface area and decreasing the solvent flow rate. For example, using 1.0 m (unswollen) of polydimethylsiloxane hollow fiber at a solvent flow rate of 0.200 ml/min, and a high turbulance chamber, a concentration factor of about 500 would be predicted for naphthalene. For the pervaporation system, a concentration factor of 4 could be acheived with 0.26 m of hollow fiber at 60 °C. This would give a total concentration factor of 2000 for the combined system. This would be superior to the batch liquid extraction of 1 L of water followed by concentration of the extract to 1 ml.

Conclusion

This work has shown that a semi-volatile organic compounds such as naphthalene can be extracted and concentrated with a hollow fiber membrane system. This system would provide continuous, real time, water concentrations without sample handling or large solvent waste. Its application to other semi-volatile compounds is currently being explored.

Literature Cited

(1) Fed. Register, 44, 233, Dec. 3, 1979.
(2) Westover, L.B.; Tou, J.C.; Mark, J.H. Anal Chem. 1974, 46, 568.
(3) Pinnick, W.J.; Lavine, B.K.; Weisenberger, C.R.; Anderson, L.B. Anal. Chem. 1980, 52, 1102.
(4) Mieure, J. P.; Mappes, G.W.; Tucker, E.S.; Dietrich, M.W. In "Identification

and Analysis of Organic Compounds from Water"; Lawrence, H.K., Ed.; Ann Arbor Science, Ann Arbor, Michigan, 1976, pp 113-133.
(5) Sodergren, A. Environ. Sci. Technol. 1987, 21, 855.
(6) Hassett, J.P.; Force, M.; Song, H.M. "Preprints of Papers, Division of Environmental Chemistry, 198th National Meeting of the American Chemical Society, Miami, FL"; American Chemical Society: Washington, DC, 1989; 496.
(7) D'Elia, N.A.; Dahuron, L.; Cussler, E.L. J. Memb. Sci. 1986, 29, 309.
(8) Prasad, R.; Sirkar, K.K. AIChE J. 1988, 34, 177.
(9) Melcher, R.G. Analytica Chim. Acta 1988, 214, 299.
(10) Melcher, R.G.; Marabito, P.L. Anal. Chem. 1990, 62, 2183.
(11) Yoshikawa, M.; Yokoi, H.; Sanui, K.; Ogata, N. J. Polym. Lett. Ed. 1984, 22, 125.
(12) Niemoller, A.; Scholz, H.; Gotz, B.; Ellinghorst, G. J. Membrane Sci. 1988, 36, 385.
(13) Cabasso, I.; Jaguar-Grodzinski, J.; Vofsi, D. J. Appl. Polym. Sci. 1974, 18, 2137.
(14) Suzuki, F.; Onozato, K. J. Appl. Polym. Sci. 1982, 27, 4229.
(15) Acharya, H.R.; Stern, S.A.; Liu, Z.Z.; Cabasso, I. J. Membrane Sci. 1988, 37, 205.
(16) Cabasso, I. I.&C.E. Prod. Res. and Dev. 1983, 22, 313.
(17) Huang, R.Y.M.; Lin, V.J.C. J. Appl. Polym. Sci. 1968, 12, 2615.
(18) Suzuki, F.; Onozato, K.; Takahashi, N. J. Appl. Polym. Sci. 1982, 27, 2179.
(19) Blanchard, R.D.; Hardy, J.K. Anal. Chem. 1986, 58, 1529.

RECEIVED May 18, 1992

Chapter 14

Real-Time Measurement of Volatile Organic Compounds in Water Using Mass Spectrometry with a Helium-Purged Hollow Fiber Membrane

L. E. Slivon[1], J. S. Ho[2], and W. L. Budde[2]

[1]Health and Environment Group, Battelle, Columbus, OH 43201
[2]Environmental Monitoring Systems Laboratory, U.S. Environmental Protection Agency, Cincinnati, OH 45268

A flow-over hollow fiber membrane interface for a mass spectrometer is described in which the interior of a silicone rubber hollow fiber membrane is continuously purged with helium. The helium purge improves permeant transport from the hollow fiber interior to the ion source of the mass spectrometer, resulting in decreased response time and sub-part-per-billion sensitivity for selected volatile organic compounds in drinking water.

Semi-permeable membranes interfaced to mass spectrometers offer the most direct means of detecting and monitoring trace concentrations of organic compounds in water while maintaining the high vacuum requirements of the mass spectrometer. Membranes in the form of hollow fibers offer a significant advantage over planar membranes because the surface of the hollow fiber is self supporting allowing the use of thin membranes having one side exposed to partial vacuum without the need for mechanical support.

Real-time monitoring of groundwater, surface water and drinking water for volatile organic pollutants by hollow fiber membrane mass spectrometry offers a sensitive and specific mechanism for detecting and monitoring regulated compounds. Rapid detection and measurement in drinking water is achievable at pollutant concentrations significantly below regulatory limits. Changes in concentration may be continuously observed allowing pollution prevention measures to be implemented in a timely manner before pollutant regulatory limits are reached or exceeded.

Previous Hollow Fiber Configurations. Multiple short lengths of silicone rubber hollow fiber membrane have been incorporated into a probe in which the interior of the fibers was exposed directly to the vacuum of the ion source *(1)*. Exposure of the fiber exterior to aqueous and vapor phase organic compounds ("flow-over" configuration) resulted in permeation of organic compounds through the fiber wall, followed by gas phase diffusion to the mass spectrometer ion source.

0097–6156/92/0508–0169$06.00/0

Similar configurations have been used for direct intravenous monitoring of halogenated compounds in the blood of rats *(2)* and direct monitoring of nitrogen trichloride formation in waste water treatment processes *(3)*. The flow-over configuration is useful for high sampling rate applications but may suffer from long response time if hollow fibers of excessive length are used.

An alternative "flow-through" configuration passes air or aqueous sample through the interior of the fiber allowing organic analytes to permeate through the fiber wall to the exterior surface, which is then exposed to the vacuum of the ion source *(4-8)*. The flow-through configuration provides greater membrane contact area for a given aqueous sample volume but imposes sampling rate limitations.

Helium-Purged Flow-over Interface. An improved flow over configuration has been described in which a silicone rubber hollow fiber is immersed into a sample of aqueous solution while the interior of the hollow fiber is purged with helium carrier onto the front of a GC column *(9)* or, in our laboratory, directly into the ion source of a conventional quadrupole mass spectrometer *(10)*. Improvements in sensitivity and response time are observed as a result of helium sweeping the permeant out of the hollow fiber interior and into the ion source. Conventional flow-over configurations may suffer in response from the accumulation of permeant in the hollow fiber interior, particularly in the case of long hollow fibers and restrictive transfer lines to the ion source. This approach allows the use of long hollow fibers (greater membrane surface area) and long permeant transfer lines which eliminates the need to bring the sample in close proximity to the mass spectrometer. This work describes the incorporation of a helium-purged flow-over hollow fiber membrane into a high capacity flow cell capable of sampling water at flow rates in excess of 1 L/min. The flow cell is preceded by a dynamic standards fortification apparatus which provides for on-line calibration or measurements by standards addition.

Standards

Individual and combined standards of vinyl chloride, benzene, carbon tetrachloride, 1,1-dichloroethene, trichloroethene and chloroform were prepared at a concentrations of 6.8 to 800 μg/mL in methanol from neat material or by dilution of individual EPA Repository Standards having indicated concentrations of 5000 or 10000 μg/mL (+ or - 10%) in methanol. The standards in methanol were then injected into a flow stream of reagent water or municipal drinking water. The drinking water flow stream was obtained from a laboratory tap.

Mass Spectrometry Instrumentation

Experiments were conducted with a Finnigan TSQ-45 tandem mass spectrometer operated as a conventional single analyzer instrument. Full scan data acquisition was conducted from 48 to 250 Daltons with a scan time of 5 seconds, and electron ionization at 70 eV, 0.25 mA emission. The electron multiplier voltage was adjusted to provide a gain of 2 x 10^5 followed by a current to voltage conversion of 10^{-7} amp/volt. Additional experiments were conducted with a Finnigan model 700 ion trap detector (ITD). Full scan data acquisition was conducted over a mass range of 40-240 daltons with a cycle time of 2 seconds. All measurements with the ion trap were conducted with version 3.0 operating software and with the Automatic Gain Control (AGC) feature enabled.

Hollow Fiber Apparatus

The flow cell containing a single hollow fiber is shown in Figure 1. The hollow fiber is Dow Corning Silastic medical grade tubing, 0.30 mm i.d., 0.64 mm o.d., 9 cm in length. Helium is continuously purged through the interior of the hollow fiber at a flow rate of 1.0 mL/min (STP) unless otherwise indicated.

The flow cell was mounted in the experimental configuration shown in Figure 2. This configuration was designed to allow sampling of reagent water or sample flow streams such as municipal drinking water, as well as fortification of either flow stream with analytical standards in methanol. Standards were introduced into the aqueous sample flow stream at rates of 4-120 μL/min with a Familic 100-N syringe pump (Japan Spectroscopic Co. Ltd.) which was modified to accept a 1.0 mL Hamilton 7001 gas tight syringe. Dynamic drinking water and reagent water fortifications were reproducibly accomplished at several analyte levels by changing the digitally selectable syringe pump delivery rate of the standard while maintaining a constant drinking or reagent water sampling rate. Fortified or non-fortified sample flow streams were then passed through a baffled tubular glass mixing chamber (1.0 cm i.d., 20 cm long) prior to introduction to the flow cell. Sample flow stream temperature was monitored to the nearest 0.1°C at the exit of the flow cell. Experiments were conducted at sampling stream temperatures between 23°C and 24°C unless otherwise noted. The flow rate of the sample stream exiting the flow cell was measured with volumetric glassware and a stop watch. Helium flow to the hollow fiber was controlled using a calibrated restrictor consisting of 53.5 cm of 0.050 mm i.d. fused silica capillary tubing. Permeant from the hollow fiber was coupled to the mass spectrometer ion source using 1 m of 0.32 mm i.d. fused silica capillary tubing maintained at 60°C.

Results and Discussion

Sensitivity as a Function of Concentration. Sensitivity was evaluated on the TSQ-45 using a 1 L/min municipal drinking water sampling stream at 24°C fortified with vinyl chloride, benzene, carbon tetrachloride, 1,1-dichloroethene, and trichloroethene. The syringe pump injection rate was varied in approximately factor of two increments while maintaining a constant drinking water sampling rate. Each challenge was conducted by activating the syringe pump at a specific standards delivery rate for a period of approximately 8 minutes, then turning off the syringe pump (challenge concentration of zero) for a comparable period of time prior to the next increment in challenge concentration.

Analyte response as a function of drinking water concentration is shown in Figure 3. The curves shown through the data for each analyte represent linear regression. Excellent response linearity characteristic of membrane interfaces is observed for vinyl chloride, benzene and carbon tetrachloride in Figure 3,

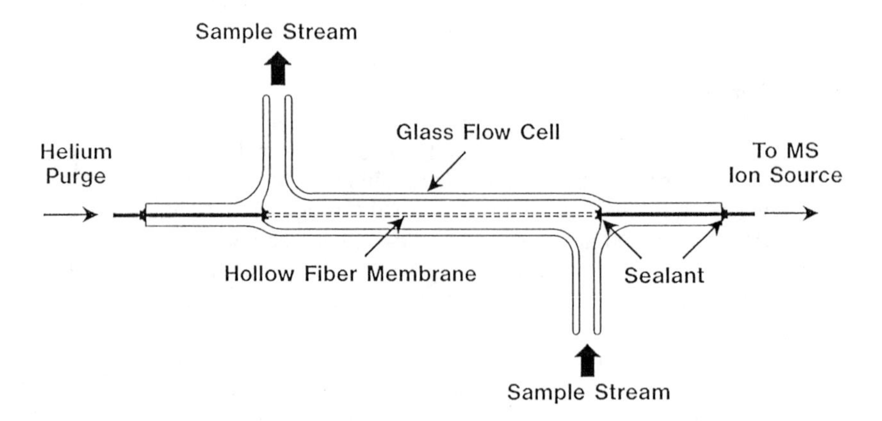

Figure 1. Continuous sampling flow cell incorporating a 9 cm helium-purged silicone rubber hollow fiber membrane.

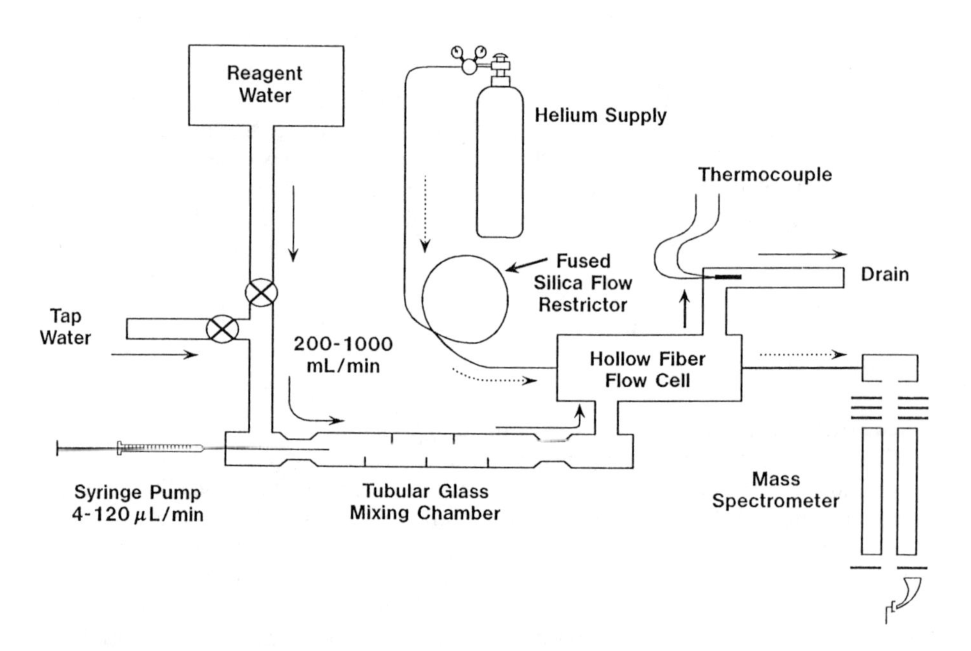

Figure 2. Experimental configuration with the hollow fiber flow cell.

including the lowest concentration of 0.18 μg/L (ppb). Similar response linearity is evident for 1,1-dichloroethene, however, the response for trichloroethene becomes non-linear at concentrations less than 0.8 ppb.

Sensitivity as a Function of Helium Flow Rate. Instrument response to 5 ppb analyte fortifications as a function of helium flow rate is summarized in Figure 4. These results, which show response maxima in the range of 0.1-0.3 mL/min helium flow, demonstrate improved sensitivity as a result of the helium purge gas, particularly for benzene. Low sensitivity in the absence of helium purge gas is the result of permeant conductance limitations imposed by the small internal diameter of the hollow fiber and the transfer line from the hollow fiber to the ion source. The slight decrease in sensitivity at helium purge rates greater than 0.3 mL/min is the result of increasing analyte dilution in the ion source by the helium purge gas.

Response Time as a Function of Helium Flow Rate. Response time as a function of helium flow rate is summarized in Figure 5. Response time is defined as the time, in minutes, for the instrument response to a specific analyte to reach 95 percent of the average steady state response. These data illustrate improved (lower) response time for the hollow fiber interface, particularly for the less volatile analytes, under helium purge conditions. The long response times in the absence of helium flow are the result of the same conductance limitations which result in reduced sensitivity.

Measurement of Chloroform in Drinking Water. The continuous measurement of trihalomethanes in drinking water represents a potential application of the hollow fiber flow cell. Chloroform is frequently observed in chlorinated drinking water supplies using conventional purge, trap and desorb techniques. Chloroform was measured in the drinking water sample stream using the method of standard addition with the hollow fiber flow cell. Two identical measurements were conducted, each at a different temperature. For each measurement, the flow cell, initially sampling distilled water, was challenged with municipal drinking water at a flow rate of 310 mL/min. After a period of 5 min, the syringe pump containing a standard of chloroform in methanol was activated for an additional 5 min period resulting in a 91.5 ppb increment in the drinking water chloroform concentration. The m/z 83 extracted ion current profiles for these two experiments, representing the base peak of chloroform, are shown superimposed in Figure 6. The response for chloroform at 39.7°C is 15 percent greater than that observed at 24.2°C, resulting from the increased diffusivity of chloroform in the membrane material at the elevated temperature. As expected, a comparable relative increase in sensitivity at elevated temperature is observed in the standard addition response, resulting in good agreement for the calculated levels of chloroform in the sampled drinking water measured at different sampling temperatures. The use of standard addition as a quantification technique is supported by the high degree of response linearity observed with the hollow fiber membrane. External standardization using fortified reagent water was not evaluated, but is less preferred because of the response dependence on sampling temperature.

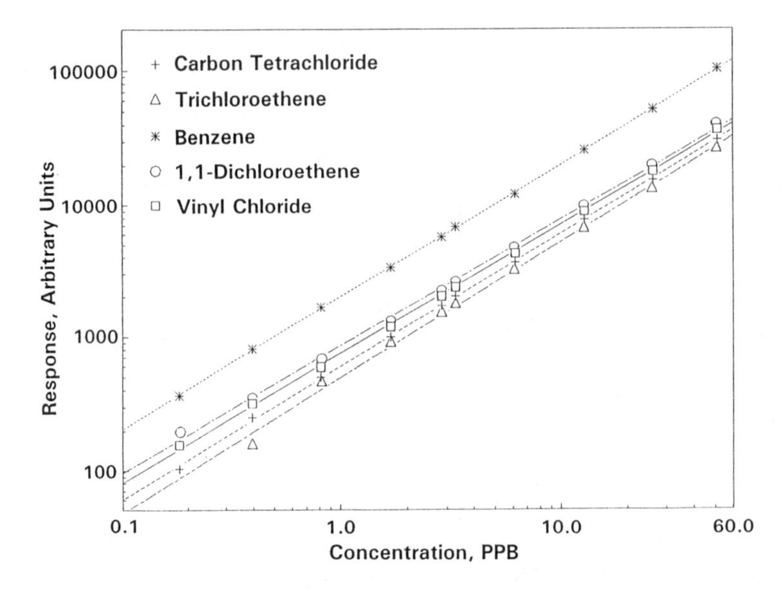

Figure 3. Flow cell response for vinyl chloride, benzene, carbon tetrachloride, 1,1-dichloroethene, and trichloroethene as a function of concentration in fortified drinking water.

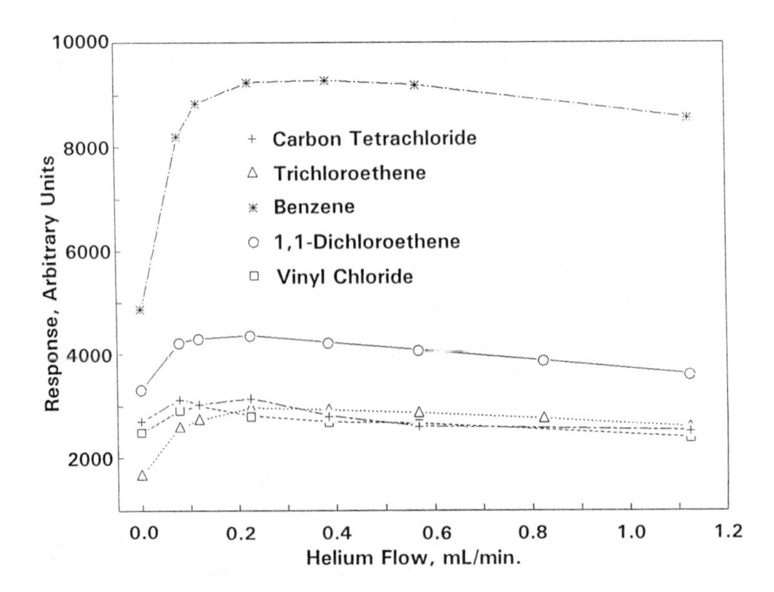

Figure 4. Flow cell sensitivity to 5 ppb of each analyte in fortified drinking water as a function of helium purge rate.

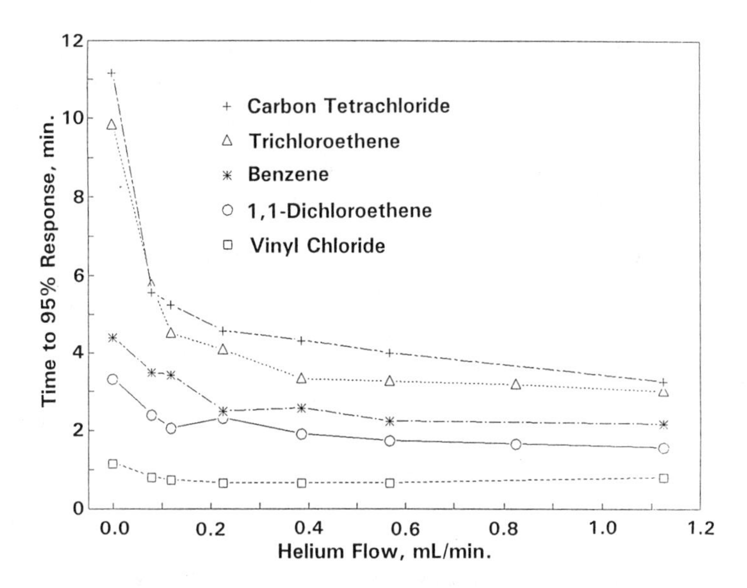

Figure 5. Flow cell response time to 5 ppb of each analyte in fortified drinking water as a function of helium purge rate.

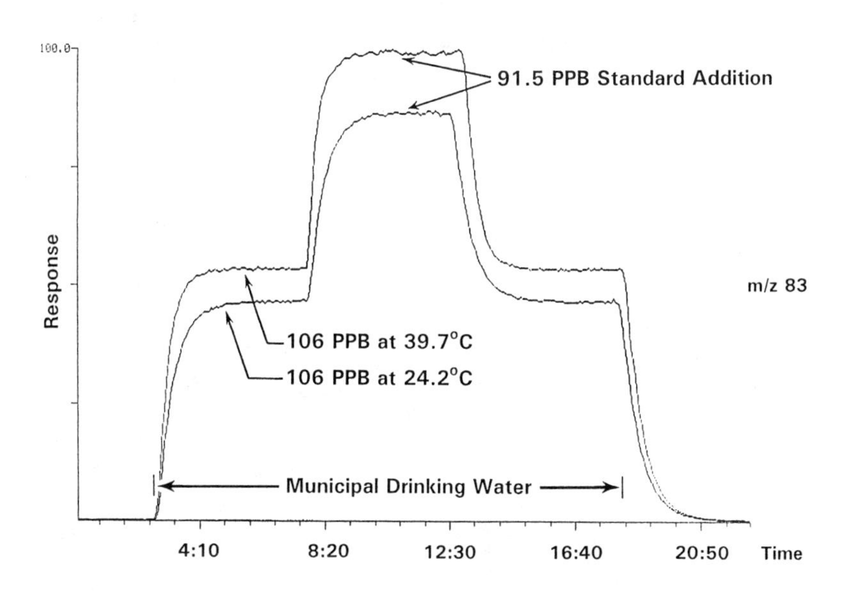

Figure 6. Comparison of standard addition measurement results for chloroform in drinking water.

Helium-Purged Hollow Fiber with an Ion Trap. Ion trap performance benefits from the presence of helium introduced into the trap as a buffer gas. Because of this operational requirement, it was anticipated that the helium-purged hollow fiber configuration combined with an ion trap would result in an ideal combination for the selective measurement of trace volatile organic compounds in water. Experiments conducted with the ion trap were similar to those described earlier with the TSQ-45 except that reverse osmosis, deionized water (reagent water) was used as the flow cell sampling stream, and the helium flow rate was reduced to 0.6 mL/min.

Vinyl chloride, benzene, and carbon tetrachloride were fortified into the reagent water sampling stream in the manner previously described for experiments involving the TSQ-45. The extracted ion current profiles for each analyte, shown in Figure 7, reveal a response at aqueous concentrations as low as 0.09 ppb. Challenge concentrations of benzene and carbon tetrachloride were the same as those indicated for vinyl chloride.

Conclusions

The use of an interior helium-purge with a flow-over hollow fiber membrane results in short response times while allowing the use of relatively long hollow fibers, having proportionately greater surface area, to improve sensitivity. The helium purge gas eliminates the need for short hollow fiber membranes close coupled to the ion source, or the need for large diameter short transfer lines between the membrane and the ion source. The helium purge improves the performance of the flow-over hollow fiber membrane by reducing the interior gas

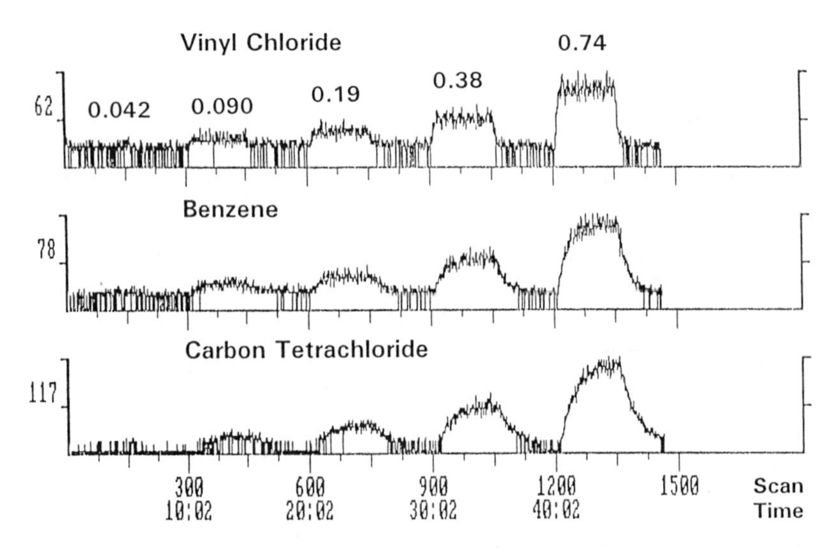

Figure 7. Ion trap extracted ion current response for vinyl chloride, benzene, and carbon tetrachloride at indicated part-per-billion concentrations of each analyte in fortified reagent water.

phase concentration of the permeant that would otherwise prevail because of conductance limitations of the small internal diameter hollow fiber and transfer line coupling to the ion source.

Comparison of the ITD results with those obtained with the TSQ-45 unexpectedly reveals comparable sensitivity between the two mass spectrometers. The absence of a significant sensitivity advantage with the ITD is likely the result of excessive permeant water vapor in the relatively closed interior volume of the ITD. In contrast, the open EI ion source of the TSQ-45 may not have permitted sufficient accumulation of water vapor in the ion source to adversely affect detection of the volatile organic compounds evaluated in this work.

Acknowledgement

The authors wish to acknowledge the helpful discussions and technical assistance provided by Battelle colleagues Mark R. Bauer and Laura Hernon-Kenny.

Literature Cited

1. Westover, L. B.; Tou, J. C.; Mark, J. H. Anal. Chem. 1974, 46, 568.
2. Brodbelt, J. S.; Cooks, R. G.; Tou, J. C.; Kallos, G. J.; Dryza, M. D. Anal. Chem. 1987, 59, 454.
3. Savickas, P. J.; LaPack, M. A.; Tou, J. C. Anal. Chem. 1989, 61, 2332.
4. Brodbelt, J. S.; Cooks, R. G. Anal. Chem. 1985, 57, 1153.
5. Sturaro, A.; Doretti, L.; Parvoli, G.; Cecchinato, F.; Frison, G.; Traldi, P. Biomed. and Environ. Mass Spectrom. 1989, 18, 707.
6. Deheandhanoo, S.; Dulak, J. Rapid Comm. Mass Spectrom. 1989, 3, 175.
7. Tou, J. C.; Rulf, D. C.; DeLassus, P. T. Anal. Chem. 1990, 62, 593.
8. Bier, M. E.; Cooks, R. G. Anal. Chem. 1987, 59, 597.
9. Nguyen, T. Q.; Nobe, K. J. Membrane Sci. 1987, 30, 11.
10. Bauer, M. R.; Slivon, L. E.; Ho, J. S.; Budde, W. L. Proc. 37th ASMS Conf. Mass Spectrom. Allied Top. 1989, 1495.

RECEIVED February 3, 1992

Chapter 15

On-Line Analysis of Liquid Streams Using a Membrane Interface and a Quadrupole Mass Spectrometer

L. A. Kephart, J. G. Dulak, G. A. Slippy, and S. Dheandhanoo[1]

Extrel Corporation, 575 Epsilon Drive, Pittsburgh, PA 15238

A permeability-selective membrane interface was used in conjunction with a quadrupole mass spectrometer for continuous monitoring of trace organic volatile compounds in water. The mechanism is known as pervaporation. The use of this interface eliminated prior sample preparation which is generally required for mass spectrometric or gas chromatography-mass spectrometric measurements. The response of the instrument to several environmentally significant chemicals was investigated. The detection limit of the instrument is in the sub-ppb range with a response time on the order of a few minutes. A linear response of the instrument to all the chemicals used in the present work was observed.

It has been recognized that Quadrupole Mass Spectrometers (QMS) have played an important role in the analysis of chemical compounds in process applications. One major advantage of the QMS is that it can be used as a general or a specific detector. Due to the low operating pressure requirement of an MS, the interfacing between the mass spectrometer and the samples has become a challenging problem, especially for process applications in which ruggedness is a main consideration. Recently, synthetic semipermeable membranes, such as Teflon, silicone rubber membranes and polyvinyl chloride have been applied for the separation of gases and volatile compounds from

[1]Corresponding author

0097–6156/92/0508–0178$06.00/0

liquid matrices (*1-8*). This type of membrane was also employed as an interface to mass spectrometers for real-time liquid sample monitoring.

The permeation of the analytes in water through the membrane is known as pervaporation or liquid permeation, i.e. the feed side of the membrane is in the liquid phase and the permeate side is in the gas phase (*9*). The permeation of the analytes is driven by the difference in chemical potentials between the feed side and the permeate side. The permeation rate increases sharply as the ratio of total pressure in the permeate side to the vapor pressure of the permeating compounds decreases (*10*). Once the pressure ratio falls below 0.4, the permeation rate is fairly independent to the pressure ratio.

Therefore, the membrane should be operated at ratios below 0.4. The low pressure in the permeate side is generally maintained by a vacuum pump. In this application the permeate side is exposed to the vacuum chamber, which houses the mass spectrometer. Pressure in the vacuum chamber is typically lower than 10^{-5} torr. Therefore, the pressure ratio is likely to be lower than 0.4 for most of the volatile organic compounds.

It was discovered that permeability is inversely proportional to the membrane thickness and the permeation rates of the analytes can be enhanced by elevating the temperature of the membrane. Although increasing the temperature of the membrane results in higher diffusivities and decreases the response time, it causes a slight reduction in the selectivity.[10] Another factor that has a strong effect on the selectivity of the membrane is the structure of permeating compounds, such as polarities and molecular dimensions. Due to its selectivity, the permeability-selective membrane serves not only as a separation septum, but also as an enrichment device. The degree of enrichment for each compound depends on its chemical and physical properties.

The pioneer work of Westover and co-workers indicated that silicone-rubber membranes were superior to other types of membranes with respect to their permeability(*1*). Furthermore, the silicone-rubber membrane is the most attractive for monitoring the trace constituents in water since the membrane is almost impermeable to polar molecules, such as water molecules.

In the present work, we explore the possibility of using a silicone rubber membrane interface (MI), in conjunction with a quadrupole mass spectrometer (QMS), for continuous monitoring of trace volatile organic compounds (VOC) in aqueous solutions. The performance of the MI/QMS instrument was assessed by testing it with several environmentally significant compounds.

Experimental Section

The instrument used in this work was designed for use in a field environment; therefore, the instrument should require minimum maintenance and should be easy to operate. A schematic of a simple silicone-rubber membrane interface is shown in Fig. 1. This diagram describes the membrane interface and the sampling system, which is computer controlled. A detail diagram of the membrane assembly is illustrated in Fig. 2. The interface consists of Silastic membrane tubing (Dow Corning, Midland, MI) of 0.037" OD and 1.5" long. The wall thickness of the membrane tubing was 0.008". The membrane tubing was located inside a 1/8" stainless steel (SS) tube. The stainless steel tubing was situated in a heating block so that the temperature of the interface could be elevated.

Sample solutions were introduced into one of the liquid vials without prior sample preparation. The other vial was used for calibration purposes. It was discovered from previous experiments that slight changes in the flow speed of the solutions through the membrane tubing caused a fluctuation in the ion signal (*8*). To provide a uniform flow of the sample through the MI, a small portion of either the sample or the calibration solution was introduced into the membrane interface by syphoning. The flow rate of sample through the membrane interface depends only on the height of the sampling port with respect to the interface. With this configuration, the sample flow rate through the membrane can be kept constant regardless of the sample flow rate through the vials. Switching sample and standard solutions can be accomplished easily by a teflon 3-way solenoid valve. Analytes that permeate the membrane were transferred into a QMS (Extrel Corp., Pittsburgh, PA) via a fused silica capillary of 250 micron ID. The use of a capillary as a means of analyte introduction into the QMS helps prevent damage to the vacuum system in cases where the membrane tubing is punctured or broken. The temperature of the SS tubing was raised to about 80°C to avoid the condensation of the analytes and reduce the response time. The actual temperature of the membrane was not measured. However, the temperature of the membrane should approximate to that of the exiting aqueous solution, which was about 40°C.

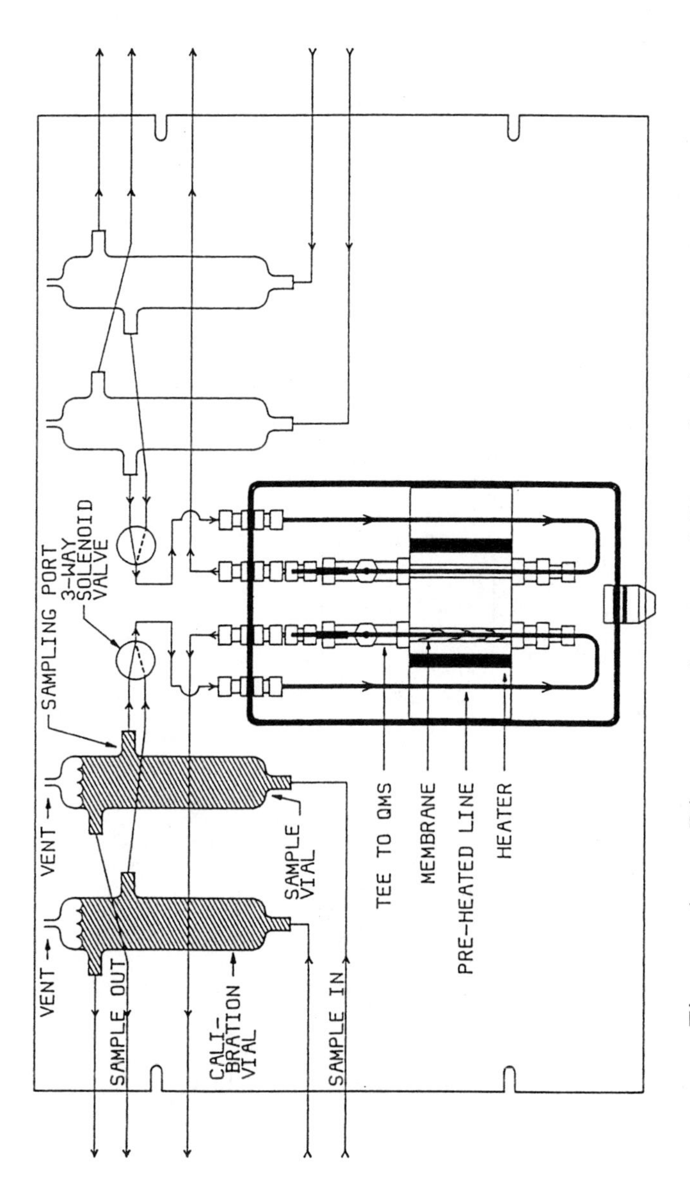

Figure 1. Diagram of permeability-selective membrane interface.

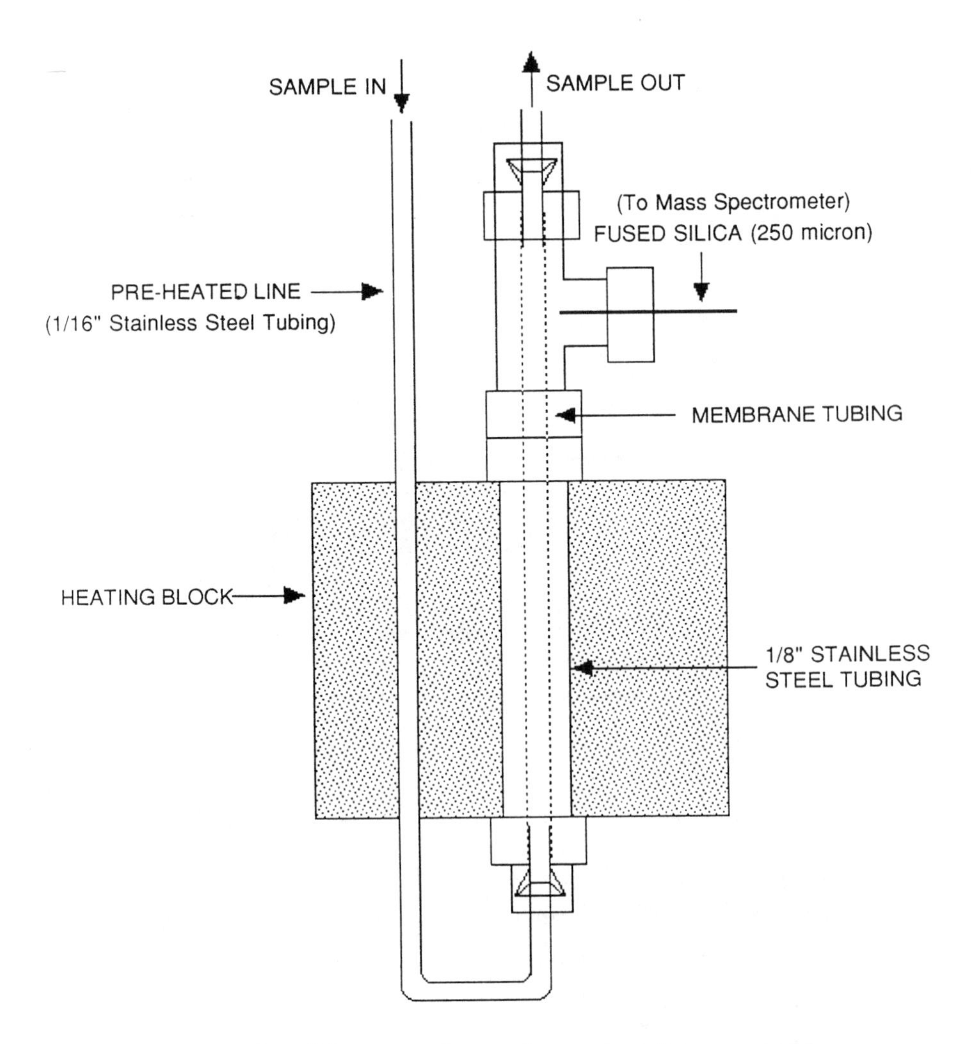

Figure 2. Diagram of membrane interface assembly.

The analytes that entered the ionizer underwent electron ionization. The energy of the electrons was set at 70 eV so that standard mass spectra could be used as a reference. Ions from the ionizer were analyzed by the QMS and were detected by an analog electron multiplier.

The chemicals used in this experiment are listed in Table I. The chemicals were obtained from various manufacturers and were used without further purification. The manufacturer of each compound is shown in parentheses. Most of these compounds are not significantly soluble in water. In such cases, the chemicals were dissolved in a small amount of methanol and then diluted with distilled water to obtain stock solutions.

The quantities of analytes were determined by monitoring the ion intensities of selected ions. The selected ions were either molecular ions or dominant fragment ions, depending on the ion yields, background intensities and interference from ions of other compounds. The background ion intensity was established by introducing distilled water into the interface and observing the monitored ions, (see Table 1), of each compound for about 10 minutes or until the ion signal was stable. Once the background intensity was determined, the distilled water was replaced by samples of known concentrations or unknown samples.

Another parameter which is important to real-time monitoring applications is the response time of the instrument. In this experiment, response time is defined as the time between the introduction of the solution and the time when the ion signal of the selected ions reaches the maximum level. Since the MI is located outside the QMS chamber, the response time of the instrument includes the permeation time and the time required for the permeated analytes to reach the ion source.

Results and Discussion

The EI spectrum of most environmentally significant compounds can be found in standard references (*11*). Thus, the selection of monitored ions for each chemical were based on the EI spectrum listed in the standard references.

The MI/QMS instrument is capable of detecting non-polar volatile organic compounds in water at levels as low as 15 parts per billion (ppb). In favorable cases, such as benzene and methylene chloride, it is likely that a detection limit in the range of a few ppb can be achieved. Due to its polarity and vapor pressure, ethylene glycol has a much lower permeability and was found to have a detection limit in the low parts per million (ppm) level. An oil was also tested in this experiment. The experimental results indicated that the oil did permeate the membrane. Although several mass peaks characteristic of oil were seen, the detection limit could not be determined because the oil compound formed an emulsion rather than a solution in water.

The MI/QMS instrument responded linearly to all the non-polar compounds used in this experiment. Figure 3 is an example of the linear response of the instrument to a few compounds. The variation in the slope of the response curve for each chemical is caused by the differences in their permeability. Due to the linear response, the instrument can be used for quantitative determination and a calibration curve for each chemical may be easily established with an appropriate standard.

Interference, i.e., the effect of the presence of one compound on the instrumental response to other compounds in the mixture, was investigated by observing the response of the instrument to three different mixed solutions of benzene (BZ), methylene chloride (MC), toluene (TO) and xylenes (XY). The first solution consisted of 200 ppb of all the aforementioned chemicals. This solution also was used for one-point calibration. As shown in Fig. 4, the first solution was introduced into the membrane for about 5 min. for calibration. The second solution contained 50 ppb of MC and 50 ppb of BZ and the third solution consisted of 50 ppb of MC and 100 ppb of BZ, TO and XY. Since the instrument appeared to respond linearly to each chemical regardless to the concentrations of the others, interference did not occur when mixtures of compounds were presented to the instrument.

Due to its selectivity, the membrane behaves as an enrichment device. The degree of enrichment is described by an enrichment factor, E. The definition of E for a chemical C in the solution can be expressed as (*6*)

$$E = \frac{C_p/W_p}{C_a/W_a} \qquad (1)$$

where C_p and W_p are number densities of component C and water in the pervaporate, respectively. C_a and W_a are number densities of component C and water in aqueous solution. For the electron ionization, the number density of molecules can be determined by ion intensity. The relationship of the number density of molecules and the ion intensity can be expressed as

TABLE I: List of chemicals which were used in this experiment

Compound	Monitored ion (amu)
Benzene (Alfa)	78
Carbon tetrachloride (Alfa)	117
Chlorobenzene (Fisher)	112
1,3-Dichlorobenzene (Aldrich)	146
1,4-Dichlorobenzene (Aldrich)	146
1,2-Dichloroethane (Fisher)	62
1,1-Dichloroethane (Alfa)	63
Ethylbenzene (Alfa)	91
Methylene Chloride (Fisher)	49
1,1,2,2-Tetrachloroethane (Aldrich)	83
Toluene (Fisher)	91
1,1,1-Trichloroethane (Alfa)	97
Trichloroethene (Fisher)	130
Trichlorofluoromethane (Freon III) (Supelco)	101
Xylenes (Alfa)	106
Ethylene glycol (Fisher)	32
Texaco Capella WF65 oil (Texaco)	47,51

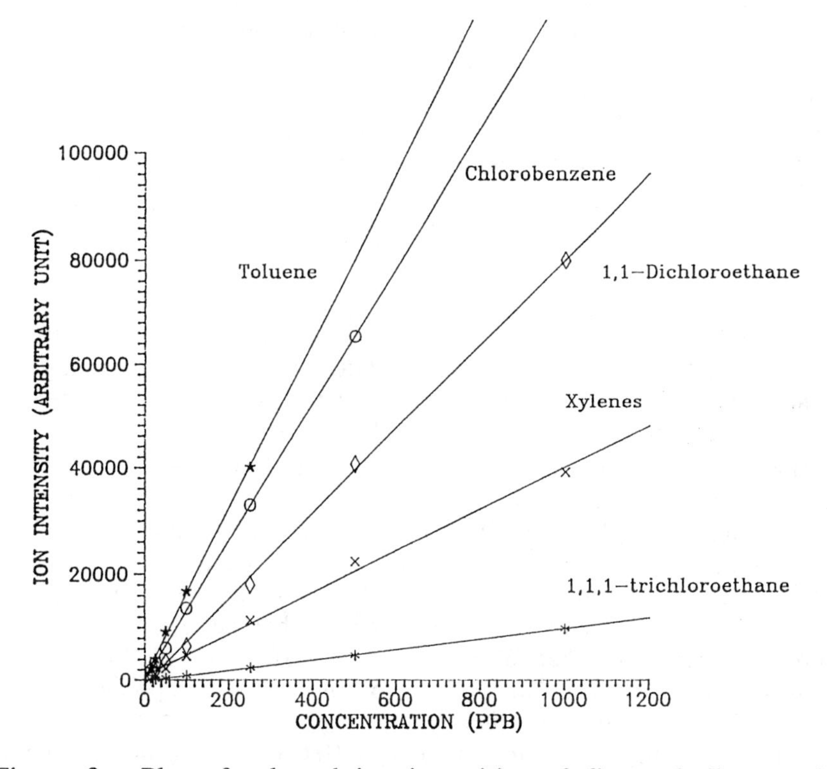

Figure 3. Plot of selected ion intensities of five volatile organic compounds as a function of concentration.

$$I = \sigma I_e d N \tag{2}$$

where I is total ion intensity, I_e is the ionizing electron intensity, σ is the total ionization cross section, d is the ionizing path length and N is the number density of molecules (*12*). This expression is valid for a large variety of molecules over a wide range of pressures. By using equation (2) and assuming that both component C molecules and water molecules experience the same ionizing electron intensity and the path length, the number density can be expressed in terms of ion intensity and equation (1) becomes

$$E = \frac{\sigma_c I_c / \sigma_w I_w}{C_a / W_a} \tag{3}$$

where Ic and Iw are ion intensities of component C and water, respectively. σ_c and σ_w represent the total electron ionization cross sections of component C and water, respectively.

Let's consider the enrichment factor of benzene. The electron ionization of water molecules yields dominant ions of masses 17 and 18 which correspond to OH^+ and H_2O^+ while benzene yields a molecular ion of mass 78. By substituting, the ionization cross sections of benzene and water, equation (3) can be rewritten as

$$E = \frac{\sigma_W}{\sigma_B} \times \frac{I_{78}}{I_{17} + I_{18}} \times \frac{W_a}{C_a} \tag{4}$$

where σ_B and σ_W are ionization cross sections of benzene and water, respectively. By using equation (4), and recognizing that the ratio of the cross section of water to that of benzene is approximately 1/5.9 (*13*) the measured enrichment factor of benzene was found to be about 1 $\times 10^4$ at about 40^oC.

The enrichment factor of chemicals increases with a decrease in solubility and/or an increase in the vapor pressure of the chemicals (*6*). The vapor pressure of ethylene glycol is about 300 times lower than that of benzene at the same temperature (*14*). Thus, the enrichment factor of ethylene glycol should be much lower than that of benzene. It is most likely that low enrichment factor contributed to the poor response of the membrane to ethylene glycol.

The response time and memory effects depend on the nature of the chemicals. According to the experimental results (see Fig.

5), the response time of the MI to volatile organic compounds was less than 2 minutes, at ~40 °C, in all cases. Response time and decay time of most of the chemicals were approximately equal.

Conclusion

The results of these experiments show that the MI/QMS instrument provides direct, virtual real-time determinations of trace constituents in water without prior sample preparation. The detection limit of the instrument can be in the sub ppb range with a response time of a few minutes. The instrument responds linearly in the ppb and ppm range. The interference effect did not appear to be a problem. Other attractive features of the instrument are its simplicity and ruggedness, which make it suitable for industrial and process control applications, such as continuous monitoring of ground water and industrial discharge water.

The MI used in the present work is located outside the ionizer, while some other workers have used MI's with the membrane inside the ionizer(*15*). The present design may cause a longer response time due to the transit time of the permeated analytes from the membrane to the ionizer. However, there are several advantages of the present design in terms of flexibility, ease of maintenance and independent control of the MI's temperature. Another advantage of an external MI is that samples from several locations may be monitored simultaneously and continuously by using several MI's coupled to a single QMS. Each MI can be used dedicate to separate stream. Permeated analytes from each membrane can be analyzed by the QMS in a user specified sequence. With this type of arrangement, the MI/QMS instrument provides a cost effective technique for industrial process control applications.

The experimental results indicate that the instrument is suitable for real-time monitoring of non-polar VOC's in liquid samples, provided that the chemicals to be monitored are known. Since a mass spectrometer cannot identify the chemicals in real time, it is not applicable for monitoring of unknown samples. Furthermore, isotope compounds normally have the same or very similar mass spectra; therefore, the QMS cannot provide isotopic information. The MI/QMS instrument is useful for rough screening for suspected contaminants in aqueous samples. However, a more specific analytical instrument, such as

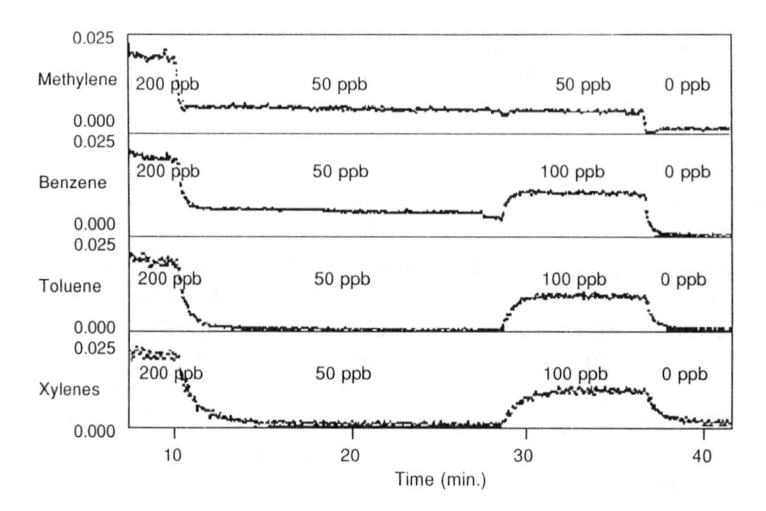

Figure 4. The response of the instrument to various concentrations of methylene chloride (MC), benzene (BZ), toluene (TO) and xylene (XY). A one point calibration with each component at 200 ppb in aqueous solutions was followed by two solutions, one containing 50 ppb of MC and BZ and the other containing 50 ppb of MC and 100 ppb of BZ, TO and XY.

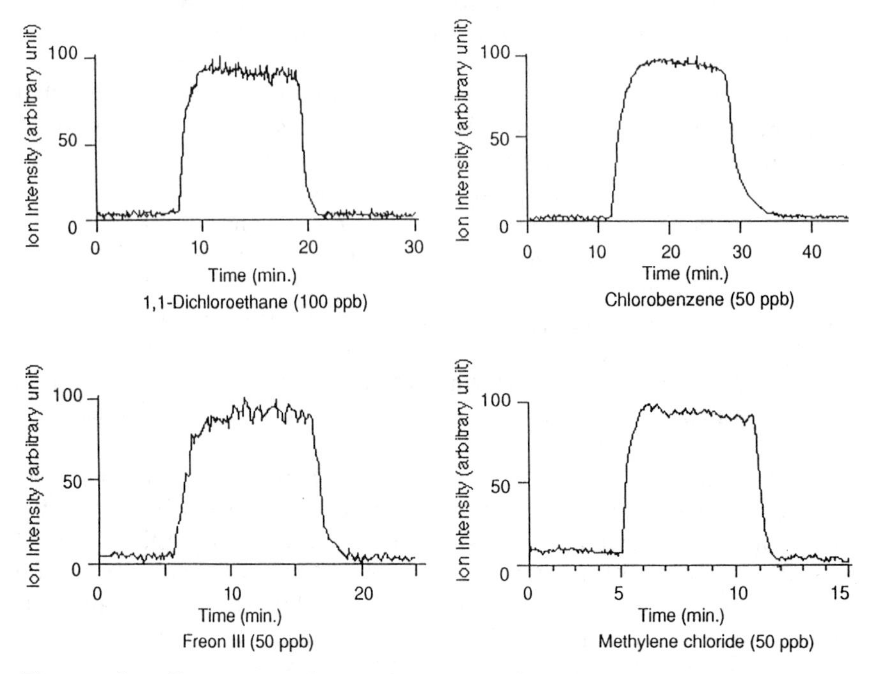

Figure 5. Response time of the MI/QMS instrument to various volatile organic compounds.

a gas chromatography-mass spectrometer (GC/MS), is needed for positive identification of the contaminants.

Acknowledgments

This work was partially supported by SBIR grant No. F40600-89-C-0013.

Literature Cited

1. Westover, L.B.; Tou, J.C.; Mark J.H. **Anal. Chem., 1974,** 46, 568.
2. Weaver, J.C.; Abrams, J.H. **Rev. Sci. Instrum.**, **1979**, 50, 478.
4. Brodbelt, J.S.; Cooks, R.G. **Anal Chem., 1985**, 57, 1153.
5. Calvo, K.C.; Weisenberger, C.R.; Anderson, L.B.; Klapper, M.H. **Anal. Chem. , 1981**, 53, 981.
6. Harland, B.; Nicholson, P.; Gillings, E. **Wat. Res., 1986**, 21, 107.
7. Cox, R.P. **In Mass Spectrom. Biotechnol. Process Anal. Control. (Proceed Workshop);** Elmar, H., Reuss, M., Eds.; Plenum, New York 1986; pp 63.
8. Dheandhanoo, S.; Dulak, J. G. **Rapid Comm. Mass Spectrom., 1989**, 3, 175 .
9. R. Kesting, R. **In Synthetic Polymeric Membranes**, John Wiley & Sons, New York (1987).
10. Bruschke, H.E.; Tusel, G. F.; Rautenbach, R. **In Reverse Osmosis and Ultrafiltration** , ed. Sourirajan, S., Matsuura T., Eds.; ACS: Washington, D.C. 1985; pp 467.
11) Eight Peak Index of Mass Spectra; Mass Spectrometry Data Center: Reading, UK (1974).
12) Fitch, W.E.; Sauter, A.D. **Anal Chem., 1983**, 55, 832.
13) Product and Vacuum Technology Reference Book; Leybold-Heraeus Inc., 1986.
14) **The Vapor Pressures of Pure Substances,** Boublik, T., Fried, V., Hala , E., Eds.; Elsevier, New York,1973.
15) Bier, M. E.; T. Kotiaho, T; Cooks, R.G. Anal. Chim. Acta, **1990**, 231, 175.

RECEIVED June 9, 1992

ANALYTICAL APPROACH TO PROCESS ANALYTICAL CHEMISTRY

Chapter 16

Electrospray Ion Mobility Spectrometry

Its Potential as a Liquid-Stream Process Sensor

C. B. Shumate[1] and H. H. Hill

Department of Chemistry, Washington State University, Pullman, WA 99164–4630

Ion mobility spectrometry is an analytical method in which ions are produced at atmospheric pressure and allowed to drift and separate in an electric field at intrinsic velocities. By monitoring the arrival time spectra of these atmospheric pressure ions, both qualitative and quantitative information can be obtained from an analytical sample. In the past, IMS has been used exclusively for gas phase analysis. By employing electrospray ionization techniques, this study evaluates IMS for use with liquid samples. Using amines as test analytes, ion mobility reproducibility was studied as a function of time, solvents, electrospray voltage, sample matrix and drift gas temperature. In addition, electrospray ion mobility spectrometry was demonstrated as a detection method for flow injection analysis, chromatography, and continuous sample stream monitoring. Detection limits as low as 5 X 10^{-15} mol/s were determined.

The ability to obtain on-line, real-time chemical information from liquid waste streams for pollution prevention and process control is a challenge for analytical chemistry. Essentially, only two analytical approaches are currently available: spectrophotometry and electrochemistry. Spectrophotometric methods take advantage of unique electronic structures in atoms and molecules by monitoring emission, absorption or fluorescence in an emerging stream. Electrochemical methods rely on unique oxidative or reductive properties of pollutants to detect their presence. In general, compounds which are not spectrophotometrically or

[1]Current address: Hamilton Company, 4970 Energy Way, Reno, NV 89520

0097–6156/92/0508–0192$06.00/0

electrochemically active must be converted to species which are, or they go undetected. Recently, in order to detect a wider variety of compounds, much more sophisticated analytical instruments such as MS, ICP, NMR, and XRF have been investigated (1). Unfortunately these instruments introduce a layer of added complexity which, although in many cases may be necessary, is undesirable for simple routine process analysis.

The objective of this research was to investigate an analytical methodology called ion mobility spectrometry (IMS) which does not rely on either the presence of chromophoric or electroactive functional groups for its analytical response. Instead, response in ion mobility spectrometry relies on gas phase electron and proton affinities. In the past, these instruments have traditionally been used for gas phase analysis although the detection of organic compounds dissolved in supercritical fluids has been reported. Comprehensive reviews of IMS for gas and supercritical fluid analysis can be found in the literature (2,3).

Initial attempts to interface IMS with liquid streams were reported by Dole and his group (4-6). The interface was accomplished by electrospraying the liquid directly into the ion mobility spectrometer. Since the electrospray process not only served to nebulize the liquid but also to ionize the sprayed droplets, no auxiliary ionization source was required. Unfortunately, in this work, they used a low temperature instrument with a bi-directional flow design and were unable to successfully stabilize solvent clustered ions before entering the drift region.

Based on Dole's work, we have recently demonstrated electrospray nebulization and ionization of liquid samples for ion mobility spectrometry using a high temperature spectrometer and a uni-directional flow design (7). These preliminary investigations showed electrospray ion mobility spectrometery (ES-IMS) response to a variety of organic compounds dissolved in a variety of solvent systems. Subsequently, this ES-IMS was interfaced to a liquid chromatograph and evaluated as a chromatographic detector (8).

Electrospray ionization is an emerging technology which is currently finding acceptance in the field of mass spectrometry (9-13). In these investigations we have focused on evaluation of the ES-IMS as a continuous process analyzer choosing test compounds which cannot be detected by traditional methods and providing more detailed information of the analytical figures of merit of the instrument.

Experimental

The Ion Mobility Spectrometer. Figure 1 represents a schematic cross-section of the spectrometer used in this work. The geometrical design of the spectrometer consists of a stack of metal rings separated by insulators. The rings are held at successively lower voltages to provide a smooth electric field down the center of the stack. Generally, first drift ring of the spectrometer was maintained at 4 to 5 kV. Drift tube voltage was doped linearly through a resistor chain which connected the drift rings. The temperature of the drift tube was kept at 150 C with a drift gas flow of 600 mL/min of air. Gate voltages were set at 60 V and the liquid flow rate was 5 uL/min of 50/50 methanol/water except where noted. Individual conditions for each experiment are noted in the figure captions. For detailed information on the construction and operation of this spectrometer the reader is referred to Shumate's Ph.D Thesis (14).

Ions were created by an electrospray ionization source. This source consisted of a metal capillary held at a high voltage (6-10 kV) through which a liquid sample was passed. A small pulse of ions is then "gated" into the drift region through a switchable orthogonal field. A fast electrometer records the arrival time of the ions at the collector. This method of operation is referred to as signal averaging as many of these arrival time spectra can be averaged to enhance the signal to noise ratio.

By placing a second gate near the collector, drift time selective detection can be achieved. This is accomplished by pulsing the first gate open and imposing a finite delay on the opening of the second gate. The second gate is then held open for a selected period to monitor ions drifting during a specific drift time window. This method of operation is useful when using the spectrometer as a tunable detector. The disadvantage of this method is the inability to concurrently record spectra. Concurrent detection and spectral acquisition can be accomplished through software using the signal averaging mode. An algorithm extracts a drift time specific ion current from each spectrum and plots the current as a function of time; this is referred to as software selective detection.

Test Compounds. Alkylamines were chosen as the test compounds. The inability of conventional liquid detection methods to respond to these compounds made them ideal for demonstrating the unique ability of ES-IMS as a detector for non-chromophoric compounds. Currently the detection of aliphatic amines relies on the chemiluminescence detection of a derivative (15). Compounds used in these studies were the following: Butyl amine, pentyl amine, hexyl amine, heptyl amine, octyl amine, decyl amine, dodecyl amine, tetradecyl amine, iso-butylamine, sec-butylamine, tert-butylamine, di-n-butylamine, tri-n-butylamine.

Operating Conditions. Flow injection, chromatographic injection and continuous introduction methods were used during the course of these investigations. In all methods, a dual piston syringe pump (MPLC micropump, Brownlee Labs Inc., Santa Clara, CA) served to deliver stable liquid flows down to rates as low as 1 uL/min.

In the **flow injection mode**, the configuration was pump-injector-detector. The injector used for the flow injection experiments was a four port, 60 nL internal volume injector (Valco C14W, Valco Ind., Houston, TX). All ion mobility spectra presented in this study were obtained from samples introduced by the flow injection method. Calibration graphs were generated by monitoring the mobility of the product ion and making three injections at each concentration level.

In the **chromatographic mode**, the configuration was pump-injector-column-detector. An isocratic separation of four alkylamines was carried out in 60:40 0.1M ammonium acetate:acetonitrile on a 2.1 mm i.d., 5 um particle cyanosilica column (Alltech Associates, Deerfield, IL.). A flow rate of 200 uL/min was used which was split approximately 50:1 after the column through the use of a tee with restricting valve on the high flow side of the split. Injections were made with a four port, 200 nL internal volume injector (C14W, Valco, houston, TX). A solution of the four trialkylamines (triethyl, tri-n-propyl, tri-n-butyl, and tri-n-hexylamine) was made at 10 parts per thousand corresponding to 2 ug of each compound on column and 100ng into the IMS after the split.

In the **continuous monitor mode**, the configuration was simply pump-detector. In this experiment the concentration of dibenzylamine was programmed from 0 to 25 ppm in methanol over 20 minutes and the current at the mobility of the product ion was monitored. This program was achieved by gradient programing the dual piston syringe pump. Syringe A was filled with pure methanol while B contained 25 ppm dibenzylamine. The gradient was programed from 0% B to 100% B over 20 minutes and then held isocratic for 20 minutes.

Results and Discussion

In ES-IMS, the ionization mechanism appears to be different from that which is observed for ES-MS. Rather than true ion-evaporation, experiments reported in these studies more closely resembled that of chemical ionization. A spectral peak existed even when no analyte was present and is referred to as the reactant ion peak. This reactant ion peak was the solvent ion background and was diminished when ion peaks from the analyte appeared. For example, we know from past experience with radioactive ion sources when water is present, the predominant reactant ions are hydronium ion clusters and that the ionization of an amine occurs by proton transfer to the amine.

$$(H_2O)_nH^+ + RNH_2 \Longleftrightarrow nH_2O + RNH_3^+$$

Data which we have generated so far with ES-IMS is consistent with this model although mass identifications of reactant and product ions must be conducted before conclusive evidence for the ionization mechanism can be established. Given that a variety of solvents were used throughout these investigations a more general equation for response can be written as

$$S_nH^+ + RNH_2 \Longleftrightarrow nS + RNH_3^+$$

where S represents the solvent molecule and, as before, RNH_2 represent the analyte. Thus, one might expect that response factor in ES-IMS will vary as a function of both solvent and analyte identity.

In addition, large quantities of solvent (several uL/min of liquid) are introduced into the ionization region of the IMS. If these solvent molecules are not efficiently evaporated and swept from the drift tube, the solvent can form cluster ions with the product ion.

$$RNH_3^+ + mS \Longleftrightarrow RNH_3S_m^+$$

These cluster ions would have different mobilities causing the ion mobility spectra to vary. Several experiments were conducted to investigate potential interferences due to solvent clustering.

Mobility Experiments. Mobility data were reported in terms of reduce mobility constants (K_o) calculated from the equation

$$K_o = (d/tE)(P/760)(273/T)$$

where d was the drift length in cm, t was the ion drift time in seconds, E was the electric field in the drift region of the spectrometer in V/cm, P was the gas pressure in the drift region in Torr and T was the temperature of the drift gas in the drift region in K. In theory, reduced mobility values are independent of instrumental conditions. In practice, they often do change with instrumental conditions if these conditions are varied over wide ranges. For normal analytical ion mobility detection, the optimal instrumental conditions do not vary greatly and the reduced mobility values should be constant from day to day.

Reproducibility. Reduced mobility constants of three compounds (tri-n-propylamine, tri-n-butylamine, and tri-n-hexylamine) were determined at three different times during the day for five days. For tri-n-butylamine the average K_0 value and standard deviation were determined to be 1.64 $\pm$.01. For tri-n-butylamine and tri-n-hexylamine the values were 1.44+-.01 and 1.17+-.01, respectively. Over this five day period the ES-IMS instrument performance was extremely reproducible.

Solvent. In a separate experiment, the effect of several solvents on mobility values were investigated. The effect that solvents have on product ion mobility was investigated by introducing tri-n-butylamine into the EIMS dissolved in five different solvents: acetone, isopropanol, acetonitrile, methanol, and water. The average K_o value was 1.44, the same as in the previous experiments, but the standard deviation of 0.02 was slightly higher than that obtained during the five day study. Nevertheless, solvent effect on mobility appeared to be minimal.

Electrospray Voltage. A third mobility experiment, in which the electrospray voltage was varied from 8000 V to 10,000 V in 500 V increments, produced an average product ion mobility value for tri-n-butylamine of 1.46 with a standard deviation of 0.02. Upon closer investigation of the data, it was found that the mobility increased from 1.44 at 8000 V to 1.50 at 10,000 V. This increase in mobility with electrospray voltage indicated that the electrospray voltage was coupling to the drift field voltage and increasing the electric field through which the ions drifted. In future designs of the ES-IMS the drift field voltage should be shielded from the electrospray voltage.

Temperature. In general, mobilities were reproducible from day to day and from injection to injection. On occasion, however, especially after the instrument had been taken apart and reassembled, mobilities could be different than previously observed. For example, the mobility of the product ion of tri-n-butylamine was measured at 50, 100, 150, and 200 C, found to be 1.36 with a standard deviation of 0.02. No definitive explanation can be given for why mobility is so much lower in this set of experiments compared to the others conducted with tri-n-butylamine. One possibility was that the spectrometer had become contaminated. In such cases, the addition of mass identified ion mobility data would aid in understanding the response and behavior of EIMS. Nevertheless, changing the temperature from 50 to 200 C was found to have little effect on the mobility. These results, coupled with those presented above indicate that the solvent was efficiently declustered from the product ion.

Matrix. Finally, matrix effects on mobility were investigated using a mixture of eight alkylamines. Figure 2 shows the complex ion mobility spectrum of the unseparated amine mixture. Eight product ion peaks could be identified. Drift times and mobility values of these peaks matched those obtained from ion mobility spectra of the individual amines. Thus, it appears that under conditions used in this study, the product ions produced

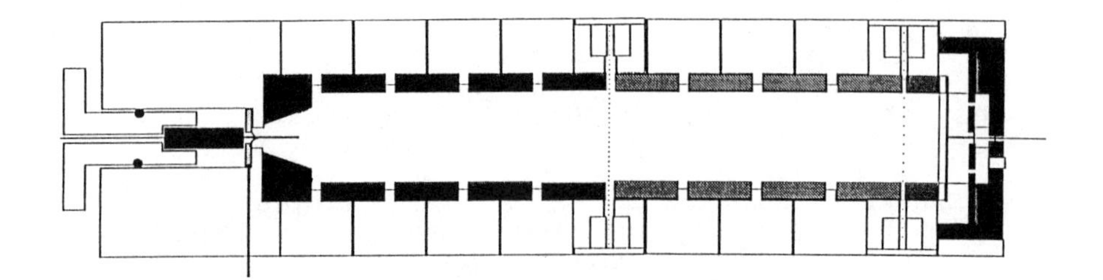

Figure 1. Schematic cross-section of electrospray ion mobility spectrometer.

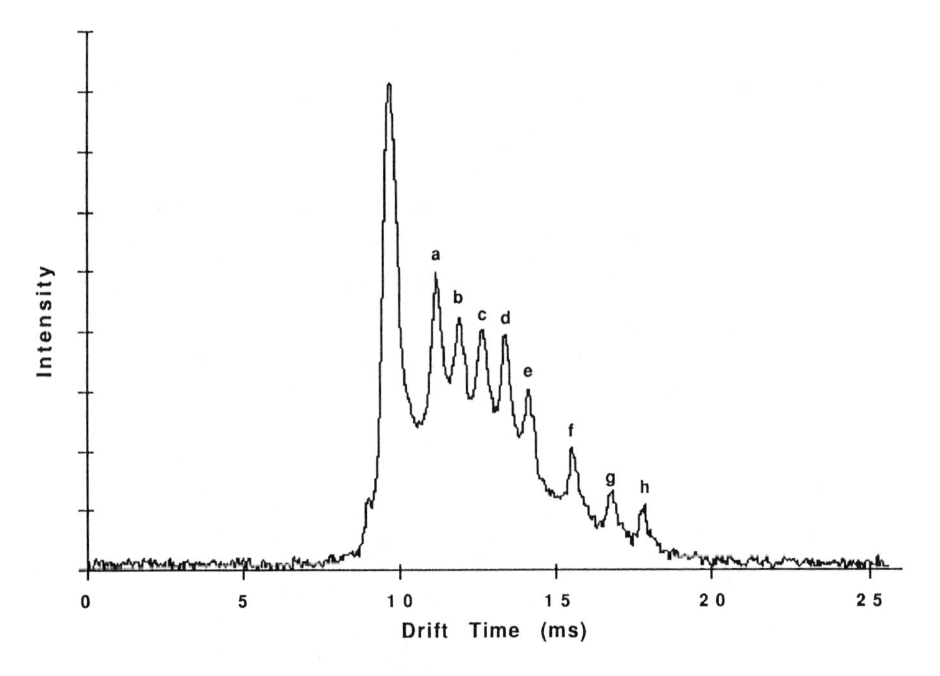

Figure 2. ES-IMS spectra of mixture of eight alkylamines. Operating conditions: needle voltage, 8000 V; drift voltage, 4000 V; temperature, 150 °C; pressure, 698 Torr; liquid flow rate, 5 μL/min of 50/50 methanol/water; drift gas flow, 600 mL/min pre-pure nitrogen; 60 nL injection of 0.1 mg/mL (each) solution.

from the test amines were stable even in the presence of other potentially interfering amines.

Flow Injection Mode. Response data were generated for the amines using the selective product ion monitoring mode. Calibration curves were obtained by discrete injections of test compounds similar to flow injection analysis. Figure 3 is a flow injection analysis trace of a series of tri-n-butylamine standards at different amplifier gains. It represents a typical calibration study produced in this way. Figure 4 shows the calibration curve generated from peak heights with a calculated (S/N = 3) detection limit of 4.5 X 10^{-15} mol/s with a dynamic range of 10^2.

Chromatographic Mode. Figure 5 shows the separation of four trialkylamines (triethyl, tri-n-propyl, tri-n-butyl, and tri-n-hexylamine) with detection by ES-IMS. The top chromatogram was produced by monitoring a large drift time window while the other chromatograms show selective product ion monitoring of the three largest compounds in the same separation.

The triethylamine produced a product ion that drifted before the reactant ion and thus appeared as a negative deflection in the non selective chromatogram. Since the monitoring window included part of the reactant ion peak, any product ion which before this window decreased the total current detected within this window.

The selective ion chromatograms demonstrate the true strength of ES-IMS as a liquid chromatographic detector. By tuning to the product ion of interest, only compounds of interest can be detected with some degree of certainty.

Continuous Monitor Mode. If the mobilities of amines are not changed as a function of the matrix and if the ES-IMS can be tuned to monitor specific ion drift times, then it should be possible to continuously monitor waste and process streams for specific amines. Figure 6 provides and example of continuously monitoring a sample in which the concentration of the test amine was continuously varied from 0 to 25 ppm. Tuned to the product ion of this test amine, the ES-IMS traced the composition change in real time. By simply shifting the drift time monitored, no response would have been seen for these changes.

Long term studies of the instrument response were conducted only for several hours at a time. Due to the prototype nature of the instrument and the high voltage requirements, the instrument could not be left unattended. For well defined parameter conditions, stability of response appeared to be good. Once the system was operating, it continued to operate in a stable manner for several hours at a time. Day to day reproducibility also appeared good as reported above. Plugging of the electrospray orifice did occur on occasion, but this could usually be traced to unfiltered solvents or injections of excessive amount of nonvolatile compounds. In gas phase

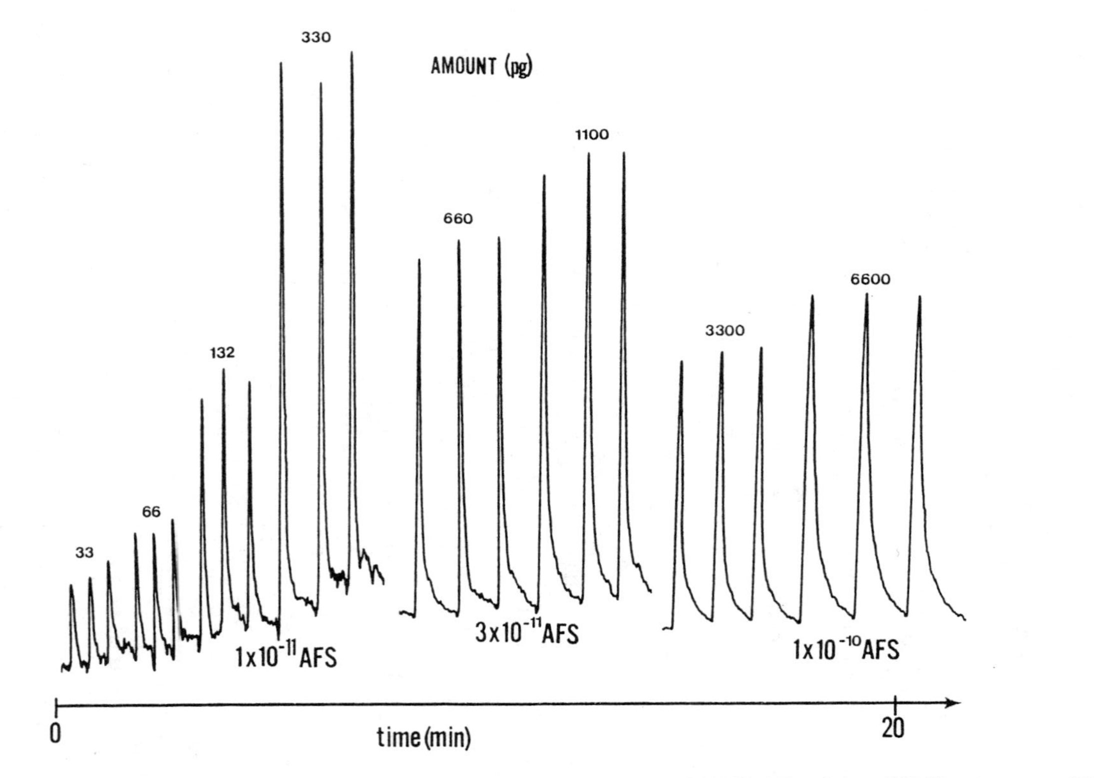

Figure 3. Flow injection analysis of tri-n-butylamine. Conditions: needle voltage, 8000 V; drift voltage, 4000 V; temperature, 150 °C; pressure, 688 Torr; liquid flow rate, 5 μL/min of methanol; drift gas flow, 600 mL/min pre-pure nitrogen; 60 nL injections; stock solution, 11 mg/mL.

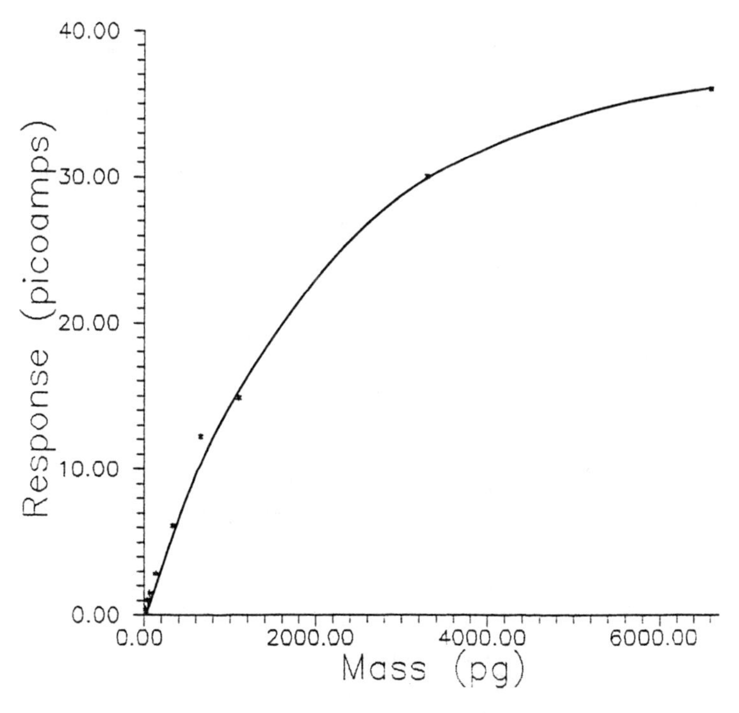

*Figure 4. Calibration curve for tri-*n*-butylamine. Conditions: 8–10 ms monitoring; electrospray voltage, 6880 V; drift voltage, 4000 V; temperature, 152 °C; pressure, 696 Torr; liquid flow rate, 5 μL/min of methanol; drift gas flow, 600 mL/min pre-pure nitrogen; 60 nL injections.*

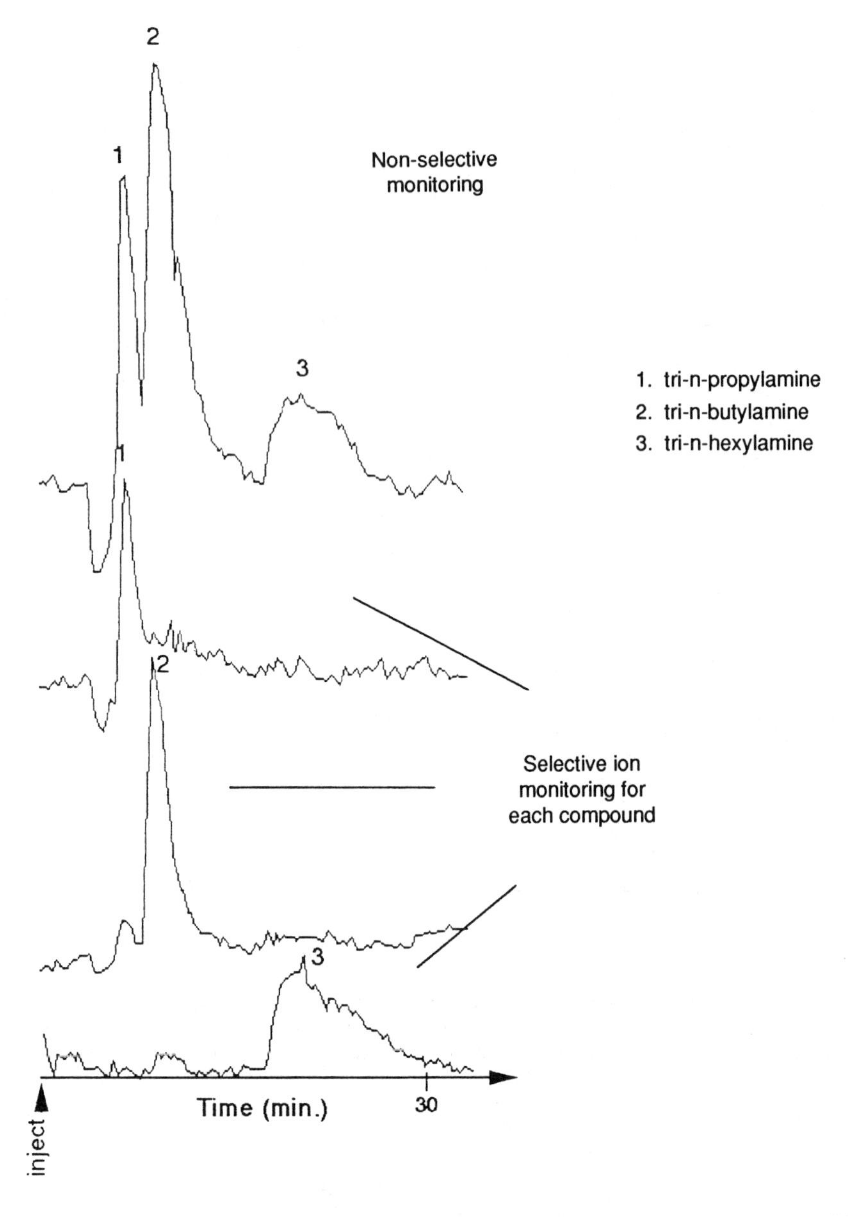

Figure 5. Non-selective and selective monitoring of a mixture of trialkylamines separated by liquid chromatography. Conditions: needle voltage, 10,000 V; drift voltage, 5000 V; drift gas flow, 600 mL/min of air; temperature, 70 °C; liquid flow rate, 200 μL/min split 50:1; mobile phase, 60/40 0.1 M ammonium acetate/acetonitrile isocratic; 200 nL injection of 100 mg/mL (each) solution.

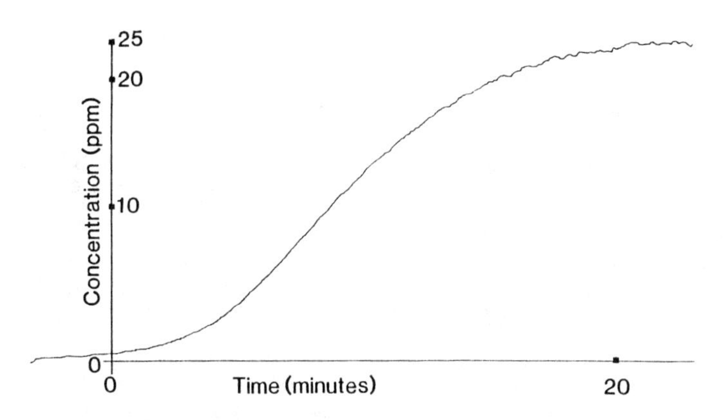

Figure 6. Constant monitoring of di-benzylamine. Conditions: 9–9.8 ms monitoring; needle voltage, 6800 V; drift voltage, 4000 V; temperature, 162 °C; pressure, 703 Torr; drift gas flow, 600 mL/min pre-pure nitrogen; liquid flow rate, 5 μL/min of A: methanol, B: methanol with DBA (0.0254 mg/500mL), gradient from 100% A to 100% B in 20 minutes.

and supercritical fluid applications, long term stability of ion mobility spectrometry is truly excellent. If guarded from contamination, IMS instruments operate for months without down time or the need for maintenance. More experience is needed, however, before such a statement can be made with respect to liquid sample introduction.

Certainly, more investigations should be undertaken with specific analyte and matrix systems in mind but coupled with the fundamental data collected during these studies these initial investigations into continuous monitor for pollution control appears promising.

Ion Abundance Experiments. In general, the response of the detector increased with increasing voltage. However, at voltages above 10 kV detector operation became unstable. For these studies, the most stable operating conditions were determined to be a voltage of 9 kV on the electrospray needle and 4.5 kV on the first drift ring of the spectrometer. Under these conditions, total ion current, noise and compound response were determined to be as follows:

Total Ion Current. First, the overall ion current produced by ES-IMS was on the order of 1 nA. This is several orders of magnitude lower than that reported by electrospray sources used with mass spectrometry. Perhaps, current is lost to the walls of the drift tube through the spraying process and could be regained by ion focusing and geometric considerations of the drift tube.

Noise. With both ion gates open and the total ion current collected at the collector, noise was commonly on the order of 6 X 10^{-11} amperes. When only the reactant ion current was monitored noise was reduced to 2 X 10^{-11} amperes. However, when the entire product ion region was monitored, but no product ions were present, the noise fell to 6 X 10^{-13} amperes and with selective product ion monitoring the noise was reduced even further to 2 X 10^{-13} amperes.

Detection Limit. From noise considerations alone, it would appear that the detection limit for the ES-IMS would appear that the reactant ion monitoring mode would have the highest detection limit, the non-selective product ion monitoring mode the next highest and the selective product ion monitoring mode the lowest. This was in fact the order we observed for the test compound tri-n-butylamine. The detection limit in the reactant ion monitoring mode was found to be 1 X 10^{-12} mol/s. In the non-selective product ion monitoring mode it was 1 X 10^{-13} mol/s. And, in the selective product ion monitoring mode it was as low as 5 X 10^{-15} mol/s.

Summary and Conclusion

Although this investigation was limited to amines as test compounds and further investigations of other classes of analytes will be necessary, the data obtained in these studies were promising. The mobility studies indicated that the detector can be reliable and reproducible over time and at various operating conditions. They also suggested that this prototype design may not be the optimal configuration. Investigations into the effects of ion focusing, shielding of the electrospay voltage from the drift field voltage, and mass identification of reactant and product ions are needed before the method can be developed to it maximum potential. Nevertheless, demonstration applications for flow injection analysis, chromatography, and continuous monitoring of liquid streams indicated that ES-IMS shows considerable potential as a detection method to compliment spectrophotometric and electrochemical methods of analysis.

Acknowledgments

This research was sponsored in part by grants from the Public Health Service, Precision Analitics, Inc., and Millipore Corporation.

Literature Cited

1. Riebe, M. T.; Daniel, J. E. *Anal. Chem.* **1990**, *62*, 65A.
2. Hill, H. H.; Siems, W. F.; St. Louis, R. H.; McMinn, D. G. *Anal. Chem.*, **1990**, *62(23)*, A362.
3. St. Louis, R. H.; Hill, H. H. *Critical Rev. in Anal. Chem.*, **1990**, *21(5)*, 321-355.
4. Gieniec, J.; Cox, H. L., Jr.; Teer, D.; Dole M. *Abstracts 20th Annu. Conf. Mass Spectrum. Allied Top.* **1972**, 276.
5. Dole, M.; Gupta, C. V.; Mack, L. L.; Nakamae, K. *Polym. Prepr.* **1977**, *18(2)*, 188.
6. Gienied, J.; Mack, L. L.; Nakamae, K.; Gupta, C.; Kumar, V.; Dole, M. *Biomed. Mass Spectrum.* **1984**, *11(6)*, 259.
7. Shumate, C. B.; Hill, H. H. *Anal. Chem.* **1989**, *61(6)*, 601-606.
8. McMinn, D. G.; Kinzer, J. A.; Shumate, C. B.; Siems, W. F.; Hill, H. H., *J. Microcolumn Separations,* **1990**, *2*, 188-192.
9. Fenn, J. B.; Mann, M.; Meng, C. K.; Wong, W. F.; Whitehouse, C. M. Science 1989, 246, 64.
10. Fenn, J.B.; Mann, M.; Meng, C. K.; Wong, S. F.; Whitehouse, C. M. Mass Spectrom. Rev. 1990, 9, 37.
11. Whitehouse, C. M.; Dreyer, R. N.; Yamashita, M; Fenn, J.B. Anal. Chem. 1985, 57, 675.
12. Smith, R. D.; Loo, J.A.; Barinaga, C.J.; Edmonds, C. G.; Udseth, H.R., J. Am. Soc. Mass Spectrom. 1990, 1, 53.
13. Smith, R. D.; Barinaga, C. J.; Udseth, H. R. J. Phys. Chem. 1989, 93, 5019.
14. Shumate, C. B. *An Electrospray Nebulization/ Ionization Interface for Liquid Introduction into an Ion Mobility Spectrometer,* Ph.D. Thesis, Washington State University, 1989.
15. Noffsinger, J. B.; Danielson, N. D. *J. Chrom.* **1987**, *387*, 520.

RECEIVED May 18, 1992

Chapter 17

Analytical-Scale Extraction of Environmental Samples Using Supercritical Fluids

Steven B. Hawthorne, David J. Miller, John J. Langenfeld, and Mark D. Burford

Energy and Environmental Research Center, University of North Dakota, Box 8213, University Station, Grand Forks, ND 58202

Analytical-scale supercritical fluid extraction (SFE) is a rapidly-developing method for the extraction of organic analytes from environmental solids including soils, sediments, air particulates, and sorbent resins. SFE virtually eliminates the need for liquid solvents, and 10 to 90-minute SFE extractions can yield recoveries of organic pollutants which compare favorably with those obtained using several hours of conventional liquid solvent extraction. Even though SFE is relatively simple to perform, the development of routine SFE methods is hampered because the physical and chemical mechanisms which control ultimate recoveries are poorly understood. To develop a quantitative SFE method, the physicochemical processes which control the partitioning of the analytes from the matrix into the extraction fluid, and the efficient collection of the extracted analytes from the supercritical fluid must be controlled. Approaches to the development of quantitative SFE methods for organic pollutants from environmental samples including soils and sediments, air particulates, sorbent resins, and water will be discussed.

The extraction of organic analytes from an environmental solid has traditionally been performed using several hundred milliliters of a liquid solvent followed by evaporation of the solvent prior to instrumental analysis. Soxhlet extraction has generally been accepted as the definitive sample preparation method, even though the technique has undergone little improvement since its introduction nearly 100 years ago (*1*). The current interest in reducing liquid solvent waste and in developing extraction methods that are rapid and have potential for field use has generated a strong interest in alternative methods to extract organics from environmental solids prior to instrumental analysis. Recent investigations have demonstrated the potential for supercritical fluid extraction (SFE) to meet

0097–6156/92/0508–0206$06.00/0

many of the idealized goals of a sample extraction/preparation technique, i.e., quantitative, rapid, inexpensive, little waste production, and potential for field-portability (*2-4*). In addition, several recent reports have demonstrated the ability to directly couple SFE with chromatographic techniques including capillary GC and SFC (*2,3*).

The investigations to date have demonstrated the potential for SFE to yield rapid and quantitative recovery of organics from environmental solids while virtually eliminating the need for liquid solvents. However, a "generic" approach to developing a successful SFE method for environmental samples is not yet available. This paper will describe some of the factors which must be considered in order to develop a useful SFE method, and will suggest approaches for testing the individual steps in an SFE method. Examples of SFE approaches and results will be presented for organic analytes from a variety of environmental matrices. Extraction recoveries, times required, and solvents used for SFE and conventional liquid solvent extraction techniques will be compared.

Steps in an SFE Extraction

A simple instrument for SFE is shown in Figure 1. In essence, SFE is performed by providing a constant pressure (typically 100 to 500 atm) of a fluid (most commonly CO_2) to the sample which is held at a known temperature (usually just above the critical temperature of the fluid, i.e., above 32°C for CO_2). Following extraction, the analytes are swept from the cell by the supercritical fluid, and then collected by depressurizing the supercritical fluid through a limiting orifice (flow restrictor). Various approaches to SFE have been described in the literature, and will not be repeated in detail here. Several commercial instruments have also recently become available, which differ primarily in the method used to depressurize the supercritical fluid (i.e., fixed versus variable restrictors) and in the method used for collecting the extracted analytes from the depressurized supercritical fluid (i.e., solvent trapping versus trapping on a sorbent resin or a cooled surface). For general descriptions of SFE techniques and environmental applications, the reader is referred to recent reviews (*2-4*).

For the purposes of developing an SFE method, it is useful to divide the SFE experiment into two conceptual steps (designated "chemical" and "plumbing"). First, the analyte must be partitioned from the matrix into the bulk supercritical fluid. This first step is controlled primarily by the physicochemical interactions of the analyte, sample matrix, and the supercritical fluid. Second, the analyte must be swept from the extraction cell by the supercritical fluid and the analyte must be efficiently collected at the outlet of the extraction cell flow restrictor. This second step depends primarily on the "plumbing" of the SFE system, i.e., the flow restrictor and analyte recovery devices. Since failure at any of these steps results in less than 100% extraction efficiencies, it is useful to evaluate each step independently during methods development.

SFE Step 1 ("Chemistry"). Partitioning the Analytes from the Matrix into the Supercritical Fluid. The first conceptual step in an SFE extraction, the partitioning of the target analytes into the bulk supercritical fluid, can be controlled by a number of physicochemical factors that are not yet well understood. However, it has been useful to divide these physicochemical factors into three areas (*5-7*); analyte solubility in the supercritical fluid, matrix/analyte interactions, and a kinetic limitation (which has previously been described by a diffusion model, ref. 5). The first requirement for good SFE recoveries is that the analyte must have sufficient solubility to be extracted in the supercritical fluid. For example, the ionic surfactant linear alkylbenzenesulfonate (LAS) cannot be extracted with pure supercritical CO_2, simply because it has no significant solubility. However, when the polarity of the supercritical fluid was increased by adding a high concentration of a polarity modifier (e.g., 40 mole % methanol in 400 atm CO_2 at 125°C), the extraction was very efficient, and recoveries averaged 99%, 91%, and 98% for the extraction of LAS from river sediment, agricultural soil, and municipal sewage sludge, respectively (*8*). The extraction of LAS also demonstrates a kinetically-limited feature to the SFE extraction rates which has been common to all environmental samples we have studied. As shown in Figure 2, the recovery of the LAS was most rapid at the beginning of the extraction, however the extraction rates slowed considerably after the first few minutes. This behavior clearly demonstrates a kinetic limitation to SFE extraction rates, and has been explained on the basis of diffusion limitations of the analytes in the sample matrix (*5*), although alternate kinetic limitations (e.g., desorption kinetics of the analytes from the active sites on the matrix) may also apply.

While the extraction of high molecular weight analytes and highly polar analytes such as LAS can be limited by solubility, the majority of SFE investigations to date have dealt with less polar and lower molecular weight species, i.e., "GC-able" organics such as PAHs, PCBs, fuel hydrocarbons, and low polarity pesticides. Although solubility data in supercritical CO_2 is not available for many of the organic pollutants of interest, sufficient data is available to indicate that analyte solubility is generally not the limiting factor for SFE of "GC-able" organics (*9*). For example, Table I lists the concentrations of several pollutants which could be extracted from a 1-gram sample using 1 mL of supercritical CO_2 at 400 atm and 50°C based on the solubility of each pollutant in supercritical CO_2 (*9*). Based on solubility considerations alone, a small volume of supercritical CO_2 should be capable of extracting extremely high concentrations of these analytes. Unfortunately, the extraction efficiencies that might be expected for such analytes are often not observed when real-world samples are extracted. For example, while Table I shows that PAHs have more than sufficient solubility to extract with CO_2, quantitative recovery of PAHs can not always be achieved in a reasonable time. As shown in Figure 3 by the recoveries of PAHs obtained from a standard reference material marine sediment (National Institute of Standards and Technology, NIST SRM 1941) using 10-minute SFE extractions at 380 atm (50°C), extraction with CO_2 did not yield nearly as high of recoveries of higher

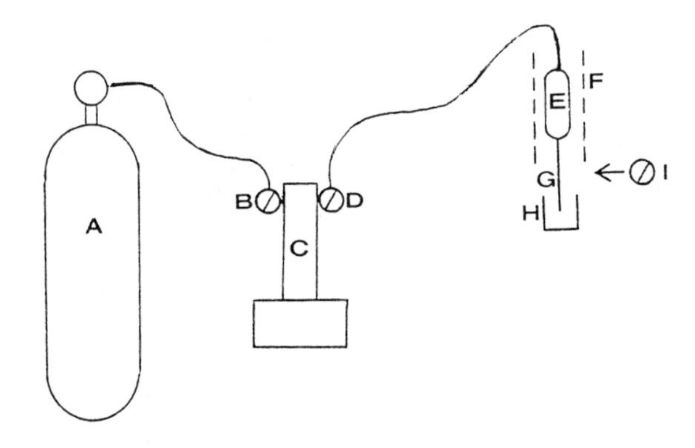

Figure 1. Components of a simple SFE instrument. The fluid reservoir (A) is connected using 1/16 inch stainless steel tubing to a shut-off valve (B) mounted on a pressure-controlled syringe pump (C). During extraction, valve (B) is closed, and shut-off valve (D) is opened to supply the pressurized extraction fluid to the extraction cell (E) which is placed in a tube heater (F) to maintain the extraction temperature. The extracted analytes then flow out of the restrictor (G) into the collection vessel (H). For static extractions, an additional shut-off valve (I) can be placed between the extraction cell and the restrictor as indicated on the figure.

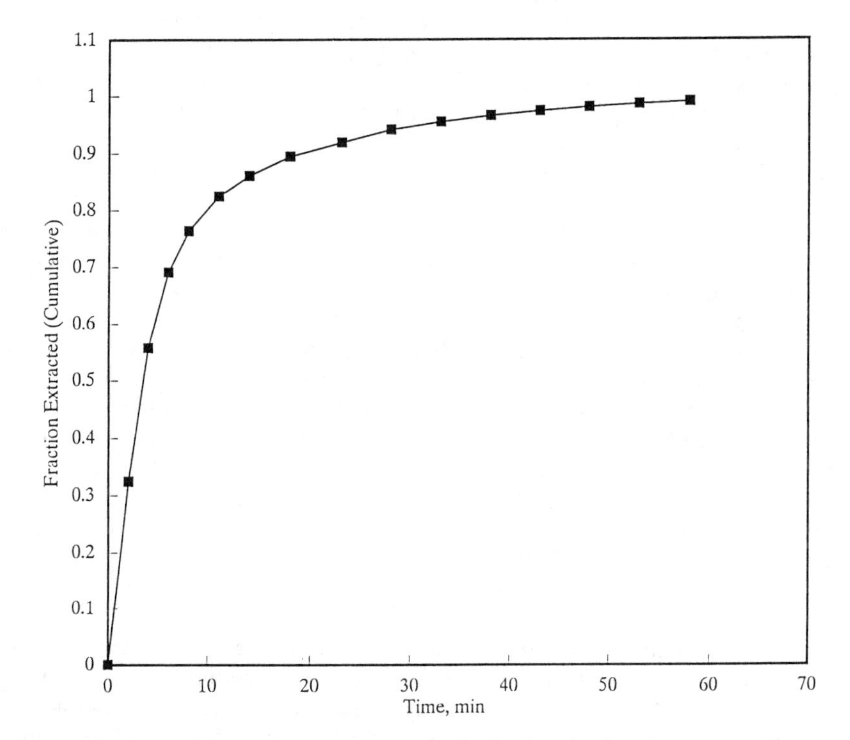

Figure 2. SFE extraction kinetics of the ionic detergent, linear alkylbenzenesulfonate (LAS) from a 1-gram sample of sewage sludge.

Table I. Solubilities of Organics in Supercritical CO_2

Species	Solubility at 400 atm, 50°C[a]
naphthalene	190000 ppm (μg/mL)
phenanthrene	13000 ppm
pyrene	2500 ppm
dibenzothiophene	11000 ppm
p-chlorophenol	137000 ppm
docosane	315000 ppm

[a] Solubilities were estimated based on the tabulations given in reference 9.

molecular weight PAHs as extraction with N_2O, although SFE with N_2O did yield good recoveries compared to the certified values (based on 16 to 20 hour Soxhlet extractions).

The results shown in Table I and Figure 3 demonstrate that high analyte solubility in the supercritical fluid is not sufficient to guarantee that quantitative recovery will be achieved, and it is clear that the interactions between the analytes and the matrix sorptive sites must be overcome to obtain good SFE recoveries. Fortunately, proper selection of SFE conditions has resulted in excellent agreement with the known concentrations of PAHs from a variety of sample matrices. As shown in Table II and Figure 4, SFE with CO_2 yields quantitative recovery of PAHs from some matrices (e.g., Tenax-GC and diesel exhaust particulates), while for other matrices recoveries are better using supercritical N_2O. While the use of CO_2 is clearly advantageous because of its low toxicity and negligible environmental impact (burning 1 gallon of gasoline generates enough CO_2 to perform several hundred extractions), the varying recoveries achieved using pure CO_2 demonstrate the need for a "stronger" extraction fluid for a generalized SFE approach. While N_2O yields better recoveries than pure CO_2 for some samples, N_2O is more hazardous for routine applications. Most likely, the use of small amounts of polarity modifiers in CO_2 will be the preferred solution in the future. For example, the use of CO_2 modified with 5% (v/v) methanol yielded quantitative recovery of the PAHs from diesel exhaust particulates in 30 minutes, compared to the 90 minutes required for pure CO_2 as shown in Table II. PCB recoveries from river sediment (NIST SRM 1939) also improved dramatically using methanol-modified (5%) CO_2 versus pure CO_2. For example, the recoveries of eleven individual PCB congeners (trichloro- to heptachloro-) averaged only 63% using pure CO_2 (40 minutes at 400 atm and 50°C) compared to the NIST 32-hour Soxhlet extraction, while the recoveries using the methanol-modified CO_2 averaged 96% compared to the Soxhlet results (*10*). These results are in agreement with similar increases in extraction efficiencies using methanol-

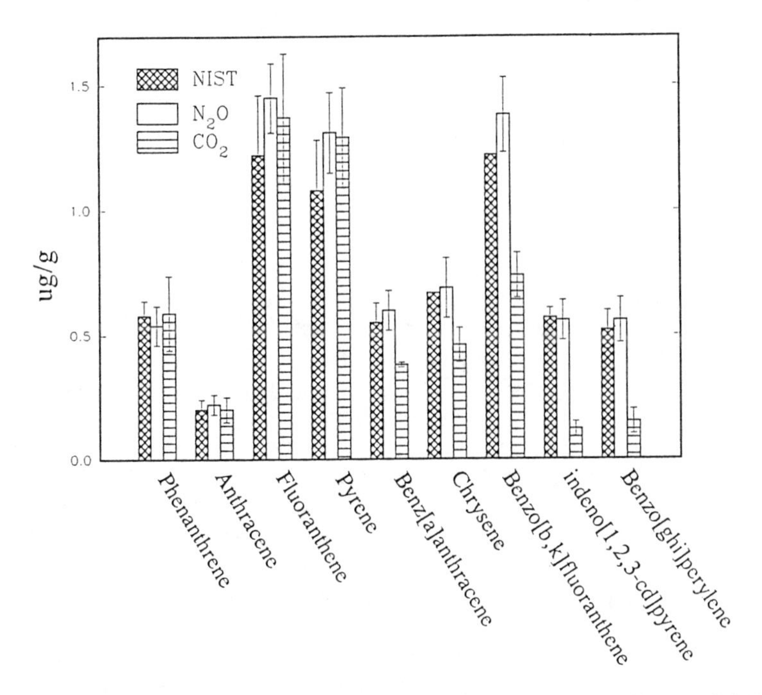

Figure 3. Comparison of PAHs extracted from marine sediment (NIST SRM 1941) using supercritical CO_2 and N_2O. Results are adapted from reference 16. (Values for perylene and benzo[a]pyrene were also certified, but were not determined in this study because of the lack of standard compounds.)

Table II. SFE Recoveries of PAHs from Diesel Exhaust Particulate (NIST 1650) Using Pure and Methanol-Modified CO_2

		SFE Recoveries[a]	
Species	Certified value (μg/g)	CO_2 (90 min)	CO_2/MeOH (30 min)
fluoranthene	51	99%	105%
pyrene	48	96%	99%
benz[a]anthracene	6.5	111%	104%
benzo[a]pyrene	1.2	112%	120%

[a] Values were adapted from reference 14 and are based on the NIST certified concentrations.

modified CO_2 compared to pure CO_2 for the extraction of chlorinated dibenzodioxins and dibenzofurans from fly ash (*11*) and pesticides including diuron, linuron, and DDT from soil (*12,13*).

The results just described, as well as a survey of the literature, make it clear that a better understanding of the interactions involving the analytes, matrix active sites, and the supercritical fluid will greatly facilitate the development of generalized SFE techniques for a variety of sample matrices.

SFE Step 2 ("Plumbing"). Sweeping the Extracted Analytes from the Extraction Cell and Efficiently Collecting the Analytes. In contrast to the matrix/analyte/supercritical fluid interactions that control the "chemical" step during SFE, the "plumbing" steps are more simple to test and understand. The first "plumbing" factor to consider is the flow rate of the supercritical fluid. When the analytes have some solubility in the supercritical fluid, but their concentrations are very high, the solubility limit of the supercritical fluid can be reached, and higher extraction flow rates would be expected to yield faster recovery rates (e.g., the extraction of fats from meats). However, as discussed above, the concentrations of environmental analytes are frequently much lower than the solubility limit of the supercritical fluid. In such cases, the effect of the extraction fluid flow rate on the recovery rates is not large unless an unreasonably low flow rate is used (i.e., the flow is too slow to efficiently sweep the volume in the cell which is not occupied by the sample matrix). For example, the extraction rate of PAHs from a 3-gram sample of railroad bed soil was essentially unchanged when the supercritical fluid flow rate was varied from 0.9 mL/min down to 0.3 mL/min as shown for phenanthrene in Figure 5 (fluid flow rates are as the liquid CO_2 measured at the pump). However, when the flow rate was further lowered to 0.15 mL/min, the recovery rates of the PAHs were lowered (Figure 5), as would be expected since the 0.15 mL/min flow rate was too slow to efficiently sweep the 1-mL void volume resulting from the spaces between the soil particles.

(a)

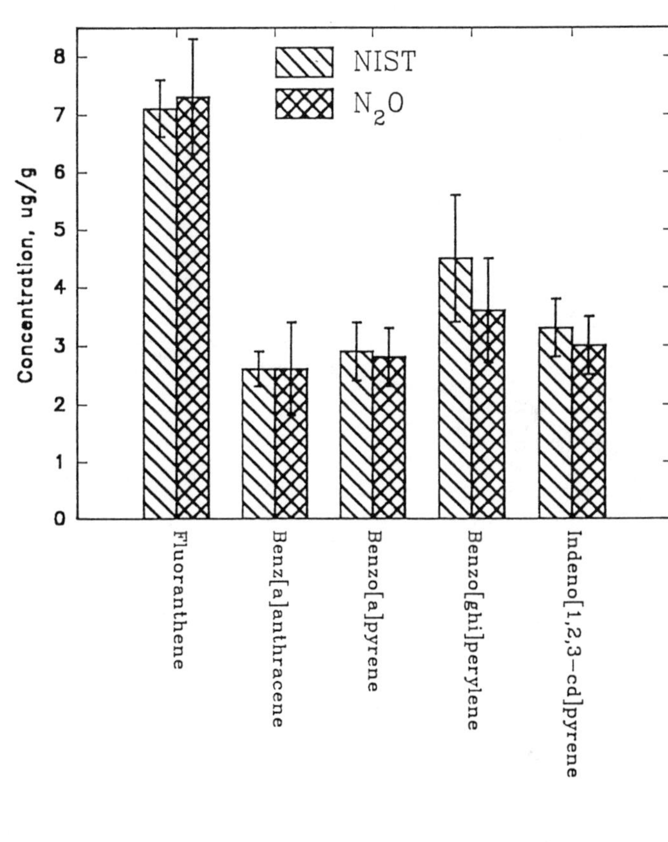

(b)

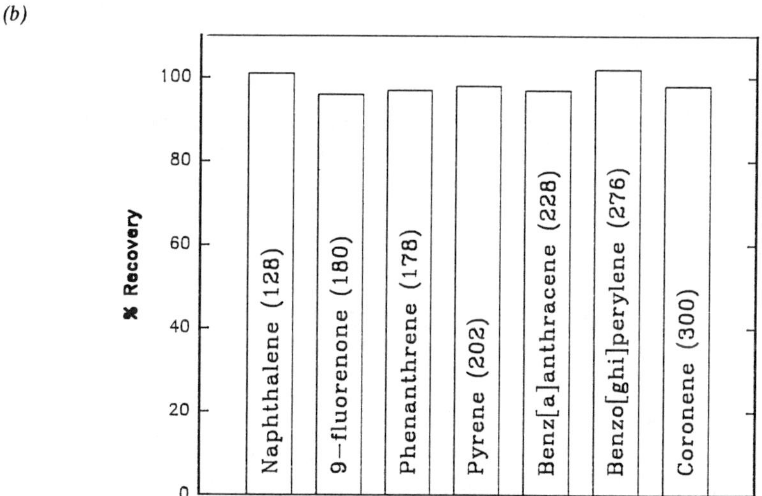

Figure 4. SFE of PAHs from NIST urban dust (air particulates, SRM 1649) using 20-minute extractions with 350 atm N_2O (a) and from Tenax-GC using 15-minute extractions with 200 atm CO_2 (b). Results are adapted from references 14 and 17.

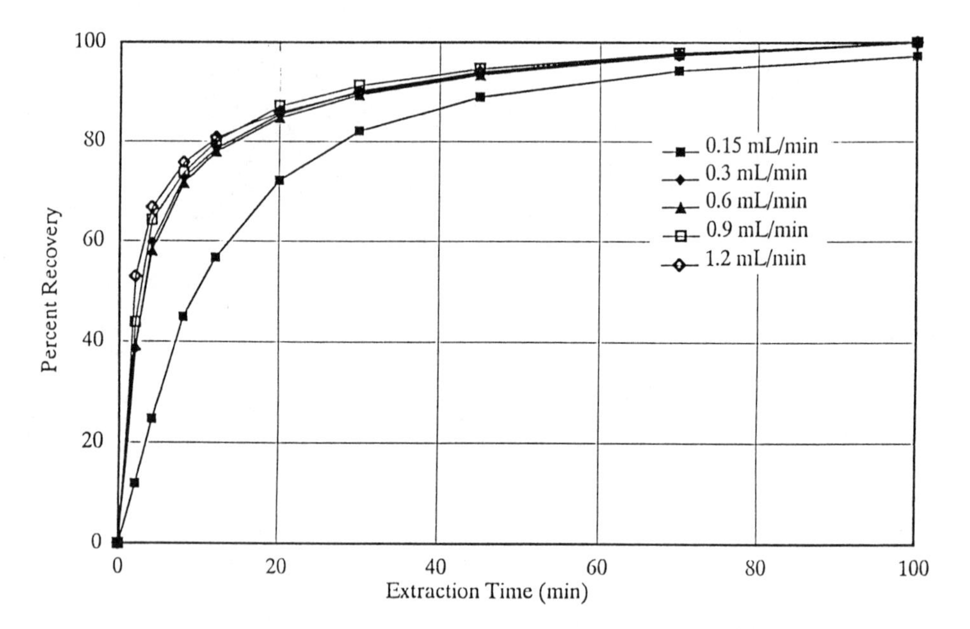

Figure 5. The effect of SFE flow rate on the recovery of phenanthrene from railroad bed soil. Phenanthrene concentration was ca. 50 ppm.

High SFE flow rates may actually yield poorer apparent recoveries because of the increased difficulty in collecting the analytes from the depressurized fluid. Supercritical CO_2 expands by a factor of ca. 500 upon depressurization to a gas, and for the most commonly-used post-extraction collection methods (discussed below), CO_2 extraction flow rates have typically been in the range of 0.5 mL to 2.0 mL/min (resulting in a gas flow rate of ca. 250 to 1000 mL/min upon depressurization). Since larger samples require higher extraction flow rates to efficiently sweep the void volume of the cell, and higher flow rates make analyte collection more difficult, analytical-scale SFE seems best applied to samples not larger than 10 to 20 grams using presently-available hardware.

The second "plumbing" factor that must be considered to obtain quantitative results from SFE is the method used to collect the analytes from the depressurized supercritical fluid. The analytes in SFE extracts are collected in two basic modes; on-line (where the SFE effluent is directly transferred to a chromatographic system and analytes are trapped in a pre-trap or directly in the chromatographic column), or off-line (where the analytes are trapped in a separate device and later subjected to analysis by chromatographic or other appropriate techniques). This paper will focus on off-line trapping of the SFE effluents since the majority of environmental applications have utilized the off-line approach, most commercial SFE instrumentation utilizes off-line collection, and off-line SFE is presently more compatible with conventional automated chromatographic instrumentation used in routine laboratories. The interested reader is referred to recent reviews for discussions about on-line SFE techniques (*2,3*).

Off-line SFE trapping methods have utilized two essentially different approaches; trapping directly in a few milliliters of liquid solvent, or trapping in a sorbent (or cryogenic) trap followed by elution of the trapped analytes with a few milliliters of liquid solvent. With both methods, the analytes are ultimately contained in a small volume of solvent which can then be stored and analyzed in a manner similar to that used for a concentrated conventional liquid solvent extract.

Since 1 mL/min of supercritical CO_2 results in ca. 500 mL/min of CO_2 gas, the collection of the extracted analytes is essentially an air sampling problem. When analytes are trapped in a sorbent or cryogenic trap, both the collection from the gaseous CO_2 stream and the method used to recover the analytes from the trap must be quantitatively efficient. Obviously, the use of cryogenic trapping is less useful as the volatility of the target analyte increases. The use of sorbent traps may also suffer from contamination by co-extracted matrix materials which may not be efficiently cleaned from the sorbent between samples. When organic modifiers are used for the SFE step, collection efficiencies on sorbent traps may also be poorer since the organic modifier may reduce the retention of the target analytes on the sorbent trap. Collection directly into a solvent appears inherently simpler than sorbent trapping, since the elution step is avoided. However, trapping on a sorbent does have the potential advantage that class-selective recoveries from the sorbent can be obtained based on the sorbent's characteristics and those of the elution solvent.

The variations in trapping conditions used (including technique variations between different operators) make prediction of the success of a proposed trapping method difficult, particularly for volatile analytes. The best collection conditions for individual analytes and samples are not always obvious, and collection conditions appear to be most often chosen based on the experience of individual researchers and (when commercially-available SFEs are used) on the hardware configuration of a particular SFE system. Because of the high flow rates of gas involved, the loss of analytes can frequently occur, and low overall recoveries from an SFE extraction will result. Since low recoveries from poor collection efficiencies can mistakenly be blamed on poor extraction conditions, it is important to test the analyte trapping system independently of the extraction step.

While it is clear that spike recovery studies are frequently not valid for determining the efficiency of the extraction step because of the inability of spikes to mimic the behavior of native analytes (the "Chemistry" step discussed above), spike recoveries are an excellent means to develop and evaluate the "Plumbing" steps, i.e., once analytes are extracted, are they efficiently swept from the cell and quantitatively collected? Regardless of whether sorbent trapping or solvent trapping is used, the development of trapping conditions that yield quantitative recovery of spiked analytes from a non-sorbtive matrix is a necessary prelude to developing the SFE conditions that are best suited to extracting the native analytes from real-world samples. The evaluation of sorbent trapping conditions can include the sorbent's polarity, volume (and capacity) and the identity and volume of liquid solvent used to recover the target analytes. For trapping in liquid solvents, it is frequently useful to test spike recoveries using solvents with different polarities. For example, when developing collection conditions for several organic pollutants from soil, spike recoveries from a non-sorbtive matrix (sand) were first determined using 2-mL portions of a variety of collection solvents. As shown in Table III, trapping efficiencies were generally better in methylene chloride than in either acetone, methanol, or hexane. Based on these results, methylene chloride was chosen as the best collection solvent, and further investigations demonstrated that all of the test species could be efficiently collected in methylene chloride if the collection vial was held at 5°C during the SFE step (Table III).

Solvent Reduction using SFE versus Conventional Extraction Methods

While the potential for SFE to yield rapid and quantitative extraction and recoveries from a broad range of environmental samples has been demonstrated by a number of researchers, the goal of this symposium was to investigate methods that are capable of reducing the use of liquid solvents and their subsequent emission (or disposal requirements). As discussed above, we have developed SFE methods that yield good agreement with NIST certified reference values for a variety of materials including PAHs from diesel exhaust

Table III. SFE Collection Efficiencies of Representative Organic Spikes from Sand Using Different Trapping Solvents

	% Recovery (RSD)[a]			
species	CH_2Cl_2	methanol	hexane	CH_2Cl_2 (5°C)
phenol	77 (2)	55 (12)	43 (2)	99 (2)
nitrobenzene	82 (4)	58 (10)	50 (2)	97 (1)
1,2-dichlorobenzene	78 (5)	58 (13)	46 (5)	93 (2)
2-nitroaniline	86 (1)	57 (5)	72 (3)	98 (7)
pyrene	90 (3)	58 (7)	80 (5)	100 (1)

[a] Recoveries are based on 10-minute extractions from sand using 400 atm CO_2 at 0.5 mL/min. RSDs are based on triplicate extractions.

particulates, urban air particulates, and marine sediment, and PCBs from river sediment, and a comparison of the extraction times and solvents required for our SFE methods versus the NIST methods based on conventional liquid solvent extraction are shown in Table IV. As can be seen in Table IV, SFE dramatically reduces both the requirements for liquid solvent usage (and disposal or emission) as well as the time required for sample extraction. In addition, on-line SFE techniques have been reported which completely eliminate the need for liquid solvents (*2,3*).

SFE of Water Samples

While the majority of investigations using SFE for the extraction of environmental samples have focussed on the extraction of solids, some initial attempts to replace the conventional methylene chloride extraction of water samples with SFE have been reported (*15*). The potential for directly extracting water samples is demonstrated in Figure 6 by the extraction of a 4-mL sample of coal gasification wastewater using CO_2 with 1 mL of methanol modifier added to the water. While such direct extractions of small water samples appears feasible, the extraction of larger (e.g., 1-liter) samples is not practical. When larger samples must be extracted to meet sensitivity requirements, the combination of solid-phase extraction with SFE may be useful as shown in Figure 7 by the extraction of wastewater components from a C-18 "Empore" sorbent disc.

Summary and Conclusions

SFE has been demonstrated to yield rapid and quantitative extractions of a variety of organic pollutants from a wide range of environmental solids, and also appears to have potential for extracting organics from water samples.

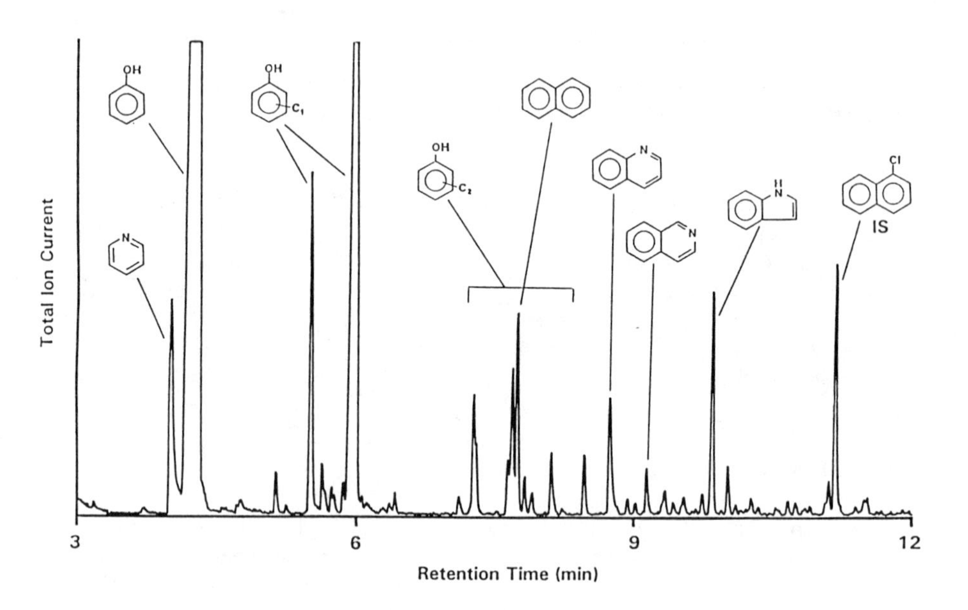

Figure 6. GC/FID chromatogram of the SFE extract from a 4-mL sample of coal gasification wastewater. Peak identifications are based on GC/MS analysis.

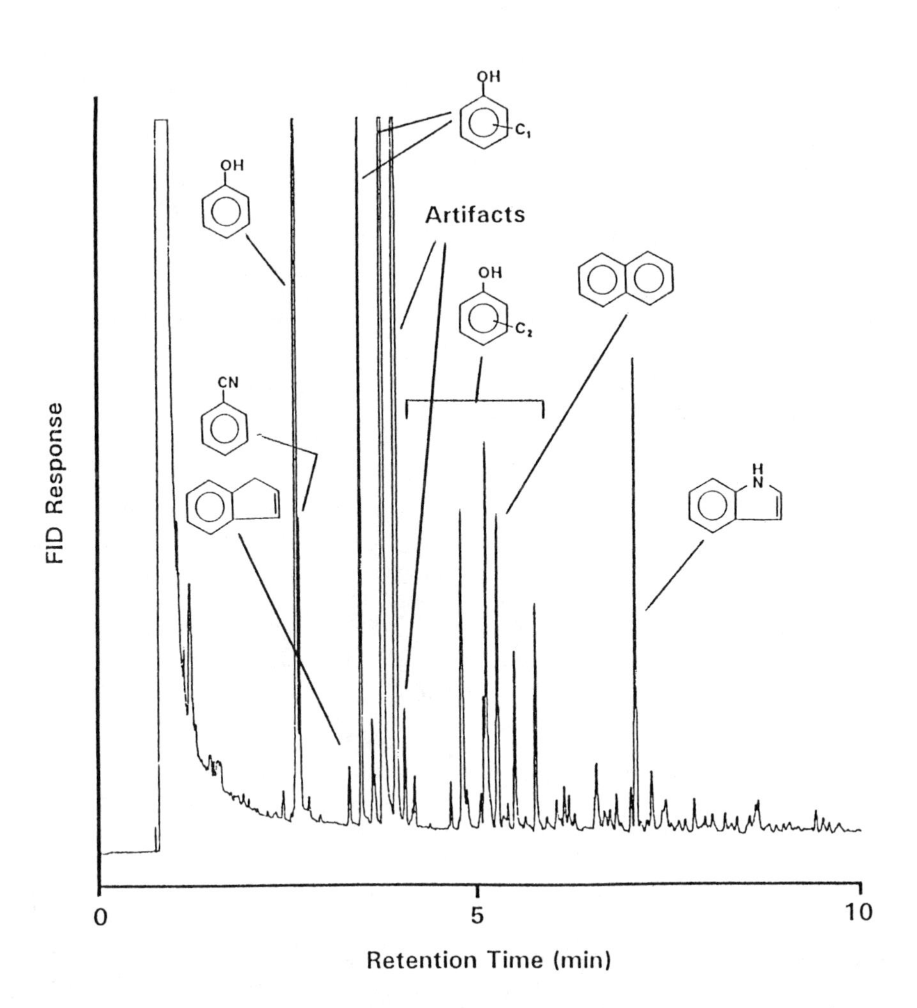

Figure 7. GC/FID chromatogram of coal gasification wastewater organics collected on an "Empore" C-18 sorbent disc. The 100 mL water sample was acidified to pH <2 before filtration through the disc. SFE was performed with 400 atm CO_2 for 10 minutes.

Table IV. Comparison of Solvent Volumes and Extraction Times Using SFE and Conventional Solvent Extraction of NIST Certified Reference Materials

	Extraction Time[a]		Liquid Solvent Volume[b]	
Sample, SRM number	NIST	SFE	NIST	SFE
PAHs, diesel particulate (1650)	--[b]	30, 90 min	450 mL	4 mL
PAHs, urban dust (1649)	48 hr	20 min	450 mL	0 mL[c]
PAHs, marine sediment (1941)	16 hr	10 min	500 mL	0 mL[c]
PCBs, river sediment (1939)	32 hr	40 min	200 mL	4 mL

[a] NIST extractions were performed with a conventional Soxhlet apparatus.
[b] The Soxhlet extraction time is not stated in the certificate of analysis.
[c] Sample was analyzed using on-line SFE-GC analysis.

Extraction times required using SFE are typically <60 minutes, and the quantity of liquid solvent required is reduced by ca. two orders of magnitude compared to conventional liquid solvent extractions. While quantitative recoveries using SFE have been demonstrated for several samples, the physicochemical mechanisms that control SFE recoveries are still poorly understood and it is likely that generally-applicable SFE methods will require the addition of organic polarity modifiers to CO_2 when either the analytes are highly polar, or the matrix/analyte interactions are strong. However, analytical-scale SFE has only recently received significant attention, and rapid advances that are now occurring in instrumentation, techniques, and mechanistic understanding should greatly increase the general applicability of SFE to environmental samples.

Acknowledgements

The authors would like to thank the U.S. Environmental Protection Agency (EMSL, Cincinnati, and the Office of Exploratory Research) and the New Jersey Department of Environmental Protection (Division of Science and Research) for financial support. Instrument loans from ISCO are also gratefully acknowledged.

Literature Cited

1. Saiki, T. *J. Biol. Chem.* **1909-1910,** *7*, 21.
2. Hawthorne, S. B. *Anal. Chem*. **1990**, *62*, 633A.
3. Veuthey, J. L.; Caude, M.; Rosset, R. *Analusis* **1990**, *18*, 103.
4. *Proceedings of the International Symposium on Supercritical Fluid Chromatography and Extraction*; Park City, Utah, January **1991**.

5. Bartle, K. D.; Clifford, A. A.; Hawthorne, S. B.; Langenfeld, J. J.; Miller, D. J.; Robinson, R. *J. Supercritical Fluids* **1990**, *3*, 143.
6. Kandiah, M.; Spiro, M. *J. Food Sci. Technol.* **1990**, *25*, 328.
7. King, M. B.; Bott, T. R.; Barr, M. J.; Mahmud, R. S.; Sanders, N. *Sep. Sci. Tech.* **1987**, *22*, 1103.
8. Hawthorne, S. B.; Miller, D. J.; Walker, D. D.; Whittington, D. E.; Moore, B. L. *J. Chromatogr.* **1991**, *541*, 185.
9. Bartle, K. D.; Clifford, A. A.; Jafar, S. A.; Shilstone, G. F. *J. Phys. Chem. Ref. Data*, in press.
10. Hawthorne, S. B.; Miller, D. J.; Langenfeld, J. J. Manuscript in preparation.
11. Alexandrou, N.; Pawliszyn, J. *Anal. Chem.* **1989**, *61*, 2770.
12. Wheeler J. R.; McNally, M. E. *J. Chromatogr. Sci.* **1989**, *27*, 534.
13. Dooley, K. M.; Ghonasgi, D.; Knopf, F. C. *Environmental Progress* **1990**, *9*, 197.
14. Hawthorne, S. B.; Miller, D. J. *J. Chromatogr. Sci.* **1986**, *24*, 258.
15. Hedrick, J. L.; Taylor, L. T. *J. High Resolut. Chromatogr.* **1990**, *13*, 312.
16. Hawthorne, S. B.; Miller, D. J.; Langenfeld, J. J. *J. Chromatogr. Sci.* **1990**, *28*, 2.
17. Hawthorne, S. B.; Miller, D. J. *J. Chromatogr.* **1987**, *403*, 63.

RECEIVED April 12, 1992

Chapter 18

Real-Time Measurement of Microbial Metabolism in Activated Sludge Samples

Anna G. Cavinato[1], Zhihong Ge[2], James B. Callis[2], and Richard E. Finger[3]

[1]D[2] Development, 1108 J Avenue, La Grande, OR 97850
[2]Center for Process Analytical Chemistry, University of Washington, BG-10, Seattle, WA 98195
[3]Municipality of Metropolitan Seattle, Seattle, WA 98104

In a wastewater secondary treatment process, where the effluent from the primary treatment is mixed with microorganisms in aeration tanks, there is no direct measurement of biological action by the added microorganisms. Current indirect correlations to changes in dissolved oxygen and pH are unreliable. In this paper a method to follow metabolic changes of activated sludge samples is presented. The changes are diagnosed by monitoring the spectroscopic properties of microbial populations in the 500–650 nm wavelength region. Measurements are performed remotely and non-invasively by means of a bifurcated fiber and a diode array spectrophotometer. The application of this technique to wastewater treatment could not only provide additional understanding of the process, but also lead to a more efficient control and protection of the plant by giving early warning signs about the health of the microbes.

In industrial and municipal wastewater treatment technology, the effluent from a primary treatment stage is mixed with microorganisms in aeration tanks. In these 'activated sludge' tanks, oxygen is consumed in the biological degradation process of organic pollutants by the microorganisms. To insure total aerobic degradation, such that the carbonaceous portion of the sludge is oxidized automatically to CO_2 and H_2O and the nitrogenous compounds are oxidized to nitrate, great effort is extended in keeping adequate levels of Dissolved Oxygen (DO) with consequent economical drawbacks (*1*). In fact, 70 to 80% of the energy required in a biological effluent treatment plant is expended for this task alone (*2*). Traditionally, control of the degradation process is accomplished by continuously monitoring the level of DO in the sludge with either polarographic or galvanic commercially available probes. However a dissolved oxygen measurement provides only indirect information of the actual biological action of the microbial population. In the worst scenario, the information may be misleading, as in the case of high values of DO because of decreased microbial activity due to toxic pollutants. Besides the intrinsic limitation in the nature of the measurement, DO probes are susceptible to fouling and present several problems related to drift, calibration, lag in the

0097–6156/92/0508–0222$06.00/0

response time, etc., thus requiring constant maintenance (*3*). Obviously, if a more direct means to monitor the physiological status of the microorganisms were available, control strategies could be developed that would supply oxygen on the base of the true metabolic rate.

Another interesting application of this concept would be used in screening for hazardous substances in waterways (*4,5*). Although the monitoring of the metabolic status in itself would not detect or identify the actual pollutants, it would provide an 'early-warning' alarm system that could trigger specific analyses of the water sample and prevent major damage to the treatment plant.

The metabolic status of the microbial population can be assessed by monitoring changes occurring in the respiratory chain at the cytochrome level. Reduced cytochromes can be identified by the positions and relative magnitude of their sharp absorption bands. Chance and colleagues (*6*) performed diffuse reflectance measurements on intact monocellular organisms in the visible region of the spectrum (400–700 nm). They showed that there is a gradient of oxidation-reduction level along the respiratory chain which is a precise indicator of the rate and nature of metabolism of the microorganisms. The application of this spectrophotometric technique to continuous monitoring of a bioreactor has been explored by Nagel (*7*).

Recently, at the Center for Process Analytical Chemistry, we developed a spectroscopic sensor able to follow real-time changes in the aerobic/anaerobic status of yeast fermentations (*8*). The measurement is performed by monitoring the cytochrome bands in the 500–650 nm region. As shown in Figure 1, metabolic variations in the fermentation broth cause large intensity changes in the cytochrome spectral features, respectively of cytochrome *c* (552 nm), cytochrome *b* (560 nm) and cytochrome aa_3, (604 nm).

In this paper a similar approach is used to monitor activated sludge samples obtained from the city water treatment plant. Preliminary data is presented that show the feasibility of continuous, on-line monitoring of the metabolic status of the microbial population by means of a non-invasive optical sensor.

Materials and Methods

Sludge. Activated sludge samples were collected from the mixed liquor channel at the Municipality of Metropolitan Seattle's Treatment Plant (Renton, WA).

Spectroscopic measurements. Spectra were collected with a Hewlett Packard 8452A photodiode array based spectrophotometer modified for use with a bifurcated fiber-optic probe. The experimental set-up has been previously described (*9*). The sludge samples were analyzed in diffuse reflectance mode by attaching the bifurcated fiber optic probe to the side of a $1 \times 1 \times 4$ cm quartz cuvette (Figure 2). Aerobic or anaerobic conditions were established by bubbling air or nitrogen and by stirring the activated sludge sample directly in the cuvette.

Data analysis. Data were acquired by a PC–AT and transferred to a DEC Station 3100 for processing and analysis. Spectra were first smoothed and a second derivative transformation was calculated using a 10 nm window with routines developed in the PRO-MATLAB environment (The Mathworks, Inc.).

Results and Discussion

A series of preliminary experiments was performed to characterize the spectroscopic features of the cytochrome system in the microbial population

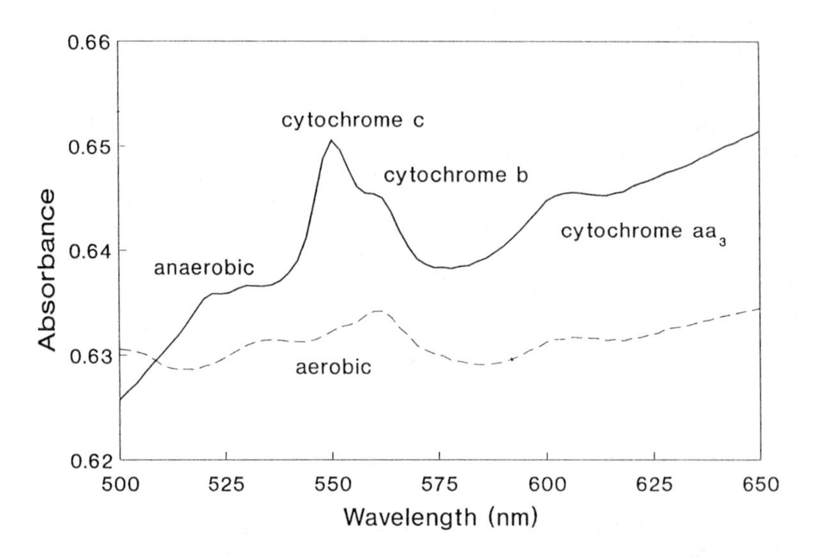

Figure 1. Absorbance spectra of cytochromes recorded in intact yeast suspensions.

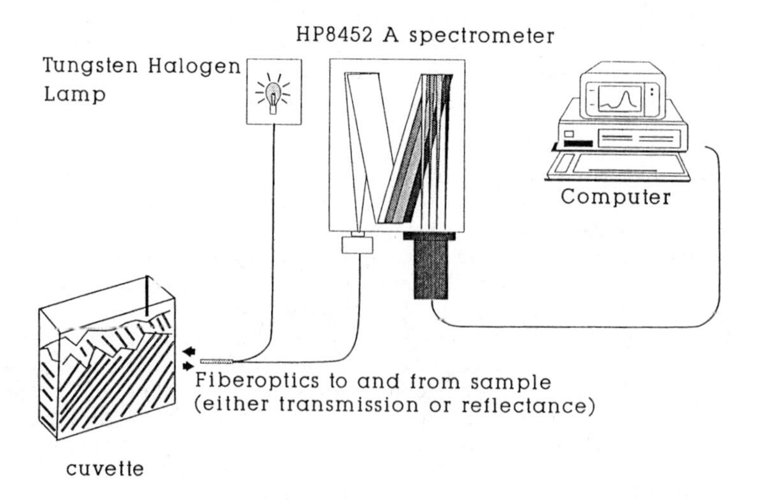

Figure 2. Diagram of instrument.

present in an activated sludge sample. Figure 3 shows characteristic spectra recorded in the 500–650 nm region under aerobic (oxidized) and anaerobic (reduced) conditions. To provide better spectral resolution and eliminate baseline offsets due to stirring and bubbling of the solution, a second derivative transformation of the spectra is performed (Figure 4).

The cytochrome bands appear not to be as well resolved as in pure cultures (*10*) due to the heterogeneity of the microbial population of the system. The spectral feature at 550 nm may be assigned to the α band of the cytochrome c. Such band tends to disappear under oxidized conditions. The cytochrome aa_3 band is also visible at 606 nm. Changes in the metabolic status of activated sludge samples were accomplished by bubbling alternatively air or nitrogen in the cuvette and were monitored by acquiring spectra at 5 min intervals. Figure 5 shows a set of second derivative absorption spectra recorded under these conditions. As the sludge is progressively aerated, the cytochrome c band tends to disappear and to shift toward the red, while the cytachrome aa_3 band remains quite stable. These changes are more obvious if cross sections of the second derivative absorption at 550 nm and 606 nm are displayed (Figure 6). In this figure the extent of the second derivative absorbance spectrum was converted to % oxidation of the metabolic status of the microorganisms. Therefore, 0% oxidation corresponds to the smallest second derivative value in the data set, while 100% corresponds to the largest one. The first point ($t = 0$) shows the degree of oxidation of the sludge sample ‘as received’ from the plant. The system appears quite anaerobic since the sludge had been kept in a bottle for transportation from the plant to the laboratory. In the next 10 minutes of the experiment, the sludge sample was kept under anaerobic conditions; after approximately 10 minutes, oxygen was bubbled in the cuvette at 5 minute intervals in different proportions, driving the system toward oxidation. After approximately 30 minutes the sludge was returned to an anaerobic status by bubbling nitrogen again. The change is clearly reflected in the cytochrome c signal.

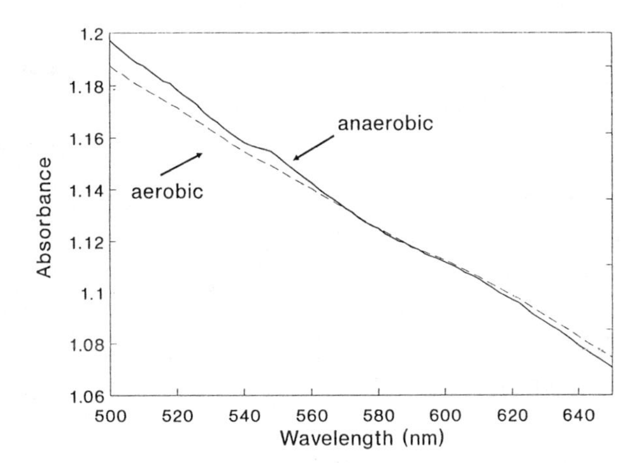

Figure 3. Absorbance spectra of cytochromes in an activated sludge sample.

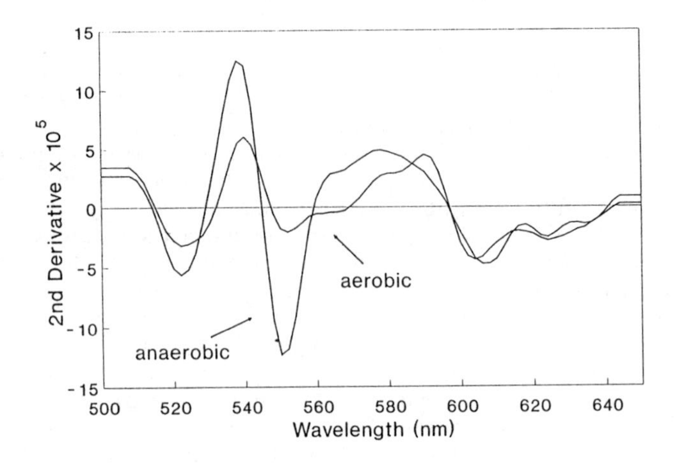

Figure 4. Second-derivative spectra of cytochromes in an activated sludge sample.

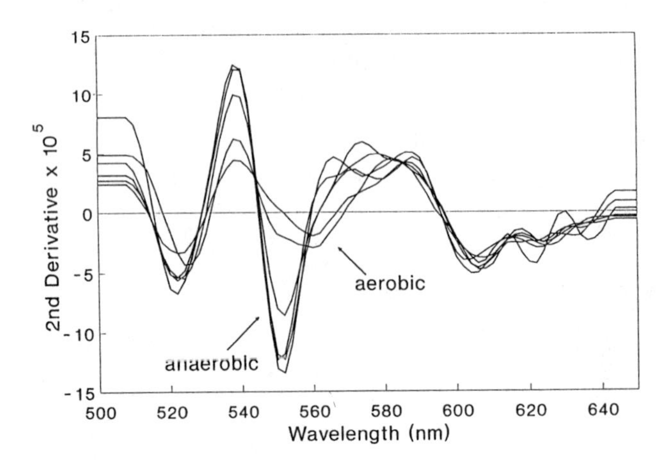

Figure 5. Second-derivative spectra of cytochromes in an activated sludge sample undergoing metabolic changes.

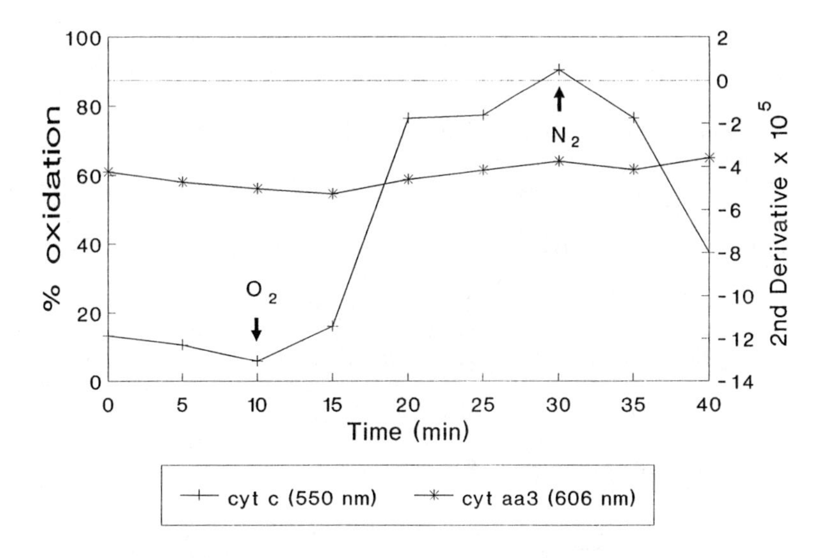

Figure 6. Redox profiles of cytochrome *c* and cytochrome aa_3 in an activated sludge sample undergoing metabolic changes.

Conclusion

These preliminary experiments suggest that monitoring of the cytochrome spectral features in the 500–650 nm region may be a valuable tool to estimate the true metabolic status of an activated sludge sample. The information obtained by this approach is similar to an Oxygen Uptake Rate (OUR) measurement and better reflects changes in the microbial metabolic status than oxygen measurements based on commercially available sensors. Based on this information, a more efficient and economic aeration system could be developed that provides the microogranisms with oxygen levels adequate for biodegradation. The same information could possibly be used to develop strategies for more efficient control and protection of the plant by 'early detection' of the presence of toxic pollutants in the wastewater stream.

Literature Cited

1. Mines, R.D.; Silverstein, J.; Sherrard, J.H.; M.K.; Weber, A.S.; *J. WPCF*, **1989**, *61*, 799–807.
2. Muller, H.G. In *Biosensors and Environmental Biotechnology*, C.P. Hollemberg and H. Sahm, Eds.; G. Fisher: Stuttgart, New York, 1988, 57–64.
3. Philichi, T.L.; Stenstrom, M.K., *J. Water Pollut. Control Fed.*, **1989**, *61*, 83–86.
4. Hansen, P.D.; Pluta, H.J.; Beeken, J.A., In *Biosensors: Applications in Medicine, Environmental Protection, and Process Control*, R.D. Schmid and F. Scheller, Eds., Weinheim: Federal Republic of Germany, New York, 1989, pp. 113–116.
5. Noack, V.; Boemisch, J.; Schneider, T. In *Biosensors: Applications in Medicine, Environmental Protection and Process Control*, R.D. Schmid and F. Scheller, Eds., Weinheim: Federal Republic of Germany, New York, 1989, pp. 243–246.
6. Chance, B.; Hess, B. *Ann. N.Y. Acad. Sci.*, **1956**, *63*, 1008–1016.
7. Nagel, B.; Bayer, C.; Iske, U.; Glomitza, F. *F. Am. Biotech. Lab.*, **1986**, *4*, 12–17.
8. Cavinato, A.G.; Mayes, D.M.; Ge, Z.; Callis, J.B. In *Frontiers in Bioprocessing II*, in press, 1991.
9. Cavinato, A.G.; Mayes, D.M.; Ge, Z.; Callis, J.B. *Anal. Chem.*, **1990**, *62*, 1977–1982.
10. Shipp, W.S., *Arch. Biochem. Biophys.*, **1972**, *150*, 459–472.

RECEIVED April 21, 1992

Chapter 19

Prompt Gamma-Ray Spectroscopy for Process Monitoring

Jennifer L. Holmes and William H. Zoller

Department of Chemistry, Center for Process Analytical Chemistry, University of Washington, BG-10, Seattle, WA 98195

Prompt gamma-ray spectroscopy (PGRS) is a noninvasive, nondestructive, analytical method. PGRS has an immediate response to allow for real time multi-elemental analysis in an industrial process or waste stream. The flow stream can be exposed to a neutron flux through the existing pipe or a modified pipe section. The prompt gamma-rays produced will be emitted within 10^{-12} seconds of neutron capture and range from 50 keV to 11 MeV (*1*). The prompt gamma-ray energy will vary as to the nuclide that produced it, and thus isotopic abundances can be determined. The high energy gamma-rays can penetrate most materials readily; however, a modified pipe or gamma-ray detection window could be constructed to allow for improved detection.

Prompt gamma-ray spectroscopy (PGRS), also refered to as prompt gamma-ray activation analysis (PGAA) and prompt gamma-ray neutron activation analysis (PGNAA), is an analytical tool able to measure some major and minor components on an isotopic and, assuming natural abundance or known isotopic ratio, elemental basis. The technique involves measuring the gamma-rays emitted by nuclei upon capture of a neutron. The nuclei formed via neutron capture have excitation energies due to the binding energy of the added neutron; this energy is released by emission of one or several prompt gamma-rays within 10^{-12} to 10^{-15} seconds of capture. The gamma-rays can reach up to 11 MeV (*1*). Commercial systems that utilize PGRS are available, such as Conac by Sia and Bulk Analyzer by Gamma Metrics. Most of these systems are designed to do analysis on dry, solid, heterogeneous material such as dry cement or coal. EG & G Ortec has recently worked with the Chinese to install PGRS process analytical systems into aluminum processing plants. This system combines elemental abundances together with other measured parameters to determine major oxide concentrations (*2*). Some other common field applications of PGRS have been borehole logging (*3*, *4* and more) and shallow water contamination monitoring (*5*).

0097–6156/92/0508–0229$06.00/0

Method

As PGRS relies upon capture of thermal neutrons, a neutron source is obviously a concern. There are three possible sources of neutrons: a nuclear reactor, an accelerator, or a radionuclide source. Of these three, a reactor is obviously too large and costly for an industrial processing plant to consider. However, with the new, more compact and safer nuclear reactor designs (*6*), and the possibility of both nuclear and coal electrical generators being operated at joint stations, the possibility of monitoring total carbon, sulfur, and nitrogen emissions from the coal generator with the neutrons generated from the nuclear reactor does exist. Several research reactors have PGRS facilities.

The PGRS facility at Los Alamos National Laboratory's (LANL) Omega-West reactor was used to do some preliminary studies (see Figures 1 and 2). The LANL PGRS facility is one of internal construction, meaning that the sample sits in the reactor core in a graphite thermal column. This design allows for a high neutron flux of 5×10^{11} neutrons/cm^2 $\times$ second (*7*) and seems to be free from some neutron scatter effects (*8*). The detector system can collect in Compton suppression and pair production mode; however, it sits six feet away from the sample. The detector distance from the sample protects it from possible neutron damage. However, this causes a poor detector geometry, leading to poor sensitivities.

The National Institute of Standards and Technology (NIST) has a PGRS facility with an external neutron beam made of collimated neutrons from the reactor core, brought up through a portal, to the room above (*9*). The detector system can collect in Compton suppression or pair production mode and is a mere 0.2 meters from the sample. This allows for a much better detector geometry than at LANL. This system has some scatter effects (*8*) and can only hold a limited sample size. The flux of thermal neutrons at the sample is about 2×10^8 neutrons/cm^2 $\times$ sec. The NIST external beam design is more similar to the PGRS processes system design discussed later. Table I shows some elemental sensitivities and detection limits at the NIST facility. Preliminary studies at NIST showed cadmium detected at 5 ppm in laboratory waste (*10*).

Another source for neutrons are accelerators. Accelerators work by bombarding a thin layer of material (deuterium, tritium, beryllium, or lithium) with deuterium, protons or alpha particles. Neutrons are then emitted in the forward direction of the bombardment. Accelerators have several advantages. The direction and energy of the neutrons can be controlled by the bombardment. The flux is more consistent and will not diminish with time, as there is no dependence on a radionuclide that decays. Also, the neutron beam can be turned on and off as desired. Accelerators can be very costly. However, leasing a radionuclide source, and continual replacement as it decays, can also be costly over time.

Another source for neutrons are radioisotopes. Californium-252 is a man-made radioisotope that is unstable and disintegrates, emitting neutrons. One milligram of ^{252}Cf can emit a neutron flux of 10^9 neutrons/sec. Neutrons are emitted multidirectionally with an energy range that has a mean value of 2.2 MeV. The flux, of course, will become weaker with time as the ^{252}Cf decays with a 2.64 year half-life. As ^{252}Cf decays significantly in a year's time, and is limited in quantity, other isotopic sources may be more desirable.

Other isotopic sources can be made by mixing a radionuclide that is a strong alpha or gamma-ray emitter with beryllium. This causes one of the following reactions:

$$^{9}Be +^{4} He \rightarrow^{12} C +^{1} n$$

$$^{9}Be + \gamma \rightarrow^{24} He +^{1} n$$

The neutron flux and energy vary with radionuclide used; possibilities include radium, plutonium, and americium.

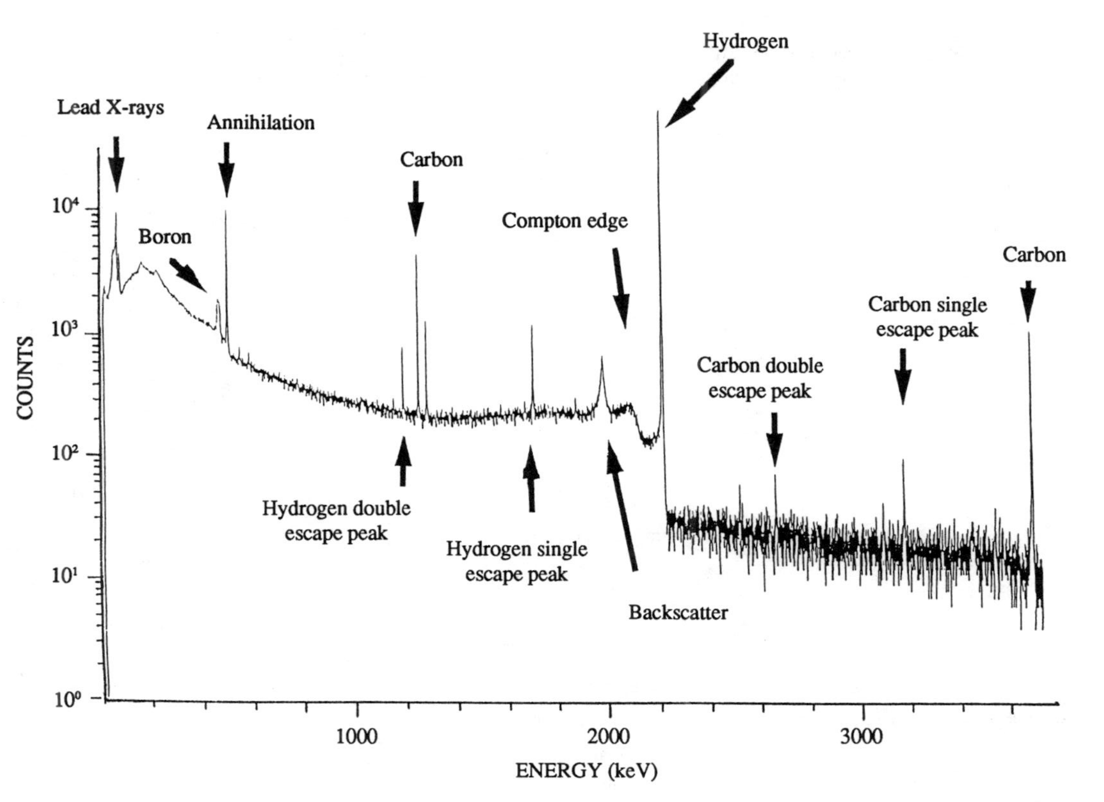

Figure 1. Prompt gamma-ray spectra of Seattle Metro sewage sludge that was analyzed at LANL. This spectra was collected in Compton suppression mode; however, note the large Compton edge due to the large concentration of hydrogen.

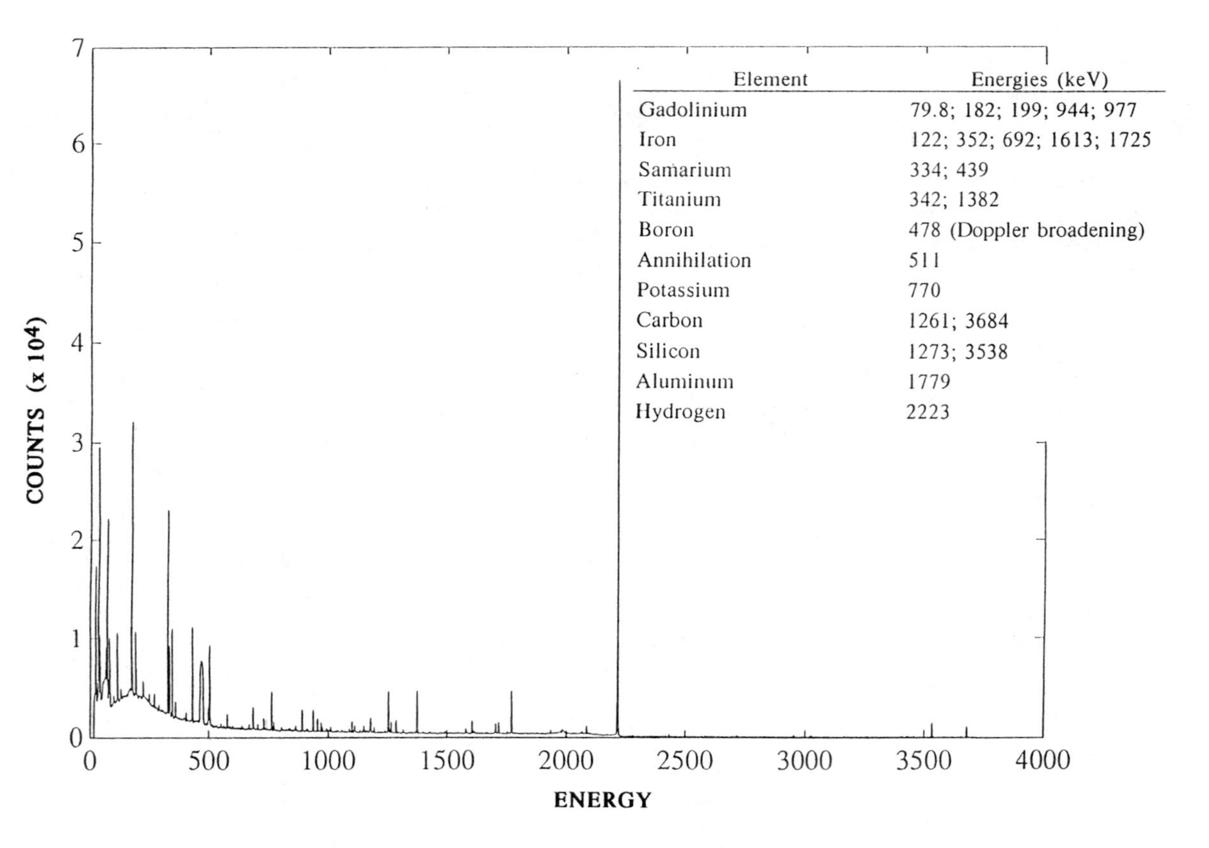

Figure 2. Prompt gamma-ray spectrum of NBS Coal Fly Ash Standard SRM1633a. Note the complexity of the spectra. Figure 1 is a log plot and this is not; notice the difference.

Any type of neutron source will require licensing from the Nuclear Regulatory Commission, and some personnel monitoring in the area where the PGRS system is operating is required. Also, a radiation safety officer must be on site; however, this position could be filled by any safety engineer, facility manager, or industrial hygienist who takes a radiation safety class. The flow stream, after its exposure to the neutron flux will remain within normal, background radiation levels, and is therefore below regulatory concern. This is due to the short amount of time that any one volume parcel is exposed to the neutron flux, as it is in a flowing system. As an example, we measured the residual radioactivity of our Municipality of Metroplitan Seattle (Metro) sewage sludge sample within twenty minutes of its removal from the reactor core at LANL, and it was not significantly above background even though it had been in the core for over four hours.

A neutron source can vary and would be chosen based on sampling constraints and elemental concerns. A detection system must also be decided.

A gamma-ray can interact with matter three ways and be detected. One, is when the gamma-ray transfers its energy totally to the electron in the detection material. This is known as the photoelectric effect. This causes the characteristic photopeaks in the spectrum (see Figure 1 and 2), which are then used to identify different isotopes. More often, the gamma-ray transfers only part of its energy to an electron in the detection material and one or several secondary gamma-ray(s) with the remaining energy are emitted. This is known as the Compton effect. This adds to the background. Its effect can vary with the sample matrix, and is not useful in identification. Compton suppression is achieved when two detectors are arranged geometrically so that one encircles the other (usually a NaI(Tl) scintillator detector surrounds a HPGe solid-state semiconductor detector), and the electronics are designed so that when both detectors detect a gamma-ray coincidentally, neither one is counted.

When a gamma-ray is of sufficiently high energy (above 1.02 MeV), a positron and electron pair can be created; this is known as pair production. As a positron will be annihilated with an electron nearly immediately, two characteristic annihilation gamma-rays of 0.51 MeV are emitted. Pair production spectra are collected with a detector geometry similar to Compton suppression; however, the electronics are such that a gamma-ray is only counted if a 0.51 MeV gamma-ray is detected in coincidence. These spectra have very low backgrounds, so an isotope that emits in this region may have lower limits of detection. Mercury and chromium both pose an environmental concern and can be difficult to measure using conventional means. Also, both emit prompt gamma-rays in this high energy region. As high energy gamma-rays can penetrate material more readily, this method of detection can be very useful.

Design

The focus of this research is to constructing a PGRS system that can measure isotopic components, in a flowing system, in real time. The isotopes will have varying sensitivities. Isotopic or elemental sensitivities, $\boldsymbol{S}$, are directly proportional to the thermal neutron crosssection, $\boldsymbol{\sigma}$, and the intensity of the photopeak, $\boldsymbol{I}$. To consider these sensitivities in a mass unit one should divide by the atomic number, $\boldsymbol{A}$. This leads to the equation (*11*):

$$\boldsymbol{S = \sigma I / A} \tag{1}$$

A list of relative sensitivities can be found in Table II. When considering actual limits of detection, one must consider the system parameters as well as isotopic characteristics to be discussed below.

Table I. Elemental sensitivities and detection limits measured at the NIST PGRS facility. The detection limits are those determined in Hawaiian Basalt. This table was adapted from tables in ref. 12. (Note that pair production collection was not reported.) The D and SE stand for double and single escape peaks

element	Eγ (keV)	sensitivities (counts/sec mg)	detection limit (mg/g) (non-suppressed)	detection limit (mg/g) (Compton suppressed)
H	2223	0.86	0.023	0.021
B	478	530	6×10^{-5}	3×10^{-5}
C	1262	3.9×10^{-4}	22	13
	3684	1.5×10^{-4}	31	16
N	1885	0.0030	2.4	1.4
	3678	0.0012	3.8	1.9
	5298	0.0017	3.6	–
F	583	0.0011	10	5
	1634D	0.0015	5	3
Na	472	0.15	5	3
	869	0.026	0.40	0.23
	1368D	0.036	6	3
Mg	585	0.0085	1.4	0.7
	1809	0.0022	3.6	2.0
	2829	0.0020	3.2	1.8
Al	1779D	0.024	0.34	0.19
	7213	0.0015	3.1	–
Si	1273	0.0036	2.7	1.6
	3539	0.0066	0.6	0.3
	4933	0.0042	1.5	–
P	637	0.0062	1.8	1.0
S	841	0.0054	0.19	0.1
	2380	0.015	0.28	0.17
	3220	0.0067	0.6	0.3
Cl	516	1.4	0.009	0.004
	785 + 788	1.2	0.013	0.006
	1165	1.0	0.009	0.006
K	770	0.13	0.085	0.045
	4869SE	0.0028	2.3	–
Ca	1943	0.022	0.32	0.18
Ti	342	0.38	0.045	0.032
	1382	0.38	0.025	0.013
V	1434D	0.33	0.022	0.013

Table I (continued)

element	Eγ (keV)	sensitivities (counts/sec mg)	detection limit (mg/g) (non-suppressed)	detection limit (mg/g) (Compton suppressed)
Cr	749	0.050	0.22	0.11
	835	0.11	0.09	0.05
Mn	212	0.46	0.07	0.06
	847D	1.05	0.009	0.004
Fe	352	0.046	0.38	0.28
	7631 + 7646	0.013	0.23	–
Co	277	112	0.017	0.014
Ni	465	0.12	0.12	0.74
Cu	278	0.12	0.17	0.13
Zn	1077	0.019	0.44	0.27
Sr	898	0.31	0.33	0.18
	1836	0.028	0.27	0.15
Cd	558	170	7×10^{-5}	3.5×10^{-5}
	651	30	3.5×10^{-4}	1.8×10^{-4}
Ba	627	0.55	0.020	0.010
	818	0.0026	1.4	0.72
Nd	618	0.55	0.020	0.01
	697	0.0026	0.008	0.004
Sm	334	640	3×10^{-5}	2×10^{-5}
	439	320	4.5×10^{-5}	3×10^{-5}
Gd	182	680	6×10^{-5}	5×10^{-5}
	944	96	1×10^{-4}	5×10^{-5}

Table II. Elemental sensitivities using equation (1). The cross-sections and intensities were adapted from ref. 13, and are reported on an elemental, not isotopic bases. The same elements and gamma-ray energies are used as in Table I whenever possible to allow for an easy comparison between these theoretical values and the measured values of Table I

element	Eγ (keV)	cross-section σ (barns)	Eγ intensity, I (%)	Atomic weight	Sensitivity (Iσ/A)	Comments
H	2223	0.33	100	1	0.33	
C	1262	0.0034	30	12	0.000085	
	3684		32		0.00009	
N	1885	0.075	22	14	0.0012	
	5298		30		0.0016	
F	583	0.0095	13	19	0.000065	
Na	472	0.4	60	23	0.01	
	869		23		0.004	
Mg	585	0.063	26	24	0.00068	
	2829		43		0.0011	
Al	7724	0.23	27	27	0.0023	a
Si	1273	0.16	16	28	0.00091	
	3539		68		0.0039	
P	637	0.18	12	31	0.0007	
S	841	0.52	76	32	0.012	
	2380		45		0.0073	
Cl	516	33	19	35.5	0.18	
	1165		20		0.19	
K	770	2.1	51	39	0.027	
Ca	1943	0.43	73	40	0.0078	
Ti	342	6.1	26	48	0.033	
	1382		69		0.088	
V	6517	5.0	18	51	0.018	a
Cr	749	3.1	11	52	0.0066	
	835		27		0.016	
Mn	212	13	8.8	55	0.021	
	26.6				0.043	b
Fe	352	2.6	12	56	0.0056	
	7631		28		0.013	
	7646		24		0.011	
Co	277	37	20	59	0.13	
Ni	465	4.4	13	59	0.01	
Cu	278	3.8	33	63.5	0.02	
Zn	1077	1.1	19	65	0.0032	

Table II (continued)

element	Eγ (keV)	cross-section σ (barns)	Eγ intensity, I (%)	Atomic weight	Sensitivity (Iσ/A)	Comments
Sr	898	1.2	28	88	0.0038	
	1836		58		0.0079	
Cd	558	2500	73	112	16.3	
	651		14		3.1	
Ba	627	1.2	12	137	0.0011	
	818		9		0.00079	
Nd	618	51	29	144	0.10	
	697		73		0.26	
Sm	334	5800	99	150	38.3	
	439		56		21.6	
Gd	199	49000	10	157	31.2	a
	944		11		34.3	

a The Eγ in reference 13 do not agree with those in Table I, and these values are listed instead.

b Table I does not list this Eγ; however, it was of higher intensity in reference 13, therefore, it was listed.

The system, matrix, and elemental characteristics are very difficult to determine; however, the following equation accounts for the significant characteristics of all these parameters:

$$\boldsymbol{R}_{ct} = \boldsymbol{N}_t \boldsymbol{\sigma} \boldsymbol{I} \Phi \Pi \boldsymbol{t} \boldsymbol{\zeta} + \boldsymbol{R}_b + \boldsymbol{R}_{co} \quad (2)$$

where,

$\boldsymbol{R}_{ct}$ = total counts in channel
$\boldsymbol{N}_t$ = number of target atoms
$\boldsymbol{\sigma}$ = neutron capture cross-section
$\boldsymbol{I}$ = intensity of gamma-ray peak
Φ = neutron flux
Π = detector geometry
$\boldsymbol{t}$ = counting time
$\boldsymbol{\zeta}$ = detector efficiency
$\boldsymbol{R}_b$ = counts in channel due to natural background
$\boldsymbol{R}_{co}$ = counts in channel due to Compton scatter.

The first product in equation (2) has both isotopic and system design characteristics. The Π and $\boldsymbol{\zeta}$ are detector parameters determined by the sample counting geometry, detector crystal, and gamma-ray energy or channel. $\boldsymbol{R}_b$ can also be thought of as a detector parameter. It is determined by natural background, amount of detector shielding, and can vary with gamma-ray energy or channel. $\boldsymbol{R}_{co}$ is the gamma-ray energy or channel count due to the Compton scatter; it can vary with matrix and gamma-ray energy or channel. Both $\boldsymbol{R}_b$ and $\boldsymbol{R}_{co}$ are considered background and fall to near zero in the high energy pair production range.

The design, as described below and pictured in Figure 3, would be used for monitoring a flow in a pipe such as a waste water feed line. The high penetrating power of neutrons and gamma-ray radiation enables the system construction to be done outside the existing piping. However, due to possible interference or possible irradiation of some materials to produce radioactive nuclides, it may be beneficial to replace the pipe with piping constructed of nonrelevant elements or poor neutron-capturing materials. A hydrocarbon-based pipe made of durable, polymeric material would be ideal. Also, a special gamma-ray window would help improve detection in the lower energy range (from 50 keV to 3 MeV). This will be discussed more when describing the detection system.

The system calls for a ^{252}Cf neutron source of 1–3 mg in size, a rather large size as compared to most borehole logging radionuclide sources. This should provide a flux of approximately 10^7–10^9 neutrons/cm^2 sec. The neutron beam must be collimated and moderated to optimize the thermal flux of neutrons into the center of the sample flow. As the source emits neutrons multi-directionally, a thin layer of neutron reflecting material, such as bismuth or beryllium opposite parabolically, encircling the neutron source, should act to optimize the flux of neutrons into the collimator and then into the sample flow. Since ^{252}Cf emits neutrons with a mean energy of 2.2 MeV, it is necessary to slow, or moderate, the neutrons to thermal energies, which is on the order of 0.02 keV. Hydrogen acts as a good moderator as it collides with the neutrons to slow them to thermal energies. Paraffin is an excellent moderator as it is a highly hydrogenated hydrocarbon. Also, ^{252}Cf emits alpha particles, and these can be easily stopped in a hydrocarbon medium. The flow stream itself could also act as a moderator if it is a hydrogenated species, such as water or oil.

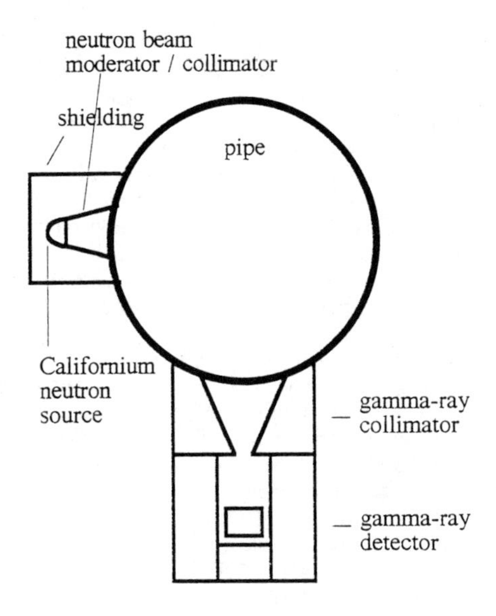

Figure 3. Schematic design of the PGRS monitoring system. Note that the neutron source and gamma-ray detectors are both completely outside the pipe. The design could be altered for different systems.

Shielding of the neutron source is required to avoid undue radiation leakage and personnel exposure. Housing the entire source assembly described above in a dense material that has a high neutron capture cross-section will effectively stop any neutron leakage. Boron carbide (B_4C) has been shown in the past to effectively stop all neutron leakage (*12*). Cadmium also has a high neutron capture cross-section, so it could also be used as shielding material. A 2–4-inch thick B_4C or CdC_2 shield would be sufficient for the ^{252}Cf source size in our design. ^{6}Li is another very effective neutron shield; however, due to cost and limited availability, it is best used sparingly. ^{252}Cf emits ten times more alpha particles than neutrons; however, as stated earlier, they are effectively stopped by the paraffin surrounding the ^{252}Cf and, therefore, do not pose a health or safety risk to personnel outside the shielded neutron source.

As mentioned earlier, there are three types of gamma-ray interaction: the photoelectric effect, which leads to the characteristic photopeak; the Compton effect, which leads to Compton scatter and pair production, which leads to the characteristic 511 keV annhilation gamma-ray photopeak. The detection scheme to reduce Compton scatter, known as Compton supression, and the detection scheme to collect for high energy energy gamma-rays or pair product, as described earlier, usually require a dual detector system. As the system designed here calls for Compton supression and pair production as the detection schemes, a dual detector system is described. However, this is not the only detection scheme available and may not be the most desirable, as explained later.

The detector system is a high purity germanium (HPGe) solid state semiconductor detector surrounded by a NaI(Tl) scintillation detector for counting in Compton suppression or pair production mode. As fast and thermal neutrons can damage the germanium crystal, it is best to set the detector system perpendicular to the neutron beam. However, neutrons will scatter in the sample flow, so additional protection is required. This is accomplished with gamma-ray collimators and a LiF or $LiCO_3$ window. Collimation not only helps to protect the detector but also helps reduce background by "focusing" the detector to the middle of the sample flow. Gamma-ray collimations is best accomplished by a material that is dense (therefore, able to shield gamma-rays), has low residual activity, and is not an interferent. Lead is an excellent collimator as it is dense and has a very low neutron capture cross section. Lead does emit low energy X-rays, as can be seen in Figures 1 and 2, so it may be necessary to line the lead with another material that does not interfere with that energy range. It may be necessary to further protect the germanium crystal with a window, something that can block neutrons but allow gamma-rays to readily penetrate it. Lithium-6 is the perfect material for such a use, as it has a large neutron capture cross section; yet is of a low density to allow gamma-rays to readily pass through it. NIST uses a window made of LiF, LANL has a window of $LiCO_3$, and both facilities have experienced insignificant neutron damage to their detection systems. However, as stated earlier, ^6Li is costly and available only in limited quantity.

HPGe detectors are costly and require low temperatures to operate; however, there are alternatives to using HPGe solid-state semiconductor detectors. Scintillation detectors are less expensive and do not require the liquid nitrogen temperatures of germanium solid state detectors; however, they have lower resolution. Solid-state semiconductor detectors operate by applying a current through a very pure, surface doped germanium crystal and measuring the resistance change due to the gamma-rays knocking the electrons from the valance to the conduction band . The electronics then sort the varying voltage shifts ($\boldsymbol{V} = \boldsymbol{ir}$), which correspond to varying gamma-ray energies, into different channels in a multichannel analyzer. Scintillation detectors vary, in that instead of measuring a voltage shift, light quanta are measured. Ion pairs produced by the interaction of scintillation materials with radiation produce secondary electrons which excite atoms in the detection medium and cause emission of light quanta. The scintillation crystal is in close contact with a photomultiplier tube which generates a current pulse, which can be converted into a voltage pulse so that it can be counted similarly to that of the solid-state semiconductor detector. Detector efficiency can vary with the Z number of the detection material and the gamma-ray energy. As iodine and bismuth have high Z numbers, they are efficient gamma-ray detection medium. Bismuth germanate is denser than sodium iodide, and thus it is a more efficient gamma-ray detection medium. 100%-efficiency HPGe solid state semiconductor (vs NaI(Tl) scintillation) detectors are available, but, as previously stated, must operate at very low temperatures and are very expensive. One may design a detector system with a bismuth germanate or a sodium iodide scintillation detector that encompasses the pipe with each half connected to its own photomultiplier tube. This geometry could collect in Compton suppression and pair production mode without such temperature constraints.

The design for any PGRS system would vary for the individual parameters and constraints. As this system's main advantage is that it can continuously monitor a flow in real time, the pipe size and flow rate must be considered. The possibility in detecting a gamma-ray is directly proportional to the number of target atoms (see equation 2), thus increasing the number of atoms, which, in turn, increases the number of gamma-rays. So it follows, the greater the volume of sample determined by the flow rate and diameter of the pipe, the greater the sensitivity. These two parameters, flow rate and pipe diameter, and their effect on detection, are considered separately.

The diameter of the pipe will govern the neutron source strength and detector geometry. A larger diameter increases the volume size, increasing the sensitivity until a maximum is reached, which will depend upon the sample density and the neutron source strength. Once the optimum size is surpassed, the neutron flux into the center of the pipe will be less than expected due to scatter and absorption in the path from the collimator to the center of flow. Also, the detection of the gamma-rays are dependent on the counting geometry, causing the detection to become poorer with distance. The increased diameter will aid in detection by increasing the sample size up to about 4–6 feet, at which time it is more likely the increased size would decrease the neutron flux and cause a poor detector geometry. If the pipe diameter is greater than the optimum size, then the design could be altered to house the neutron source in the center of the pipe. This design would require less neutron collimation, and would take advantage of the multi-directional emission of neutrons; however, the radionuclide would greatly increase the gamma-ray background, as it too is a gamma-ray emitter.

The statistics of the concentrations are related to the total number of counts, or gamma-rays, as it is a direct count. The total number of counts would be proportional to not only the total number of atoms, or volume, but also the amount of time the counts are collected. Therefore, a trade-off between maximum flow rate and optimum counting statistics exists. The flow rate itself would not be of much concern in the system design. Both the flow and the monitoring are continuous, so one could design a computer program (usually a computer is used as the multichannel analyzer) to continuously scan and integrate energies or channels, corresponding to elements of interest. By comparing the areas of these regions to previously measured areas (of the same counting time or time corrected), a change in concentration of the element(s) would be apparent. The statistics of the concentration are based on the count, as it is a direct count, so optimizing the integration time would be governed by the detector count rate and the flow rate.

In the case of this system being used to monitor waste water into a treatment plant, an increase in harmful elements, such as cadmium or mercury, would trigger a warning system so that the sewage flow could be diverted into a holding tank. There, these harmful elements could be precipitated or plated out before being introduced into the bacteria active treatment ponds. This type of system would also ensure that the sediment associated with this treatment facility would meet criteria to be safely used in agriculture rather than needing to be disposed of as hazardous waste. In industrial processing, such real-time information on flow streams could prove to be valuable not only as waste monitoring but also for processing parameters.

Conclusion

PGRS has been used both in the field with borehole logging and at reactors for major and minor elemental abundance data. The studies we have conducted so far are preliminary; however, all matrix types—aqueous, organic, liquid or solid—can be analyzed by PGRS. Processing plants are very complex systems; real-time multi-elemental abundance data of flow streams can be valuable for catalyst protection or monitoring mixing parameters. Increased environmental concerns and an ever-increasing demand for industry to monitor their waste emissions calls for a flow-monitoring system. PGRS appears to be one analytical tool capable of noninvasive, real-time, continuous, elemental monitoring of a flowing system.

Acknowledgements

We would like to thank the Center for Process Analytical Chemistry at the University of Washington for funding and computer support. We would also like to thank Dr. Joseph Edwards for supplying us with the Metro sewage sludge sample, and Dr. Robert Vandenbosch for helpful discussions.

Literature Cited

1. Failey, M. P.; Anderson, D. L.; Zoller, W. H.; Gordon, G. E. and Lindstrom, R. M., "Neutron-Capture Prompt Gamma-ray Activation Analysis for Multielement Determination in Complex Samples," *Anal. Chem.* **1979**, *51*, 2209–2221.
2. Seymour, R., Ortec EG&G, personal communication.
3. Miksell, J. L.; Dotson, D. W.; Sentle, F. E.; Zych; R. S., Kogger, J. and Goldman, L., *"In-Situ* Capture Gamma-Ray Analysis of Coal in an Oversized Borehole," *Nuclear Instruments and Methods* **1983**, *215*, 561–566.
4. Tanner, A. B.; Moxham, R. M.; Sentfle, F. E. and Baicker, J. A., "A Probe for Neutron Activation Analysis in a Drill Hole Using ^{252}Cf and a Ge(Li) Detector Collected by a Melting Cryogen," *Nuclear Instruments and Methods* **1972**, *100*, 1–7.
5. Chung, C. and Tseng, T. C., *"In-Situ* Prompt Gamma Ray Activation Analysis of Water Pollutants Using a Shallow 252Cf-HPGe Probe," *Nuclear Instruments and Methods in Physics Research* **1988**, *A267*, 223–230.
6. Stinson, S. C. "Nuclear," *C& News* **1991**,*69(24)*,. 36–40.
7. Minor, M., Los Alamos National Laboratory, personal communication.
8. Mackey, E. A.; Gordon, G. E.; Lindstrom, R. M. and Anderson, D. L., "Effects of Target Shape and Neutron Scattering on Element Sensitivities for Neutron-Capture Prompt Gamma-Ray Activation Analysis," *Anal. Chem.* **1991**, *62*, 288–292.
9. Anderson, D. L.; Failey, M. P.; Zoller, W. H.; Walters, W. B.; Gordon, G. E. and Lindstrom, R. M., "Facility for Non-Destructive Analysis for Major and Trace Elements Using Neutron-Capture Gamma-Ray Spectrometry," *J. of Radioanal. Chem. 1981*, *63*, 97–119.
10. Anderson, D. L., Food and Drug Administration at the National Institute of Standards and Technology, personal communication.
11. Henkelmann, R. and Born, H. J., "Analytical Use of Neutron-Capture Gamma-Rays," *J. of Radioanal. Chem.* **1973**, *16*, 473–481.
12. Failey, M. P., "Neutron-Capture Prompt Gamma-Ray Activation Analysis: A Versatile Nondestructive Technique For Multi-Element Analysis of Complex Matrices," Ph.D. Thesis, University of Maryland, College Park, MD, 1979.
13. Lone, M. A.; Leavitt, R. A. and Harrison, D. A. "Prompt Gamma Rays from Thermal Neutron Capture," *Atomic Nuclear Data Tables* **1981**, *26*, 511–559.

RECEIVED April 23, 1992

Chapter 20

Trace Analysis of Organic Compounds in Groundwater

On-Column Preconcentration and Thermal Gradient Microbore Liquid Chromatography with Dual-Wavelength Absorbance Detection

Leslie K. Moore and Robert E. Synovec

Department of Chemistry, Center for Process Analytical Chemistry, University of Washington, BG-10, Seattle, WA 98195

Thermal gradient microbore liquid chromatography with dual wavelength absorbance detection is evaluated as a tunable on-site analyzer. Gradient programming allows the selectivity of the analyzer to be adjusted thus potentially identifying chemical interferences in a process or an environmental upset. The dynamic range for typical microbore liquid chromatography analytes is extended by concentrating the sample on-column. Analytes are concentrated 500 fold from 500 μL injection volumes without introducing appreciable band broadening following gradient elution. The relative change in capacity factor with temperature is found to be slightly higher for highly retained peaks, thus temperature programming allows analysis of highly retained compounds such as high molecular weight polycyclic aromatic hydrocarbons (PAHs). The effects of high temperature on band broadening was evaluated. For a given capacity factor, temperature improved chromatographic efficiency over room temperature, isocratic elution.

In the last decade there has been increasing interest in on-site chemical analysis for process and environmental monitoring (*1*). On-site analysis and feedback control enables efficient utilization of resources and minimizes waste. To date, a number of in-situ spectroscopic and electrochemical methods have been developed for continuous real time monitoring (*2*). Available sensors are adequate for well characterized systems, however they lack the broad selectivity necessary to characterize constituents of an unforeseen process upset. Many devices also lack the flexibility to monitor extreme changes in the concentration of expected analytes. It would be useful to have an analyzer that responds to many different compounds and a wide range of concentrations. Furthermore, because a majority of process and environmental problems are aqueous based, it is of interest to develop an on-site analyzer for aqueous samples.

Gradient microbore liquid chromatography (μLC) with dual wavelength absorbance detection has been proposed as an on-site tunable liquid analyzer (*3*). Chromatography provides greater selectivity than single sensors and has the potential to identify interferences in a process or an environmental upset.

0097–6156/92/0508–0243$06.00/0

Microbore chromatography is well suited for remote analysis because the system requires a minimal amount of solvent preparation and storage. Thermal gradient microbore liquid chromatography (TG-μLC) optimizes analyzer selectivity, and solves the problem of applying gradients to small volume systems (*4*). It is much easier to control the temperature of the column than it is to uniformly mix two mobile phases with the precision necessary for optimal separation. Both mobile phase and temperature gradients improve selectivity by changing elution conditions with time. However separations at elevated temperatures are potentially more efficient than mobile phase gradient separations as increased temperature facilitates mass transfer of the analyte to the stationary phase (*5*). A dual wavelength absorbance detector with a fiber optic based *in-situ* flow cell is incorporated to optimize detection. The 1.2 μL flow cell and the column are both heated, so the analytes are separated and detected at the same temperature. The detector simultaneously measures the absorbance at two wavelength regions. The first is a wavelength region of interest for measuring the absorbance of analytes. The second is a reference wavelength region where little or no absorbance is expected. The detector produces two signals. The difference signal is the difference between the intensities at the two wavelength regions, and the sum signal is the sum of the intensities at the two wavelength regions. The difference signal corrects for baseline drift due to the temperature gradient and minimizes correlated noises associated with source fluctuations and changes in ambient temperature (*6*). The sum signal possesses information regarding column temperature changes. Two dual wavelength detectors are used in this study. A position sensitive detector (PSD) based dual wavelength detector has been described previously (*7,8*). The difference signal from this detector stabilizes baseline drift 15 fold over a conventional single wavelength absorbance detector. Typically, the difference signal also has low baseline noise characteristics of 1.1×10^{-4} AU ($3 \times$ root mean square noise level). A second dual wavelength detector is a two diode detector. The two diode detector is operationally the same as the PSD detector, except there are two diodes instead of one continuous photoelectric surface. Other multiwavelength detectors such as photodiode array detectors and charge coupling devices could also be used to correct for refractive index aberrations due to temperature by employing the two wavelength subtraction technique.

The dynamic range of the analyzer is extended by preconcentrating dilute solutions before the chromatographic separation. On-line concentration techniques include liquid-liquid (*9*) and membrane extraction (*10*), however solid phase extraction (SPE) is commonly employed because it is a rapid way to extract and concentrate analytes from relatively clean matrices (*11*). A SPE cartridge often consists of the same packing material as the analytical column, however the mobile phase conditions are chosen so the analytes are highly retained. It is clear that the broadening associated with eluting the analytes from a SPE cartridge could be eliminated by extracting and separating the analytes on one column. Guinebault and Broquaire have demonstrated on-column preconcentration and separation on analytical sized columns (*12,13*) and Slais has shown large volumes can be concentrated onto microbore columns (*14*). In this work we optimize the parameters associated with on-column preconcentration for microbore columns. We will also investigate the potential of thermal gradient elution in conjunction with dual wavelength absorbance detection for on-site analysis. Sample volumes of 500 μL will be injected on a microbore column enabling trace analysis of dinitrophenylhydrazine (DNPH) derivatized aldehydes and ketones in air samples, and PAHs in water samples. The benefits of temperature programming in terms of increased mass transfer and improved chromatographic efficiency will also be reported.

Experimental

Chromatographic System: A schematic diagram of the chromatographic system is illustrated in Figure 1. A syringe pump (ISCO, LC-500, Lincoln, NE) was utilized to deliver the mobile phase at a flow of 50 μL/min in all cases. Injections of 1 μL were accomplished with a microbore injection valve (Rheodyne 7520 Cotati, CA.). The 500 μL injections were introduced with a six port injection valve (Rheodyne 7125 Cotati, CA) and a 500 μL loop made in-house. The aldehyde and ketone samples were run on a low pressure (<500 p.s.i.) 1 mm × 200 mm 10 μ PRP-1 C18 derivatized polystyrene-divinylbenzene based column (Keystone Scientific Inc., Bellefonte, PA.). The PAH data was obtained with a 1 mm × 250 mm 5 μ C18 derivatized vinyl alcohol co-polymer based column (Keystone Scientific, Bellefonte, PA). The polymer based packing materials were used because they are thermally stable to 150°C. Bonded silica based columns have been shown to be stable to 160°C (15), but experiments in our labs indicate column failure when bonded silica based columns are heated above 100°C in aqueous based solvents. Reports indicate the solubility of bare silica in water rises exponentially with temperature (*16*). In addition, C18 derivatized polymer columns posses greater surface area coverage than C18 derivatized silica based columns, making them ideal for on-column sampling and preconcentration (*17*). The column and the flow cell rested on a heating strip (Wellman, SS2181, Shelbyville, IN) which was controlled by a temperature control unit (FIAtron Lab Systems, TC-55, Oconomowoc, WI). An independent thermocouple monitored the temperature of the column and was recorded on a strip recorder (Fischer 5000, Pittsburgh, PA). The column maintained contact with a 1 cm × 5 cm × 15 cm aluminium block that was affixed to the heating strip. The column, aluminium block and flow cell were covered with a heat sink compound (Rawn Co., Spooner, WI) and were wrapped in fiberglass insulation to promote uniform heating. A 100 psi pressure restrictor (Upchurch, Oak Harbor, WA) was necessary to maintain the eluent in a liquid phase at all times.

Detection and Data Collection: The lamp, fiber optics, 5 mm pathlength flow cell (1.2 μL volume), monochromator and position sensitive detector (PSD) have been described in a previous report on the dual wavelength PSD detector (*7,8*). All detection specifications are the same as reported except the PSD is configured with 5×10^7 V/A resistors at the outputs A and B. Wavelengths of 378 nm and 443 nm were used in the analysis of aldehyde and ketone dinitrophenylhydrazine (DNPH) derivatives and the difference signal was set to an electronic null by a slight adjustment of the mask. A dual wavelength detector utilizing two UV enhanced photodiodes (UV-100BQ, EG&G Solid State Products Group, Montgomery, PA) was constructed to detect polyaromatic hydrocarbons (PAHs) in the UV region of the spectrum. Wavelengths of 285 nm and 346 nm were used for the analyses. With both detectors, the difference signal was amplified and collected on a personal computer (Leading Edge, Model D, Canton, MA) equipped with a 12 bit PC analog to digital converter (DACA, IBM, Boca Raten, FL). The software was written in house. For these experiments the signal was averaged from 1000 points to 1 point on the fly for a reading of approximately

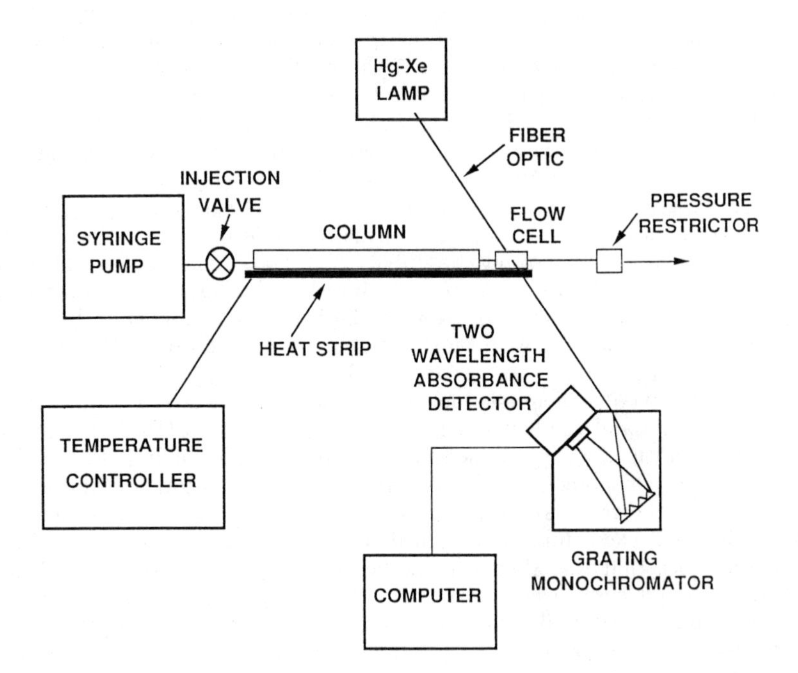

Figure 1. Thermal gradient microbore liquid chromatography analyzer instrumentation.

1 point per second. The sum signal was collected simultaneously on a chart recorder (Fischer 5000, Pittsburgh, PA).

Reagents: The mobile phase consisted of high purity acetonitrile (Burdick and Jackson UV grade, Muskegon, MI) and water (distilled, deionized with a 18 M-ohm Milli-Q water system, Millipore Corp., Bedford, MA). The aldehyde and ketone DNPH derivatives were dissolved in either water or the mobile phase. The preparation of the DNPH derivatives has been described in earlier work (*7*). The DNPH derivatized air sample was a gift of Dr. William Zoller, University of Washington. The PAH primary standards were prepared from solid standards in either acetonitrile or a mix of acetonitrile and tetrahydrofuran (J.T. Baker, Phillipsburg, NJ). Working standards were made in acetonitrile from the primary standard solutions. The soil sample was a gift of Friedman and Bruya, Inc. (Seattle, WA). The injected sample was prepared by shaking 10 mg of the soil sample with 50 μL methanol. The methanol extract was diluted 500 fold in water (20 μL in 10 ml) and filtered with a 0.02 μ syringe filter (Anatop syringe filters, Alltech, Deerfield, IL) before injection.

Results and Discussion

On-column preconcentration and thermal gradient reverse phase microbore liquid chromatography was demonstrated for two environmentally interesting groups of analytes. The first group is atmospheric aldehydes and ketones. These compounds are key intermediates in the oxidation of hydrocarbons (*18*) and indicators of urban pollution. The compounds were derivatized with dinitrophenylhydrazine (DNPH) enabling absorbance detection in the UV-visible region of the spectrum. In Figure 2 the isocratic, isothermal elution of a 1 μL sample of aldehyde and ketone DNPH derivatives is compared to on-column preconcentration and thermal gradient elution of a 500 μL aqueous dilution of the same sample eluted with a weaker solvent. The temperature program was from 28°C to 100°C in 10 minutes. The gradient program adequately resolved and eluted the analytes in half the time of the isothermal, isocratic run. The technique provided a 500-fold improvement in concentration detection limits for the DNPH derivatives. Figure 2 also illustrates that TG-LC separations are limited by the relatively poor efficiency of polymer-based columns. The popularity of techniques using these columns will encourage the development of higher performance polymer-based stationary phases.

To optimize the preconcentration of analytes onto the stationary phase, peak broadening in the loading step must be minimized. This broadening depends on how strongly the analytes are retained in the loading step. While previous theories have been developed for predicting the retention of analytes onto a stationary phase (*19*), we estimate the effective volume (V_{eff}) or the preconcentrated volume of analytes on the head of the column to be

$$V_{eff} = V_{inj}/(k'_L + 1) \tag{1}$$

where V_{inj} is the volume of sample injected and k'_L is the capacity factor of a given analyte when loaded on the column (*3*). The capacity factor (k') is calculated from $k' = (t_r - t_o)/t_o$, where t_r is the retention time of the analyte and t_o is the dead time or the retention time for a nonretained analyte. To optimize the analysis, k'_L for a representative aldehyde-DNPH was determined. A plot of log k'_L

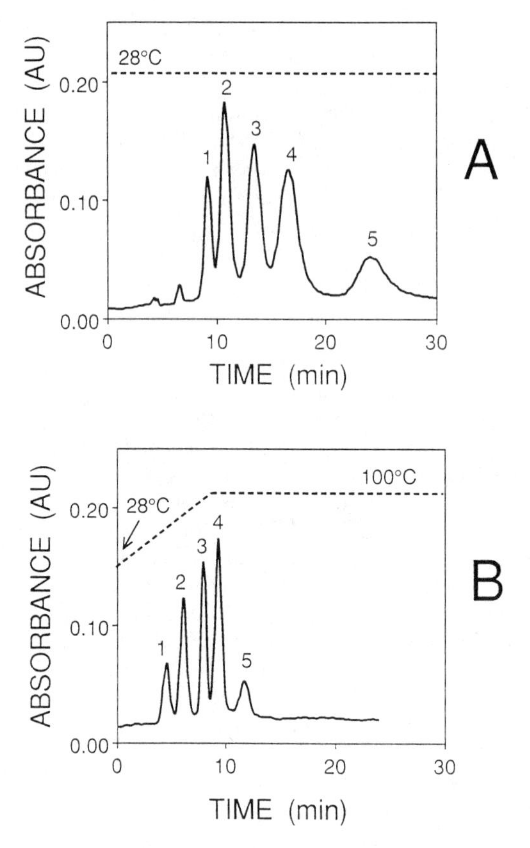

Figure 2. (A) Isocratic, isothermal separation of 5 aldehyde and ketone DNPH derivatives. 1 μL injected; mobile phase: 75% acetonitrile, 25% water; flow: 50 μL/min; T = 28°C. 10 parts per million (ppm) each. 1 – Formaldehyde-DNPH, 2 – Acetaldehyde-DNPH, 3 – Propanal-DNPH, 4 – Methyl ethyl ketone-DNPH, 5 – Benzaldehyde-DNPH. (B) On-column preconcentration and thermal gradient elution of a 500 μL aqueous sample. Mobile phase: 65% acetonitrile, 35% water; flow = 50 μL/min; T = 28°C to 100°C in 10 minutes. Analytes as above, 30 parts per billion (ppb) each.

vs. percent water in the mobile phase (v/v) at room temperature established a best fit linear regression of

$$\log k'_L = 0.045W - 0.154 \quad (2)$$

where W is the percent water, for data collected over a range of 0% to 20% water. Bitteur and Rosset have shown that a linear extrapolation of such data to conditions of 100% water is proper for most C18 separations (*17*). Guiochon *et al.* used similar plots to estimate overloading of short columns (*19*). The room temperature data enables one to predict the capacity factor of loading (k'_L) at very weak mobile phase compositions. For example, at room temperature, benzaldehyde-DNPH in a mobile phase of 35% water, 65% acetonitrile has a $k'_L = 27$. To estimate how temperature changes analyte retention, thermal gradient elutions of benzaldehyde-DNPH were performed with a temperature program of 28°C to 100°C in 3 minutes in various mobile phase combinations. A plot of $\log k'_g$ vs. percent water in the mobile phase (v/v) established a best fit linear regression of

$$\log k'_g = 0.044W - 0.801 \quad (3)$$

where W extends from 20% to 40% water and k'_g is the capacity factor of an analyte eluted with a thermal gradient program. k'_g is defined in the same way as k'_L in this study for comparison purposes. Consequently, benzaldehyde-DNPH in a mobile phase of 65% acetonitrile 35% water is loaded at $k'_L = 27$ and the thermal gradient program elutes it at $k'_g = 5$ relative to the start of the gradient. A $k'_L = 27$ is not ideal for preconcentration while minimizing peak broadening. From equation 1 it is clear that the on-column preconcentration and minimal band broadening are optimized by maximizing the loading capacity factor (k'_L), which requires increasing W in equation 2.

To increase the k'_L of an analyte, the sample was injected in water. This technique has been reported in the literature as a way to reduce band broadening in reverse phase chromatography (*12–14*). Aqueous sample volumes on the order of four times the void volume of the column establish a transient mobile phase so the injected sample, in 100% water, will become a very small volume when loaded on the column. From equation 2, the k'_L for aldehyde and ketone derivatives in 100% water is estimated to be to approximately 10,000. Thus, following from equation 1, the effective injection volume is potentially decreased by a factor of 10,000 relative to the injected sample volume. After the water based sample is loaded, the mobile phase following the sample lowers the k' of the analytes. At this time the thermal gradient program is initiated and the analytes elute with a $k' < 10$ relative to the start of the temperature program. It is clear from Figure 2B that the preconcentration technique does not introduce peak broadening. Figure 2B shows that 500 μL of injected sample elutes with peak widths less than 50 μL. Figure 3 demonstrates this technique for an environmental air sample in which the constituent aldehydes and ketones have been derivatized with DNPH. Formaldehyde-DNPH, acetaldehyde-DNPH and propanal-DNPH can be determined at the parts per billion (ppb) level by comparing the sample to the five component standard in Figure 2B. The limit of detection for this analysis is 1 ppb for the aldehyde and ketone DNPH derivatives.

The TG-μLC analyzer was also used to detect PAHs in aqueous samples. Figure 4 compares room temperature, isocratic elution of a 1 μL PAH standard in 100% acetonitrile to the preconcentration and thermal gradient of the same

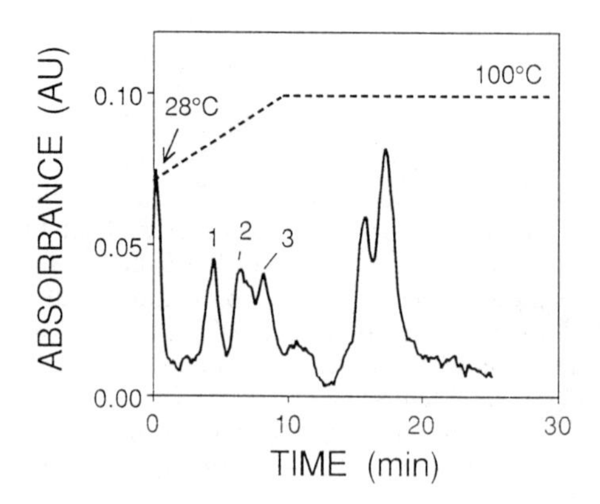

Figure 3. On-column preconcentration and TG elution of an environmental air sample derivatized with DNPH. 500 μL injected; mobile phase: 65% acetonitrile, 35% water; flow: 50 μL/min; T = 28°C to 100°C in 10 minutes. 1 – Formaldehyde-DNPH, 2 – Acetaldehyde-DNPH, 3 – Propanal-DNPH. Limit of detection is 1 ppb for the analytes.

compounds in a 500 μL aqueous sample. The thermal gradient run was performed in a much weaker mobile phase of 70% acetonitrile, 30% water. The analytes are eluted in approximately the same amount of time in both runs but the TG-μLC shows the benefit of gradually changing the solubility of the analytes in the mobile phase with time. Temperature gradient programming separates early eluting peaks such as napthalene (NAP) and acenapthalene (ACE) without broadening the later eluting peaks like cornene (COR). The width of the cornene peaks is approximately 2.5 times smaller when preconcentrated and eluted with a thermal gradient program, even though the volume injected was 500 times greater than the isocratic, isothermal run. In Figure 4 it is apparent that perylene and benzo(g,h,i)perlyene co-elute and have a difference signal opposite to the signal of the other PAHs. For perylene and benzo(g,h,i)perylene the reference wavelength region has a greater absorbance than the analytical wavelength region creating a negative peak. Negative chromatographic peaks are Gaussian and can be integrated in the same way as positive peaks. The dual wavelength absorbance detector provides two channel spectral information much like two pixels on a photodiode array. This additional information may aid in performing complex spearations. Figure 5 is a chromatogram of a water sample prepared from a coal tar/soil sample. 500 μL of the aqueous sample was preconcentrated on-column, and the analytes were eluted with a thermal gradient program of 28°C to 150°C in 20 minutes. Napthalene, pyrene and benzo(e)pyrene at ppb levels can be identified in the sample by comparison with the seven component standard shown in Figure 4B.

It was of interest to compare the effects of temperature gradients and mobile phase gradients on analyte retention. The temperature dependance of k' can be expressed as

$$\ln k' = \Delta H/(RT) - \Delta S/R + \ln f \tag{4}$$

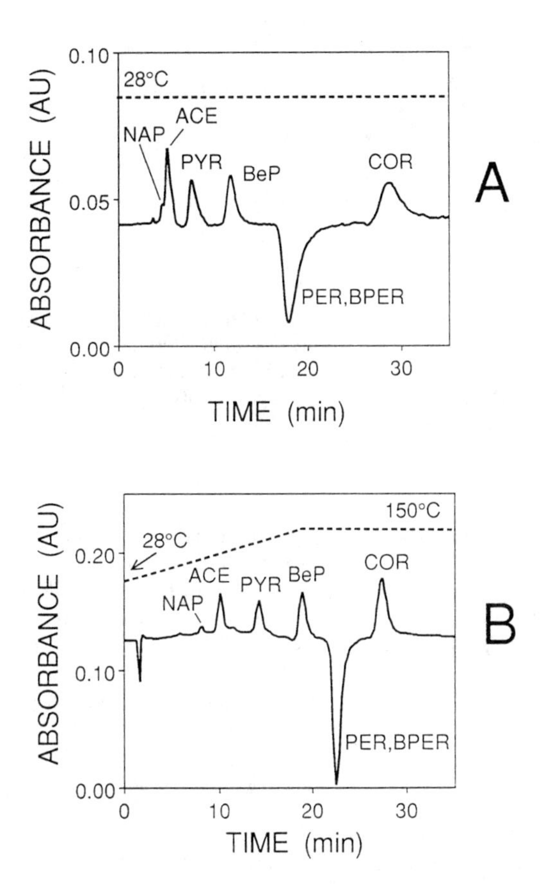

Figure 4. (A) Isocratic, isothermal separation of seven PAHs. 1 μL injected; mobile phase: 100% acetonitrile; flow: 50 μL/min; T = 28°C. NAP = naptha-lene, ACE = acenapthalene, PYR = pyrene, BeP = Benzo(e)pyrene, PER = Perylene, BPER = Benzo(ghi)perylene, COR = Cornene. 30 ppm each. (B) On-column preconcentration and thermal gradient elution of the same seven PAHs. 500 μL aqueous sample; mobile phase: 70% acetonitrile, 30% water; flow: 50 μL/min; T = 28°C to 150°C in 20 minutes; 100 ppb each.

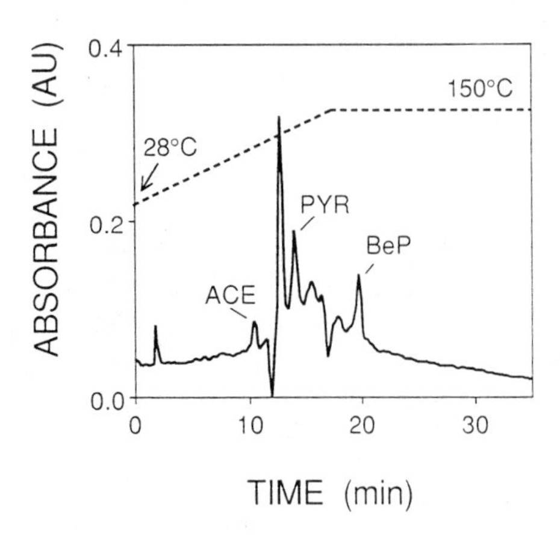

Figure 5. On-column preconcentration and thermal gradient elution of an aqueous based coal tar/soil sample extract. 500 μL injected; mobile phase: 70% acetonitrile, 30% water; T = 28°C to 150°C in 20 minutes. ACE = acenapthalene, PYR = pyrene, BeP = Benzo(e)pyrene. Limit of detection is 2 ppb for the analytes.

where ΔH and ΔS are the enthalpy and entropy, respectively, of an analyte partitioning from the stationary phase to the mobile phase, R is the universal gas constant, T is the temperature in kelvin, and f is the phase volume ratio of the stationary to mobile phase. Enthalpy can be determined experimentally by constructing a van't Hoffs plot, ln k' vs $1/T$, from equation 2. The slope of the line is $\Delta H/R$. To quantitate the change in capacity factor with temperature for an analyte, the ratio of capacity factor for the analyte eluted under two different conditions is considered. The capacity factor ratio is k'_1/k'_2,

$$k'_1/k'_2 = \exp(\Delta H(T_2 - T_1))/(RT_1T_2) \tag{5}$$

where k'_1 and k'_2 are capacity factors for an analyte at temperatures T_1 and T_2, respectively. Equation 5 predicts temperature to have a greater effect on those analytes with large enthalpies. Figure 6A characterizes the change in k' with mobile phase for seven PAH analytes with this system. The graph indicates that aromatic rings from two rings to seven rings can be eluted with 100% acetonitrile. This separation is shown in Figure 4A, where poor resolution for analytes of low k' was observed as well as severe peak broadening at high k'. At 70% acetonitrile 30% water the analytes of low k' are fully resolved, but not all the analytes are eluted in a reasonable amount of time. In 70% acetonitrile, 30% water cornene is estimated to have a $k' = 240$ and a retention time of 8.5 hours. Figure 6B demonstrates that in 70% acetonitrile 30% water the application of a TG program from 28°C to 150°C in 20 minutes will elute all seven PAHs, including cornene. The data indicates thermal gradient elution is more effective than pure acetonitrile in lowering the capacity factors of highly retained peaks. The increase in solubility of analytes at high temperature extends the range of detectable analytes to very highly retained species. The change in mobile phase necessary to match the

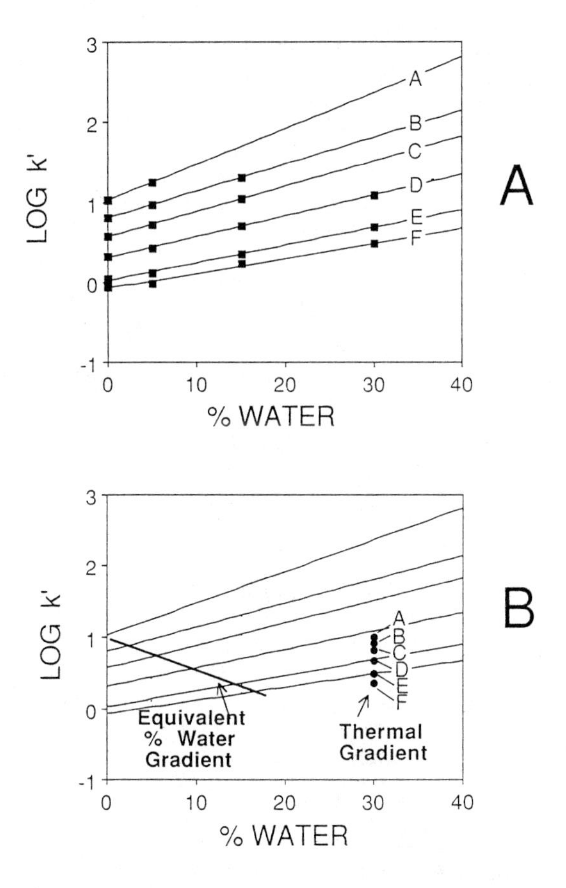

Figure 6. (A) Capacity factors of seven PAHs in various mobile phase conditions. A = Cornene, B = Perylene and benzo(ghi)perylene coelution, C = Benzo(e)pyrene, D = Pyrene, E = Acenapthalene, F = Napthalene. (B) Same as above with the capacity factors associated with a thermal gradient elution of the seven PAHs superimposed. T = 28°C to 150°C in 20 minutes. The equivalent mobile phase gradient equivalent from 70% acetonitrile to 100% acetonitrile is shown for comparison.

thermal gradient program is indicated by the line in Figure 6B. The dependence of k' on mobile phase modifier is linear and k' is generally considered to double, with a 10% decrease in organic modifier (*20*).

Finally, it is of interest to investigate the benefit of thermal gradient programming on chromatographic efficiency. Temperature has been shown to narrow peaks (*21,22*). Anita and Horvath propose an increase in the diffusivity of an analyte with temperature, thus the mass transfer contribution to band broadening is reduced (*5*). The relationship of temperature and diffusivity is described through the Stokes-Einstein Law for translational diffusion of molecules,

$$D = (kT)/(6\pi r\eta) \tag{6}$$

where k is a constant, T is temperature, r is the radius of the molecule and η is the viscosity. The van Deemter equation predicts chromatographic efficiency as defined by the height of one theoretical plate (H) (*23*). The van Deemter equation is

$$H_{total} = H_1 + H_2 + H_3 \tag{7}$$

where, H_1 is the multipath and extra column band broadening term, H_2 is the longitudinal diffusion term, and H_3 is the mass transfer term. For liquid chromatography, the longitudinal diffusion term is essentially zero and the mass transfer term is the largest contributor to H_{total}. Thus, the mass transfer term (H_3) can be defined as follows,

$$H_3 = H_{total} - H_1 \tag{8}$$

The mass transfer term can also be defined as

$$H_3 = (C_s d_p^{\,2})/D \tag{9}$$

where C_s is a constant, d_p is the thickness of the stationary phase or the particle diameter, and D is the translational diffusion coefficient for the analyte in the mobile phase. A five fold increase in D has been estimated for aqueous based separations of large molecular weight analytes at 125°C when compared to room temperature separations of the same analytes (*5*). Temperature also increases the solubility of the analyte in the mobile phase, so analytes elute with lower capacity factors.

It was of interest to conduct experiments to measure the effects of temperature on band broadening. Temperature can narrow peaks by lowering the k' of an analyte. In these experiments, the mobile phase composition and column temperature were selected so peak variance could be compared at constant k'. By expressing the data in this way, chromatographic efficiency at high temperature is associated with increased mass transfer of the analytes (*24*). Variance (σ^2) was calculated from the standard deviation of a Gaussian peak (σ), where $\sigma = w_{1/2}/2.354$ and $w_{1/2}$ is the width of the peak half way between the peak maximum and the average baseline noise (*23*). Peak variance is related to the height of one theoretical plate through

$$H_{total} = \sigma^2/L \tag{10}$$

where L is the length of the column. Three conditions were studied: (A) room temperature with changes in mobile phase, (B) changes in temperature with constant mobile phase composition, and (C) high temperature with changes in mobile phase but the mobile phase is relatively weak. Each data point represents the average of at least three duplicate runs. Figure 7 shows that the variances

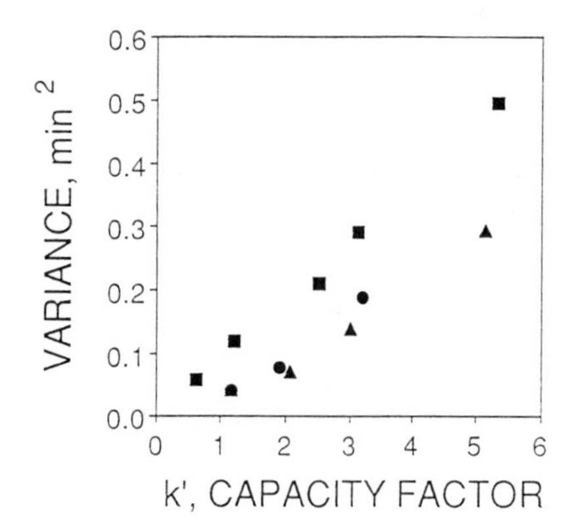

Figure 7. Band broadening study: variance of a benzaldehyde-DNPH peak as a function of k', at three different conditions. (■) T = 28°C; mobile phase changed from 100% acetonitrile to 80% acetonitrile, 20% water in 5 steps. (●) Mobile phase maintained at 80% acetonitrile, 20% water; T = 28°C, 50°C, 75°C and 100°C. (▲) T = 100°C; mobile phase changed from 80% acetonitrile, 20% water to 65% acetonitrile, 35% water in four steps.

associated with high temperature (100°C) separations were consistently smaller than the variances associated with the room temperature separations in a stronger solvent. Furthermore, at a given k', the variance associated with changes in temperature (only) were found consistently lower when compared to variances associated with changes in mobile phase (only). The data implies that for a given k', temperature has a greater effect on improving efficiency than mobile phase. By combining equations 8, 9 and 10,

$$(H_{RT} - H_1)/(H_{HT} - H_1) = ({\sigma_{RT}}^2 - {\sigma_1}^2)/({\sigma_{HT}}^2 - {\sigma_1}^2) = D_{HT}/D_{RT} \quad (11)$$

and using the data in Figure 7, the mass transfer term (H_3) and diffusivity of analytes (D) can be estimated at two different temperatures. H_3 and D for benzaldehyde-DNPH (B-DNPH) at high temperature (HT) and room temperature (RT) can be calculated from the peak variance (σ^2), and the broadening associated with multipath and extra column band broadening effects (${\sigma_1}^2$) shown in Figure 7. ${\sigma_1}^2$ was obtained by interpolating the data to the y-intercept, and was estimated to be 0.02 minutes^2. For a $k' = 3$, the σ^2 of B-DNPH in weak solvent at high temperature was about 0.25 minutes^2, while the variance associated with the room temperature separation in strong solvent was about 0.12 minutes^2. Consequently, the diffusivity of B-DNPH at high temperature was about twice that of the room temperature separation. The efficiency was also about two times better at high temperature. Examining band broadening at constant k' allows the effects of temperature on chromatographic efficiency to be evaluated independent of enhanced solubility of the analyte in the mobile phase. The study indicates that high temperature separations minimize band broadening by facilitating mass transfer of the analytes.

Conclusion

Thermal gradient microbore LC with dual wavelength absorbance detection has potential as a tunable analyzer for on-site chemical analysis. Thermal gradient programming is easily implemented in the field, and is ideal for application to small columns. The single fiber optic based dual absorbance detector eliminates problems associated with high temperature detection and minimizes correlated noise. The dynamic range of the analyzer is extended by employing on-column preconcentration. Thus, the analyzer is well suited for trace analysis of analytes in aqueous samples. Further development and evaluation of chromatographic columns suitable for high temperature must be done before the applicability of thermal gradient programming will be completely realized.

Literature Cited

1. Callis, J.B., Illman, D.L. and Kowalski, B.R., *Anal. Chem.*, **1987**, *59*, 624A–637A.
2. Murray, R.W., Dessy, R.E., Heinman, W.R., Janta, J., Seitz, W.R., *ACS Symposium Series*, V. 403, 1989.
3. Sulya, A.W., Moore, L.K. and Synovec, R.E., *SPIE Proceedings: Chem. Bio. and Env. Fiber Sensors*, **1991**, V. 1434, in press.
4. Renn, C.N. and Synovec, R.E., *Anal. Chem.*, **1991**, *63*, 568–574.
5. Anita, F.D. and Horvath, C., *J. Chromatogr.*, **1978**, *167*, 41–65.
6. Synovec, R.E., Renn, C. N. and Moore, L.K., *SPIE Proceedings: Chem., Bio., and Env. Fiber Sensors*, **1989**, *1172*, 49–59.
7. Renn, C.N. and Synovec, R.E., *Anal. Chem.*, **1990**, *62*, 558–564.
8. Renn, C.N. and Synovec, R.E., *Anal. Chem.*, **1988**, *60*, 1188–1193.
9. Atallah, R.H., Ruzicka, J., Christian, G.D., *Anal. Chem.*, **1987**, *59*, 2909–2914.
10. Melcher, R.G., *Anal. Chim. Acta*, **1988**, *214*, 299–313.
11. Van Horne, K.C., Sorbent Extraction Handbook, *Analytichem. International*, Harbor City, CA., 1986.
12. Guinebault, P.R. and Broquaire, M., *J. Liq. Chromatogr.*, **1981**, *4(11)*, 2039–2061.
13. Guinebault, P. and Broquaire, M., *J. Chromatogr.*, **1981**, *217*, 509–522.
14. Slais, K., Kourilova, D. and Krejci, M., *J. Chromatogr.*, **1983**, *282*, 363–370.
15. Ashraf-Khorassani, M., Taylor, L.T., *Anal. Chem.*, **1988**, *60*, 1529–1533.
16. Sosman, *Phases of Silica*, Rutgers University Press, New Brunswick, N.J., 1965.
17. Bitteur, S. and Rosset, R., *J. Chromatogr.*, **1987**, *394*, 279–293.
18. J.J. Zieman, "Tropospheric Carbonyl Compounds at Three Contiental Sites: Importance of Biogenic Hydrocarbons and Implication or the Oxidative Chemistry," Ph.D. Thesis, 1991, University of Washington.
19. Guiochon, G. *et al.*, *J. Chromatogr. Sci.*, **1983**, *21*, 27–33.
20. Katz, E.P., Lochmuller, C.H., Scott, R.P.W., *Anal. Chem.*, **1989**, *61*, 349–355.

21. Knox, J.H., Vasvari, G., *J. Chromatogr.*, **1973**, *83*, 181–194.
22. Grushka, E., Kitka, E.J., Jr., *Anal. Chem.*, **1974**, *46*, 1370–1375.
23. Karger, B.L., Snyder, L.R., Horvath, C., *Introduction to Separation Science*, Wiley-Interscience, N.Y., 1973.
24. Basak, S. and Compans, R.W., *J. HRC&CC*, **1981**, *4*, 300–302.

RECEIVED April 24, 1992

Chapter 21

Multicomponent Vapor Monitoring Using Arrays of Chemical Sensors

W. Patrick Carey

Department of Electrical Engineering, Center for Process Analytical Chemistry, University of Washington, FT-10, Seattle, WA 98195

Multicomponent analysis of vapor mixtures is performed by combining arrays of partially selective sensors and multivariate statistics. Sensors in an array instrument give differential responses to multiple analytes which allows the resolution of their combined effect. This array-analyte pattern which distinguishes between other analytes in the sample is similar, but with less information content, to identification of analytes in spectroscopy. Examples of quartz crystal microbalances and metal oxide semiconductor arrays are presented to show how they can be used in process analytical chemistry and pollution prevention. Calibration is performed with both linear and nonparametric regression techniques.

Analytical instrumentation, when applied to a process environment, can have a large impact on the real time monitoring of chemical components in a variety of different applications. When the process is defined as the environment, analysis of the atmosphere or groundwater systems can produce information about the dynamics of natural or polluting chemicals. More commonly as in this symposia, the process is defined as a chemical manufacturing process where real time analysis plays a role in increasing the process efficiency and quality or by decreasing unwanted reaction pathways that may occur. All of these sensing applications have a direct impact on the monitoring and reduction of pollution generated by anthropogenic sources. If one can create an efficient process based on tight control of the chemical dynamics, the by products of the process can be minimized.

The more common types of process instrumentation on the market consist of gas chromatographs, mass spectrometers and optical spectrometers (UV-VIS, IR, NIR) converted or designed to handle process streams. Although the degree of information about the process from these types of instruments is substantial, they are expensive to buy, operate and in some instances too bulky for application. An alternative approach that solves these application deficiencies is the use of chemical microsensors. Chemical microsensors have a robust quality due to their small size, versatility and cost, but the actual implementation of microsensors in chemical applications is limited due to device limitations. Several sensors such as glass pH and

0097–6156/92/0508–0258$06.00/0

ion specific electrodes have earned a favorable reputation and are used routinely in both laboratory and field applications. However, the literature is full of researched sensors with comparable figures of merit, but they are not accepted due to a variety of inherent problems. The problem of analyte selectivity is one that plagues almost every sensor on and off the market. Since only complex samples exist outside the laboratory, microsensors have to address this problem in order to become reliable and rugged analysis instruments. One approach to counter the problem of selectivity is the use of the sensor array concept. In this concept, multiple sensors are used, and their only selection criteria is that they have differential selectivity to the analytes of interest and any interferences. The advantages of this approach are a multicomponent determination capability and fewer sample preparation procedures to eliminate interferences, but the drawbacks are calibration and maintenance of numerous sensors instead of one.

This presentation describes two types of array instruments developed for vapor sensing for both ambient air and process analytical monitoring. The first is an array of quartz crystal microbalances where each sensor in the array is coated with a thin film organic structure such as a wax or a silicone polymer. This instrument was first evaluated for its ability to resolve chlorocarbon mixtures (*1*) and was later evaluated for an industrial drying process (*2*). The second instrument consists of an array of metal oxide semiconductors based on Taguchi gas sensors for the analysis of organic vapors or classes of compounds including aromatics (toluene, benzene, and xylenes), methylene chloride, chlorocarbons and alcohols. Additionally, one cannot address the application of sensor arrays without addressing the issue of calibration. In this study, two types of calibration techniques are used in order to calibration sensors which behave linearly (quartz crystal microbalances) and nonlinearly (solid state semiconductors).

Multivariate Analysis Techniques

Detection and quantitation of analytes in multicomponent samples with partially selective sensors requires the use of multivariate calibration. The two types of multivariate calibration techniques used in this paper are ones based on linear models such as multiple linear regression (MLR) (*3*), principal components regression (PCR), and partial least squares (PLS) (*4*) and nonparametric techniques such as multivariate adaptive regression splines (MARS) (*5*).

Linear Models. The multivariate capability of array instruments is established by the use of mathematical methods such as MLR, PCR, and PLS. These techniques take advantage of the response pattern produced by a sensor array for a given analyte to both identify and quantitate that component. The requirements for these methods include the existence of a unique response pattern for each analyte, linear additivity with respect to analyte concentration, and that all components in an analysis are represented in the calibration. An analytes' array response must be unique from other analyte response patterns so that linear combinations of two or more patterns do not equal a third pattern of an analyte in the system. This requirement is usually met by having more sensors in the array than analytes in the calibration. The prediction of a component's concentration, $\hat{c}_i$, in an unknown sample is performed by the following equation:

$$\hat{c}_i = \mathbf{r}_{un}^T \mathbf{R}^+ \mathbf{c}_i \tag{1}$$

where $\mathbf{r}_{un}^T$ is the response pattern (column vector) transposed from the array for an unknown sample, $\mathbf{R}^+$ is the pseudoinverse of the known calibration response patterns, and $\mathbf{c}_i$ is the vector of calibration concentrations for analyte *i* corresponding to the response patterns in **R**. The pseudoinverse of **R** is calculated by its singular value decomposition, $\mathbf{R} = \mathbf{USV}^T$, and recalculating $\mathbf{R}^+$ by using the inverse of the singular values in **S**, $\mathbf{R}^+ = \mathbf{VS}^{-1}\mathbf{U}^T$. When all the singular values are used, MLR results; when only the relevant singular values are retained, PCR is performed. The diagonal elements of **S** contain the square roots of the eigenvalues of **R**, and **U** and $\mathbf{V}^T$ contain the eigenvectors of $\mathbf{RR}^T$ and $\mathbf{R}^T\mathbf{R}$, respectively. For PLS, $\mathbf{R}^+$ is calculated in a more complex way taking into account the correlation between **R** and $\mathbf{c}_i$ (*6*).

Nonlinear Models. The MARS algorithms was developed from nonlinear regression methods to be a more robust technique for handling multivariate nonlinear data sets. One of the disadvantages of the above multivariate models such as MLR and PCR is that they are based on the linear least-squares estimator. PLS can attempt to correlate nonlinear trends into its model by including a polynomial fit in the inner relation (*7*). However, nonparametric techniques, which do not assume any underlying model for the data, can approximate nonlinear correlations much more effectively. As with any model building scheme, this means a lot of calibration samples since there is no linear assumptions. MARS is a new method based on adaptive computation of continuous spline functions in multiple dimensions. The function takes the form as follows:

$$\hat{f}(x_1, \ldots\ldots, x_n) = \sum_{m=0}^{M} a_m \prod_{k=1}^{K_m} b\,(x_{v(k,m)} \mid t_{k,m}) \qquad (2)$$

where a_m are the coefficients of the expansion, b is the univariate spline basis functions, x are the original independent variables characterized by the knots, t. The adaptive nature of the algorithm is present since the variable set $v_{(k,m)}$ and $t_{(k,m)}$ are determined by the data. Since methods based on nonparametric techniques are good at modeling data, the robustness necessary for prediction of future samples is unknown for MARS in chemical applications and is currently under study.

Chemical Sensor Arrays

Quartz Crystal Microbalance Sensor Array. The quartz crystal microbalance is a precise tuning device used in electronic circuits that is capable of oscillating from 5 to 15 MHz with an accuracy of better than a tenth of a hertz. The crystal itself is a thin AT-cut quartz disk or square about 1 cm across with electrodes deposited on both sides forming a capacitor region. When used in sensor applications, the surface of the crystal is modified with a thin film that can partition a vapor onto its surface. The crystal, being an harmonic oscillator, changes frequency whenever the surface mass changes and has a theoretical sensitivity in the nanogram range. The function relating mass and frequency is as follows:

$$\Delta F = -2.3 \times 10^{-6}\, F^2\, \Delta M/A \qquad (3)$$

where F is the fundamental frequency, M is mass in grams and A is the area of the sensor surface in cm^2. The use of this sensor was first described by King (*8*) in 1964, and various applications in the past decades have been summarized (*9-10*).

Since the sensor design of quartz crystal microbalances for specific applications involves the application of an adsorptive coating, total selectivity for a given analyte is usually a problem. Using arrays of crystals with different coating materials, the response pattern from the array can be used to identify and quantitate an analyte in the presence of other interfering compounds. Because the response of each sensor is a linear function of mass, eq 3, calibration can be performed by using linear models such as partial least-squares regression.

The first application of mixture resolution and quantitation was investigated on a system containing mixtures of 1,1,2-trichloroethane (TCA), 1,2-Dichloropropane (DCP), and m-dichlorobenzene (DCB). Nine sensors were used with the following coating materials: di(2-ethylhexyl)sebacate, ethylene glycol phthalate, quadrol, octahexyl vinyl ether, 1,2,3-tris(2-cyanoethoxy)propane, silicone SE-54, silicone DC-710, dioctyl phthalate, and silicone OV-225. The objective of this study was to determine whether this set of sensors could resolve similar molecules such as chlorocarbons even though their interaction with the stationary phases might be similar.

Figure 1 shows the response patterns from the sensor array of the three analytes tested. The degree of effectiveness to which multivariate regression performs in multicomponent analysis depends on the uniqueness of the sensor array response pattern of the pure analytes. The portion of an individual analyte's response pattern which is unique from all other patterns in the mixture is the part of the pattern used for identification and quantification. The array response patterns for the components in each sample set are plotted to show the degree of similarity or collinearity. As the degree of collinearity increases, error is amplified in the analysis by the propagation of sensor array response error. Additionally, as similarity increases, the limit of determination of each component rises due to the decrease of the effective signal-to-noise in the response pattern. Formulas developed to express the figures of merit such as sensitivity, selectivity, and limit of determination were given by Lorber (*11*).

Figure 2 shows the dynamic range plots of TCA and DCB to the array. Plotted on the y-axis is the summation of all nine sensor responses. TCA being a lighter molecule has a lower response intensity, but it is linear throughout the concentration range of 250 to 3250 ppm, whereas, DCB is not linear since saturation of the stationary phase is starting about 1500 ppm. During calibration and test sample preparation, the concentrations of both analytes were kept within the linear response regions.

Two Component Mixture. The results of MLR and PLS regression for this two component case are reported in Table I. The calibration data for the PLS model was autoscaled (subtraction of mean and scaled by variance for each column) and only two latent variables were used. For the first sample set of DCB and TCA, the PLS predictions were better in all cases. The average relative error was 21.3% for MLR and 5.7% for PLS for DCB and was 26.1% for MLR and 4.77% for PLS for TCA. The precision of each sensor was previously reported and confirmed in these experiments to be approximately 3-6% relative to the magnitude of the response (*12*). However, since the array yields nine responses for each sample and each sample was analyzed in triplicate, the prediction results can be better than this 3-6% due to signal averaging. The propagation of error introduced by the response error affects the prediction properties of MLR more than PLS by a factor of four to six. These results indicate the superior ability of PLS to deal with this response error and the instabilities created by collinearity.

Three Component Mixture. A second sample set composed of three components, DCB, TCA, and DCP, was analyzed by the same array to demonstrate the effect of severe collinearity in multivariate quantitation. When MLR was used to predict the concentrations of the analytes in the four test samples, Table II, large errors were realized for DCP. In fact, the trend in the actual concentrations of DCP in samples 1 through 4 increased while the predicted concentrations decreased. This is due to both the collinearity problem associated with MLR and the low sensitivity of DCP to the array. Improved results in this highly collinear case were obtained from

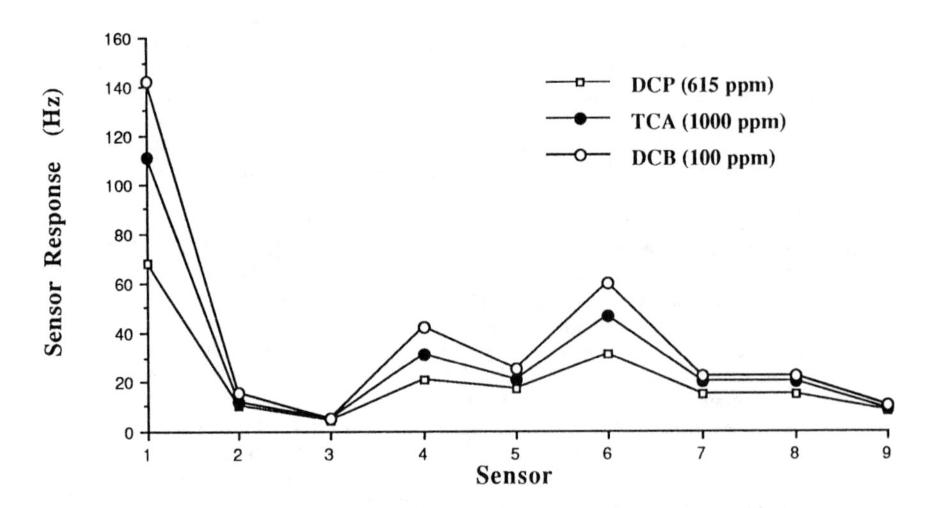

Figure 1. Chlorocarbon response patterns from the nine element quartz crystal sensor array. (Reproduced from reference 1. Copyright 1987 American Chemical Society.)

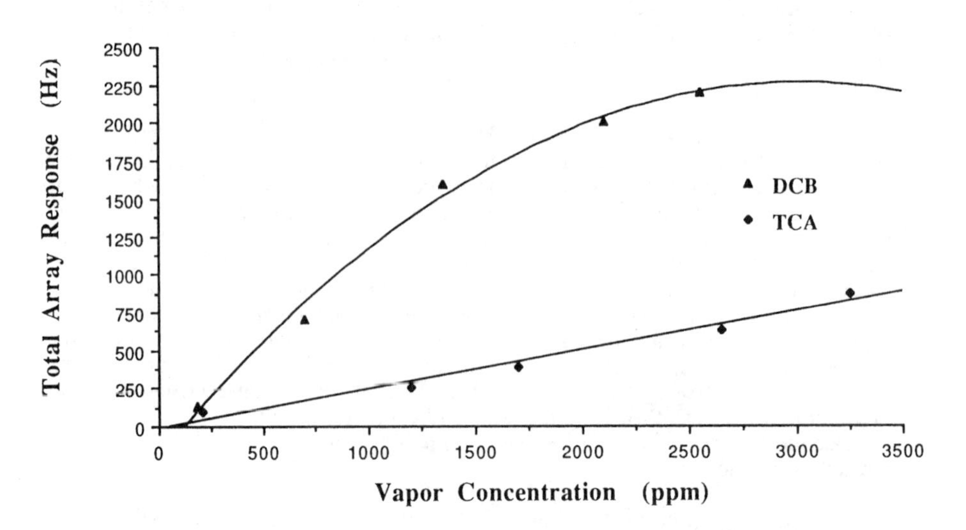

Figure 2. Dynamic range plots from DCB and TCA with the nine element quartz crystal microbalance array. (Reproduced from reference 1. Copyright 1987 American Chemical Society.)

Table I. Multivariate prediction results for the two component case

	actual concn, ppm		*PLS Predicted concn, ppm*		*rel error, %*		*MLR predicted concn, ppm*		*rel error, %*	
	DCB	TCA	DCB	TCA	DCB	TCA	DCB	TCA	DCB	TCA
1.	85	1720	83.3	1714	2.00	0.35	79.9	1746	6.00	1.51
2.	170	1075	170	1144	0.00	6.42	202	940	18.8	12.6
3.	119	1505	112	1563	5.88	3.85	160	1269	34.5	15.7
4.	510	1075	448	1168	12.2	8.65	574	821	12.6	23.6
5.	850	1505	784	1574	7.76	4.58	554	2662	34.8	76.9
Av					5.57	4.77			21.3	26.1

SOURCE: Reprinted with permission from ref. 1. Copyright 1987.

Table II. Multivariate prediction results for the three component case

	actual concn, ppm			*PLS predicted concn, ppm*			*rel error, %*			*MLR predicted concn, ppm*			*rel error, %*		
	DCB	DCP	TCA	DCB	DCP	TCA	DCB	DCP	TCA	DCB	DCP	TCA	DCB	DCP	TCA
1.	102	718	968	113	804	860	10.8	12.0	11.2	120	1043	802	17.7	45.3	17.2
2.	102	923	753	104	748	791	1.96	19.0	5.05	106	810	785	3.92	12.2	4.25
3.	136	923	860	119	841	907	12.5	8.88	5.47	140	779	901	2.94	15.6	4.77
4.	136	1025	860	123	867	944	9.56	15.4	9.77	164	408	1017	20.6	60.2	18.3
Av							8.71	13.8	7.87				11.3	33.3	11.1

SOURCE: Reprinted with permission from ref. 1. Copyright 1987.

PLS when each analyte was calibrated and predicted individually. This technique allows the maximum amount of response information to be used in the model. The average relative prediction error for PLS was 10.1% and for MLR was 18.6%. It must be emphasized that this is a worst case example of collinearity and that a limitation exists for all multivariate techniques to perform well with highly collinear data. This also provides evidence that selectivity still plays an important role in sensor development.

Process Monitoring. The monitoring of an industrial process with an array of microsensors is an important step in the evaluation of their capability and future as process instruments. One process that can be used to carry out this evaluation and for which the piezoelectric sensor is ideally suited is bulk drying of materials. In this process, in which large amounts of materials are dried, energy intensive dryers and rotating columns operate on a time scale of days. The degree of dryness is checked periodically by manually taking a sample and measuring the solvent content. Since these checks are performed at intervals, the real time knowledge of the dryness is unknown and the inefficient process control procedures result in wasted energy, raw materials, and off gases. The use of the piezoelectric crystal sensor array provides the unique ability to simultaneously monitor in-line the evaporated gases in order to determine the degree of dryness of the sample. This information can then be relayed to the operators, who can make efficient adjustments to the process and limit the energy used to run the process.

This study is focused on the evaluation of a six-element chemical sensor array of piezoelectric crystals in monitoring the evaporation of common solvents. This evaluation is performed on a laboratory rotary evaporator system developed to simulate a actual plant drying unit. The sample contained water, hexane, and methylene chloride as solvents. Multicomponent prediction of solvent concentrations in the exhaust gases is used as an indicator of the the degree of dryness of the products.

The detector cell is comprised of six coated 9-MHz quartz crystals housed in a plexiglass block with sealed wire leads where the crystals are arranged horizontally with respect to the flow path. The coatings used for the six crystals are 1: Di (2-ethylhexyl) Sebacate, 2: Ethylene Glycol Phthalate, 3: 1,2,3-Tris (2-cyanoethoxy) propane, 4: DC-710, 5: OV-225, 6: SE-54.

In order to correlate the drying process to the detector signal, a calibration scheme based on the measurement of the solvent content of the wet material by gas chromatography was established. During the drying process, samples are drawn from the chamber and analyzed. This analysis is performed by extracting the solvent from 100 mg of the sample with dimethylformamide (1 mL). The extracting solvent contains an internal standard that is used to ratio the peak areas of the sample analysis to a standard mixture.

The calibration method used in this study is PCR as denoted in eq 1. This method is the primary choice since it performs well with collinear data (*13*). An alternate method of partial least squares regression was first attempted but ruled out due to the lack of calibration samples used in creating the model. The means of calibration were to use the sensor array response patterns measured when the pure solvents were tested in the apparatus without any solid sample. Using this data, a calibration model that best relates the degree of dryness of the sample to the sensor responses of the array was established. The main difficulty in this type of analysis is that there is no relationship between the sample dryness and the amount of solvent vapors reaching the sensors. During a drying process, the evaporation rate of a solvent is constant giving a steady signal at the detector. However, the solvent content is decreasing at the same rate as the evaporation, but the detector in the gas phase does not respond to this decrease in concentration, only the constant flux of vapor reaching it. Therefore, the response patterns at each of the array scans of the exhaust vapor during a material drying run were used for predicting the vapor concentrations of the solvents based on the calibration response patterns of the pure solvents. This information gives the solvent concentration values in the gas phase and can then be related, if desired, to the calibration data from the gas chromatographic analysis to confirm the drying progression, Table III.

Table III. Calibration of Drying Process by Gas Chromatography

Time, min	*Water, w/w %*	*Hexane, w/w ppm*	*Methylene Chloride, w/w ppm*
4.2	9.3	94.1	87.8
14.7	10.3	76.4	79.3
33.6	5.2	68.4	73.6
63.0	3.5	57.3	74.1
105.0	4.6	59.6	73.8
125.6	7.8	47.5	43.1

SOURCE: Reprinted with permission from ref. 2. Copyright 1988.

This calibration included the three pure response patterns for the solvents, and the predicted concentration profiles are presented in Figure 3. In this analysis the solvent-sample interaction effect is evident as well as the distillation effect mentioned above. By examination of the solvent content in the sample, Table III, the trend of drying follows closely with the drying profiles. Methylene chloride, since its boiling point is much less than hexane and water, evaporates rapidly and completely by 13 minutes. After 13 minutes the sample uses the heat flux input to start evaporating the hexane and water. This is exactly the same process that occurs in a simple distillation. The hexane and water concentrations follow a similar drying course corresponding to the degree of dryness. The water content is relatively constant for 15 minutes denoting no evaporation until the methylene chloride was released. When 63 minutes is reached, the concentrations of the two solvents in the sample are relatively constant indicating the drying process is complete.

There is a lack of fit at some portions of the drying profiles, where one of the components is dried, and another component is evaporating. An example of this is where methylene chloride shows negative concentration values after it has been removed from the sample. This is due to calibration error in creating a model that does not exactly represent the actual gas phase exhaust and changes in the sensing environment over time. Several causes of this are the absence of the sample material being present when the pure solvents are analyzed for the calibration, the error present in the sensor responses that occurs from slight thermal drift due to changes in vapor heat capacities (the sensors were not corrected for temperature changes), random noise and the interferences from possible unknown vapors coming off of the samples themselves.

Metal Oxide Semiconductors Arrays. Metal oxide semiconductors sensors are composed of n-type binary metal oxides which provides a catalytic surface for oxidation/reduction reactions to occur. The principle of operation is that when heated to approximately 200-400° C, oxygen is chemisorbed into the semiconductor surface and ionizes. This ionized oxygen causes an increase in the semiconductor resistance and is very reactive with any vapor species it contacts on the surface. When a reaction does occur, the resistance decreases due to a released electron into the conduction band. The equation describing this operation is as follows:

$$\frac{R}{R_o} = P_{O_2}^{-\beta} \left(1 + P_i^{n_i}\right)^{\beta} \quad (4)$$

where R and R_o are the semiconductor measured and initial resistances respectively, P is the partial pressure of oxygen and of the vapor i, n_i is an equilibrium adsorption coefficient of the vapor and β is a term describing the surface energy of the semiconductor (*14*). In normal operation, the oxygen partial pressure is constant, so it is eliminated from the equation. From this equation a nonlinear relationship exists between vapor concentration and sensor response.

In order to obtain selectivity for a given compound, the type of metal oxide and a catalyst which is doped into the oxide are selected. Generally, two types of oxides, tin oxide and zinc oxide, are used as the semiconductor material. Catalyst are used to lower the barrier energy of adsorption of both oxygen and the gas component to be measured. Common catalyst used are platinum and palladium, but other metals are available depending upon the type of vapor to be sensed.

Using arrays of commercially available sensors, the objective is to perform multicomponent vapor analysis of hazardous or process important compounds. Each sensor of the array is a tin dioxide ceramic substrate with dopants added to alter the selectivity. The response of an array of Taguchi gas sensors (TGS, Figaro) to toluene and benzene is shown in Figure 4. The most noticeable effect of these type of sensors is the nonlinear responses as concentration increases as presented by eq 4. Initially our research is oriented toward resolving toluene and benzene mixtures. With these two compounds having similar structure, the role of the catalyst will be important in order to distinguish them. Alternatively, classes of organic vapors could be resolved and quantitated based on their similar interactions.

The array consists of eight TGS sensors in which vapors are passed through from a vapor generation system involving bubblers (-15° C) and mass flow controllers. Seventy 2 component mixtures of toluene and benzene were generated in the range of 50 - 400 ppm of each analyte.

Table IV. Calibration Error for Various Regression Methods

	RMS_{res} (ppm)	
Method	*Toluene*	*Benzene*
PLS1	36.2	78.0
PLS1-CV	39.9	86.5
poly-PLS	33.7	83.8
MARS	19.2	64.8

Results for the calibration of this data set using three techniques is presented in Table IV. PLS, PLS with a polynomial inner relation (poly-PLS) and MARS were used. Using PLS with five latent variables the standard deviation of the calibration error (RMSres) is 36.2 ppm for toluene and 78.0 ppm for benzene. Poly-PLS provided no significant decrease in calibration error. In this attempt, five latent variables were used with a third order polynomial fit in the inner relation. Using

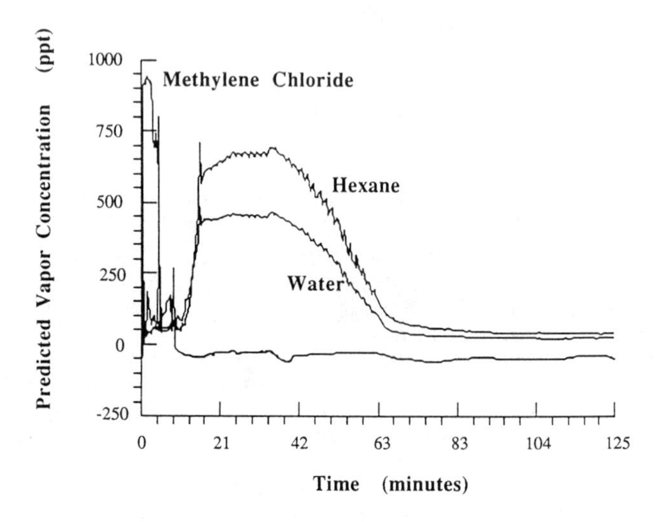

Figure 3. Drying profile of a sample containing three solvents. Solvent vapors are detected by a six element quartz crystal microbalance array. (Reproduced from reference 2. Copyright 1988 American Chemical Society.)

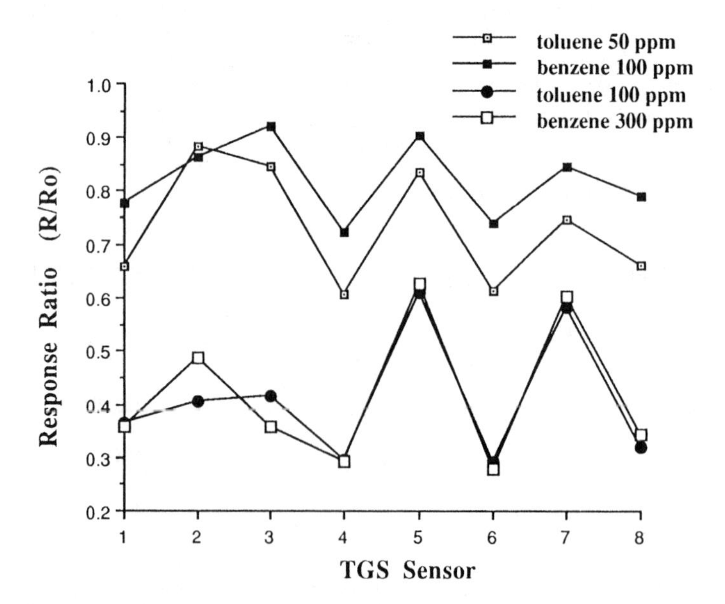

Figure 4. Response patterns for toluene and benzene at different concentrations from an eight element Taguchi gas sensor array.

MARS, however, the nonlinear trends in the data can be approximated and the calibration error dropped to 19.2 ppm for toluene and 64.8 ppm for benzene. This is a significant improvement when tested using a F-ratio comparison of errors. The important topic of robust prediction qualities is not clear with MARS at this time. Further work in this area is currently being pursued.

Conclusion

The use of sensor arrays provides a unique robustness to the chemical sensor. It allows one to performed analyses on complex samples that may cause the individual sensors some interference problems. The application of the quartz crystal microbalance array provided evidence that resolving similar compounds (in two and three component mixtures) can be done at a minimum cost of precision. The use of nonlinear calibration algorithms will expand the usefulness of the array concept to a variety of different sensors since the majority of solid-state approaches have nonlinear functions. In addition, the solid-state approach can eliminate problems with bulky apparatus since arrays could be integrated onto a single silicon chip (i.e. chemfets, and metal oxide semiconductors).

Acknowledgments

This research was supported by a grant from the Center for Process Analytical Chemistry, a NSF-Industry cooperative research center.

Literature Cited

1. Carey, W.P.; Beebe, K.R.; Kowalski, B.R. *Anal. Chem.* **1987**, *59*, 1529.
2. Carey, W.P.; Kowalski, B.R. *Anal. Chem.* **1988**, *60*, 541.
3. Draper, N. R.; Smith, H. *Applied Regression Analysis;* Wiley: New York, NY, 1981; Chapter 4.
4. Geladi, P.; Kowalski, B. R. *Anal. Chim. Acta* **1986**, *185*, 1.
5. Friedman, J. H. *Ann. of Stat.* **1991**, *19*, 1.
6. Lorber, A; Wangen, L.E.; Kowalski, B.R. *J. Chemo.* **1987**, *1*, 19.
7. Wold, S.; Wold, N.K.; Skagerberg, B. *Chemom. Intelligent Lab. Sys.* **1989**, *7*, 53.
8. King, W.H. *Anal. Chem.* **1964**, *36*, 1735.
9. Suleiman, A.A.; Guilbault, G.G. *Anal. Chem.* **1984**, *56*, 2964.
10. McCallum, J.J. *Analyst* **1989**, *114*, 1173.
11. Lorber, A. *Anal. Chem.* **1986**, *58*, 1167.
12. Guilbault, G. G. *Intern. J. Environ. Anal. Chem.* **1981**, *10*, 89.
13. Mandel, J. *J. Res. Nat. Bur. Stan.* **1985**, *90*, 465.
14. Clifford, P. in *Homgeneous Semiconducting Gas Sensors: A Comprehensive Model;* Seiyama, T., Ed.; Analytical Chemistry Symposia Series; Elsevier: 1983, Vol. 17; pp 135-146.

RECEIVED February 3, 1992

Chapter 22

Development of Fiber-Optic Immunosensors for Environmental Analysis

T. Vo-Dinh[1], G. D. Griffin[1], J. P. Alarie[1], M. J. Sepaniak[2], and J. R. Bowyer[2]

[1]Oak Ridge National Laboratory, Advanced Monitoring Development Group, Health and Safety Research Division, Oak Ridge, TN 37831–6101
[2]University of Tennessee, Department of Chemistry, Knoxville, TN 37996

A review of the principle and applications of immunofluorescence spectroscopy to the development of antibody-based fiberoptics sensors is presented. Special focus is devoted to antibody-based fiberoptics fluoroimmunosensors developed to detect important pollutants such as the carcinogenic polyaromatic compounds or to aflatoxin. The fiberoptics sensor utilizes antibodies covalently bound to its tip or encapsulated on the probe tip. The usefulness of fluorimmmunosensors for environmental monitoring will be discussed.

The combination of fiberoptics technology and advanced optical sensors promises to open new horizons in medical, clinical and environmental monitoring applications. In the area of human health protection against environmental pollutants, there is a strong need for sensitive and selective instrumentation to analyze complex samples. Two environmental pollutants of interest are polycyclic aromatic compounds and aflatoxin. Polycyclic aromatic compounds (PACs) are ubiquitous environmental pollutants that represent the largest class of suspected chemical carcinogens *(1)*. Aflatoxins are the metabolites of the fungal species Aspergillus flavus and aspergillus parasiticus *(2)*. Thcy occur naturally in foodstuffs such as peanuts, corn, milk, and animal feed. Due to the extremely toxic nature of aflatoxins and their metabolites, many countries have introduced regulations and legislation limiting maximum levels permitted in food. This in turn has sparked interest in the development of simple, sensitive and rapid methods for the analysis of aflatoxins.

Use of Antibodies for Chemical Monitoring

Antibody

Antibodies consist of hundreds of individual amino acids arranged in a highly ordered sequence (Figure 1). The antibodies are actually produced by immune

0097–6156/92/0508–0270$06.00/0

system cells (B cells) when those cells are exposed to substances or molecules, which are called antigens. The antibodies produced following antigen exposure have recognition/binding sites for specific molecular structures (or substructures) of the antigen. The way in which antigen and antigen-specific antibody interact may perhaps be understood as analogous to a lock and key fit, in which specific configurations of a unique key enable it to open a lock. In the same way, an antigen-specific antibody "fits" its unique antigen in a highly specific manner, so that hollows, protrusions, planes, ridges, etc. (in a word, the total 3-dimensional structure) of antigen and antibody are complementary. This unique property of antibodies is the key to their usefulness in immunosensors; this ability to recognize molecular structures allows one to develop antibodies that bind specifically to chemicals, biomolecules, microorganism components, etc. One can then use such antibodies as specific "detectors" to identify an analyte of interest that is present, even in extremely small amounts, in a myriad of other chemical substances. Another antibody property of paramount importance to their analytical role in immunosensors is the strength or avidity/affinity of the antigen-antibody interaction. Because of the variety of interactions which can take place as the antigen-antibody surfaces lie in close proximity one to another, the overall strength of the interaction can be considerable, with correspondingly favorable association and equilibrium constants.

Antibody-Antigen Interaction in Chemical Detection

Since understanding antibody-antigen interactions is critical in utilizing antibodies as analytical devices, the forces involved in the antigen-antibody interaction will be briefly discussed. The antibody-antigen complex is not held together by covalent bonds; nevertheless, the strength of the interaction can be gauged from the often strikingly high antigen binding constants. It is generally agreed that there are four factors involved in the antigen-antibody interaction: (1) electrostatic forces; (2) hydrogen bonding; (3) hydrophobic attractions; and (4) Van der Waal's interactions *(3)*. Electrostatic or Coulombic interactions occur between opposite electrical charges, and thus involve ionized sites (e.g., -COOH and -NH_2 groups on amino acid side chains) or less strongly attracting dipoles. Hydrogen bonds involve interaction of a hydrogen atom, covalently bonded to a more electronegative atom and thus having a partial positive charge, with an unshared electron pair of a second electronegative atom. The interaction of water molecules is a classic example of hydrogen bonding; a variety of amino acid functional groups (particularly -OH and -NH_2) could be involved in such interaction *(2)*. The hydrophobic interactions occur as a result of a strong tendency for apolar atomic groups (e.g., side chains of valine, leucine, phenylalanine, proline, and tryptophan) to associate with one another in an aqueous environment; thus, their net interaction with water is decreased. Van der Waal's attractions are the result of external electron clouds of atoms forming dipole attractions, the dipoles being induced in a given atom by the very close approach of another atom which has a fluctuating dipole *(3)*. This last force becomes increasingly stronger as the interatomic distances decrease. In fact, the forces of hydrogen bonding, hydrophobic interaction and Van der

Waal's attraction are all relatively weak in binding strength, and only become significant upon close approach of the pair of molecules. Studies of the few antigen-antibody interactions which have been dissected at the atomic level by X-ray crystallography have indicated that the predominant attractive forces in the antigen-antibody bond arise from a large number of hydrophobic and hydrogen bond interactions, as well as Van der Waals forces arising from the close approach of the antibody to the antigen *(3,4,5)*. Thus the overall remarkable strength of the interaction is due to the extremely close fit of the molecular surfaces (1-2Å) *(3)* the exquisite complementarity between antigen and antibody, and the formation of a large number of individual weak interactions, which become significant en masse.

Polyclonal or Monoclonal Antibodies for Sensors

It is important to know whether the antibodies desired for a particular application are to be derived from a polyclonal source or by monoclonal technology in sensor development. Monoclonal antibodies are antibodies which are produced by the daughters of a single "B" cell. Since all the daughters are producing exactly the same antibody as the parent cell, a monoclonal antibody is completely homogeneous with all antibody molecules being the same type of immunoglobulin and binding to the antigenic determinant *(6)*. Polyclonal antibodies, by contrast, are antibodies circulating in serum of animals immunized with a specific antigen. Because these antibodies have arisen from clones of a number of separate "B" cells, the antibodies are heterogeneous and different antibodies in this mixture react with different antigenic determinants. Therefore, polyclonal antisera will always be a mixture while monoclonals are not a mixture (unless deliberately mixed in the laboratory).

Polyclonal antibodies are relatively easy to develop (by immunizing an appropriate experimental animal and then taking an appropriate quantity of blood). There are two major disadvantages in regard to polyclonal antibodies. First, these antibodies will always have multiple specificity (even if it is possible to purify only the antibodies to a single antigen from the antisera, if the antigen has more than one epitope, the antibodies will be a mixture of antibodies recognizing different epitopes), and hence can never provide the monospecificity of a monoclonal antibody *(7)*. Second, because the antibodies arise from bleeding an immunized animal at some point in the immunization protocol, different batches of antisera taken at different times will inevitably have a somewhat different antibody composition in terms of specificity and avidity *(7)*.

Monoclonal antibodies can, at least theoretically, provide the solution to the problems of polyclonal antisera listed above. Kohler *(6)* lists the following advantages associated with monoclonal antibodies: (1) each hybrid cell line produces only one unique antibody; (2) there is potentially an unlimited antibody supply; (3) immunization with an impure antigen can still lead to an antibody against only the antigen of interest; (4) potentially all specificities (i.e., antibodies against all antigenic epitopes) can be obtained; (5) it is possible to manipulate monoclonal antibodies by genetic engineering techniques; (6) the technique is very general, in terms of what antigen can be used, and the desired properties of the antibody. Since each successful hybridoma is the result of a fusion of a myeloma cell with a "B" cell reacting with one epitope of the antigen, it can be seen that an antigen preparation that contains a number of impurities can still provide good results *(8,9)*. This fact alone recommends the monoclonal antibody technology for antigens which can only be obtained in very small amounts and/or in an impure state.

Different Designs of Immunosensors

The type of immunoassay (homogeneous or heterogeneous assays) and the choice of the detection technique determine the design of an fluoroimmunosensor (FIS) device. The instrument development is also based on the selection of the fiberoptics sensor design (e.g., light transmission onto the distal end with covalently bound antibodies or microcavity, or excitation and collection of light via the evanescent-field method). This section illustrates the different combinations of FIS designs and immunoassay procedures.

Immobilized Antibodies

Several strategies can be used to retain the antibody at the sensing probe. Whatever procedure is involved, one requirement is that the antibody, as much as is possible, retains its antigen-binding activity. Perhaps the easiest and most satisfactory procedures enclose the antibody in solution, within a semi-permeable membrane cap which fits over the end of the sensor *(10,11)*. The analyte solution is kept separate from the antibody by the semi-permeable membrane, through which the analyte of interest diffuses, and then interacts with the antibody. Obviously, such an arrangement only works for relatively small analytes (antigens) which can diffuse through the semi-permeable membrane (whose pores must not allow the antibody to pass through). Other potential problems could arise from diffusion limitations or adsorption on the membrane. Nevertheless, the authors have found the "membrane-drum" type sensor to perform well for detection of the metabolite of benzo(a)pyrene, BPT (r-7,t-8,9,c-10-tetrahydroxy-7,8,9,10-tetrahydrobenzo(a)pyrene) at ultra-levels in aqueous solution *(10,11)*.

A wide variety of procedures may be used to adsorb/link the antibody to the fiber itself. Although simple adsorption on quartz/glass (or better, plastic) is possible, most investigators prefer a more firm anchorage, particularly when multiple washes may be envisioned. A variety of covalent linkages may be utilized - the important caveats being: (1) to avoid denaturing the antibody during linkage, so it does not lose antigen-binding activity, and (2) to avoid linking at the antigen-binding site, because such linkage may provide stearic hindrance to antigen binding.

Comparative studies using several different procedures to attach antibody to silica beads has previously been completed *(12)*. The beads are first chemically derivatized with 3-glycidoxypropyltrimethoxysilane (GOPS); GOPS can also be used to derivatize quartz optical fibers. The use of this reagent introduces diol groups on the surface of the spheres. After this initial treatment, different techniques were utilized to attach antibody. In one method, HIO_4 was used to oxidize the diols to aldehyde groups, and upon addition of antibody, covalent coupling occurred through formation of the Schiff base with free primary amino groups present in the antibody protein (e.g., ϵ-amino of lysine). Obviously the site of attachment on the antibody cannot be controlled. The Schiff base linkage is subsequently reduced with sodium cyanoborohydride to stabilize the linkage. In another procedure, the GOPS-derivatized beads were treated with 1,1'-carbonyldiimidazole (CDI), followed by antibody. The linkage was again through a free primary amino group on the antibody. (Note that cyanogen bromide and N-hydroxy-succinimide are also frequently used as coupling agents for binding through primary amino groups).

Alarie et al *(12)* used free-SH groups on the antibody molecule. To generate these, $F(ab')_2$ fragments were prepared and the S-S bonds in the hinge region were reduced with dithiothreitol. Silica beads were derivatized with GOPS, activated with 2-fluoro-1-methylpyridinium toluene-4-sulfonate, and subsequently reacted with the reduced Fab fragments. The linkage of antibody, in this case, occurs at the SH groups in the hinge; the antigen binding site should therefore be unhindered. Alarie et al *(12)* also investigated a procedure where antibody is linked to the beads through Protein A binding. Silica beads having Protein A on the surface were incubated with antibody, and the resulting complex was stabilized by cross-linking the antibody covalently to the protein A with dimethylsuberimidate. In this case, Protein A is known to bind antibody in the Fc region, so again the antigen binding site should be free. For all different immobilization procedures, the total amount of antibody immobilized and the amount of active immobilized antibody was determined, using two antigen-antibody systems.

Linkage via Protein A was found to preserve antibody activity, although the amount of antibody bound was rather low in comparison to other procedures. Somewhat surprising was the fact that CDI produced large amounts of antibody bound and reasonable retention of antibody activity. The linkage via the SH group of the Fab fragment was approximately equivalent to CDI coupling. The direct linkage via GOPS and HIO_4 activation was least satisfactory as there were large losses of antibody activity. A conclusion that might be drawn from this study is that random linkage on the antibody, while unattractive on theoretical grounds, may in actual practice be acceptable, probably because there are many available amino groups on the antibody surface, a large proportion not being in the antigen-binding site. It may still be worthwhile, however, to attempt coupling in the Fc region, simply because one should be assured of retaining the bulk of the antibody activity.

Affinity-Avidity Considerations

Immunosensors are affected strongly by the affinity/avidity of an antibody for its antigen. The affinity of an antibody will determine the overall sensitivity (i.e., limit of detection will increase with increases in affinity) and specificity (specificity increases with larger differences in antibody affinity for specific and non-specific antigen) of an analytical system based on this antibody *(7)*.

There is a continuous process of association and dissociation between antigen and antibody, during which antibody and antigen may become separated since the forces holding antibody and antigen together are non-covalent. This fundamental reversibility of the antibody-antigen interaction must be grasped, i.e., that there is a constant separation and reattachment of antibody and antigen molecules as the two species interact in solution. This reaction can be written (where Ab = antibody and Ag = antigen): $Ab + Ag \rightleftharpoons Ab{\cdot}Ag$. The law of mass action as applied to this reversible reaction produces the following equation for the equilibrium constant:

$$K = \frac{[Ab{\cdot}Ag]}{[Ab][Ag]}$$

This equilibrium constant K is the affinity constant. If there is strong interaction between Ab and Ag, the equilibrium will favor the [Ab·Ag] complex, the affinity constant will be relatively large, and the antibody can be said to show strong (or high) affinity. Conversely, a smaller affinity constant will mean a shift toward greater concentrations of free antibody and antigen, and the antibody can be said to show a correspondingly lower affinity.

$$Ab + Ag \xrightarrow{kass} Ab{\cdot}Ag$$
$$Ab{\cdot}Ag \xrightarrow{kdiss} Ab + Ag$$

$$\text{Therefore } K = \frac{kass}{kdiss}$$

An understanding of the significance of the inter-relationship of the k_{diss} and the equilibrium constant is important when developing an immunosensor. As the sensing device is washed during various stages of the procedure, free antigen is removed, and the Ab·Ag complex will dissociate to some extent to re-establish equilibrium conditions. The extent of this dissociation will have important effects on the sensitivity of the device. Tijssen *(7)* presents theoretical data on the effect of multiple washes on the extent of antibody saturation, as a function of affinity constant, K. For antibodies with high K (10^9), 2 washes reduce 90% saturation to 80%. Starting with the same initial saturation, antibodies with $K = 10^7$ show a reduction to 20% saturation, while antibodies with $K = 10^6$ are reduced to $<1\%$ saturation. For these low affinity antibodies, therefore, there is little relation between the initial antigen concentration and what remains associated with the antibody after several washes.

The affinity/avidity of an antibody not only determines the strength of the antigen-antibody interaction but has important consequences with regard to antibody specificity. Antibodies are often spoken of as having cross reactivity. What this essentially means is that the antibodies (or a subpopulation of antibodies in a polyclonal antiserum) react with an antigenic epitope other than the one which induced their formation *(13)*. This situation is particularly apt to occur when the antibody preparation is heterologous, with many antibodies of differing combining site geometry (i.e., an antiserum). The cross-reactive antibodies identify and bind to antigenic sites structurally similar to the antigenic determinant used for immunization. Because these cross-reacting antigenic sites are only <u>similar</u> to the antigenic epitope the antibody was raised against, it is to be expected that there will be less complementarity between the antibody site and the similar antigen, and therefore the affinity constant will be lower than for the antibody-immunizing antigen interaction. (There is a theoretical possibility that antibodies with low affinity for the immunizing antigenic epitope may have higher affinity for other antigenic epitopes). Thus the specificity of an antibody preparation can be defined in terms of its affinity constant for the immunizing antigen, compared to its affinity constant for cross-reacting antigen epitopes. Although cross-reactivity is often thought of as a problem for polyclonal antisera, monoclonal antibodies can also demonstrate cross-reactivity. Also, polyclonal antisera can contain highly specific antibodies as well as subpopulations of less specific or cross-reactive antibodies.

Environmental Monitoring

Detection of the Carcinogen Benzo(a)pyrene (BP)

The instrument for BP detection consists of an optical fiber having antibodies immobilized at the sensor tip (Figure 2); excitation radiation from a He-Cd laser (325 nm) is sent through a beamsplitter onto the incidence end of the optical fiber *(14,15)*. The laser radiation is transmitted inside the fiber onto the sensor tip, where it excites the analyte molecules (e.g., BP) bound to the antibodies. The excited antigen fluorescence is collected and retransmitted back to the incidence end of the fiber, directed by the beamsplitter onto the entrance slit of a monochromator, and recorded by a photomultiplier. The intensity of the fluorescence signal measured is proportional to the amount of antigen bound to the sensor tip. In this study the measurement of BP molecules bound to the antibodies on the fiberoptics tip involved a three-step procedure: (1) the fiber-optics sensor tip was immersed into a 5-μL drop of sample solution containing BP; during the incubation time, set at 10 min, the BP molecules, which diffused towards the sensor tip, were bound to the antibodies immobilized on the sensor tip. (2) Following incubation the sensor tip was removed from the sample and rinsed with a phosphate-buffered saline (PBS) solution; this operation took about 10 s. (3) To conduct the measurement, laser excitation radiation was directed to the sensor tip by opening a shutter; the fluorescence from the BP molecules excited by the laser radiation was measured for a few seconds. The detection limit of the FIS device for BP is 1 femtomole.

Detection of Biomarkers of Environmental Exposure (DNA-Adducts)

With membrane-drum sensors, measurements can be obtained using a sequential and a stepwise procedure *(10,11)*. Sequential measurements were performed by filling the membrane sensor tip with antibody solution (typically 0.3 mg/mL) for each sample. After each measurement the sensor head was refilled with fresh antibody. Each FIS was incubated in 1-mL stirred antigen solution for a given time interval and rinsed in PBS solution for about 5 min. During this period, fluorescent antigen diffused across the membrane and was conjugated to its specific antibody. When the sensor was rinsed in PBS, unbound antigens and/or

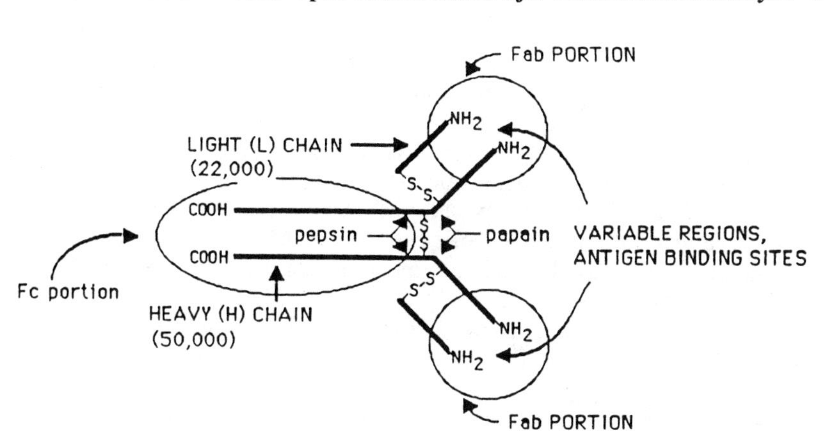

Figure 1: Structure of IgG Antibody

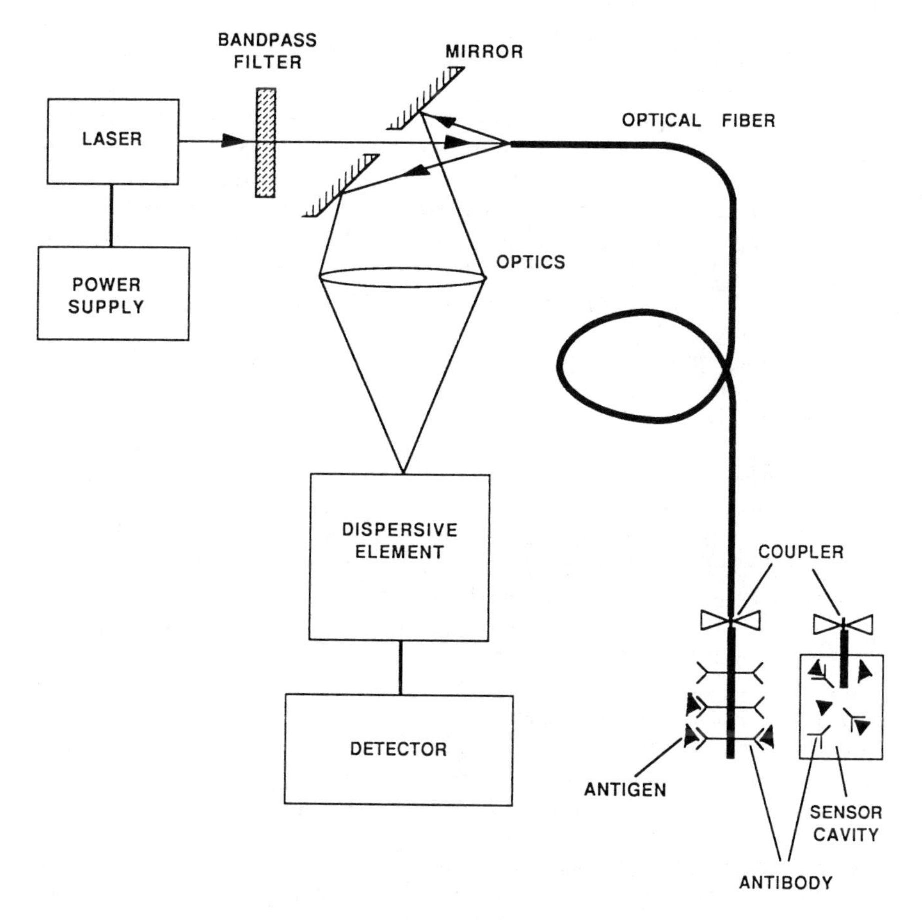

Figure 2: Fluoroimmunosensor (FIS) with a Sensing Tip Having Covalently Bound Antibody

interfering substances were dialyzed out of the sensing tip. Antibody-bound material remained, confined by the membrane to the fiber's viewing region. Signal was obtained either from the slope of the signal rise or from the difference between pre- and post-incubation signals in blank PBS solutions. Signal-to-noise (S/N) values were calculated using the peak-to-peak noise of blank PBS solutions. Since the lowest detectable antibody-bound benzopyrene-tetrol (BP) concentration visible to the fiber is 1×10^{-9} M, the limit of detection (LOD) using 15-min incubations is simply determined by 1×10^{-9} M/2 or $\sim 5 \times 10^{-10}$ M BPT. Lower LODs are attainable with longer incubations due to the fact that longer incubations allow the sensor to retain more Ab:BPT complex within its viewing region. The absolute LOD is 40 attomoles.

Microscale Regenerable Environmental Sensors

Figure 3 is a schematic diagram of the signal acquisition and sensor reagent delivery systems developed for direct fluoroimmunoassay measurements using a regenerable fiberoptic-based sensor. The signal acquisition system is shown in the upper portion of the figure and is used in the following manner to obtain a fluorescence signal for the analyte. A He-Cd laser (λ_{ex} = 325nm) was used as the excitation source. The beam was passed through a bandpass filter and then through a micro-hole in the center of the mirror. The beam was then directed on the distal end of an optical fiber which transmitted the light to the micro-chamber of the sensor. The resultant fluorescence of the analyte was detected and transmitted back along the optical fiber. The signal was then collimated by a lens and reflected off the mirror to a second lens. This lens focused the optical signal through a second filter (designed to minimize the noise due to scattering and reflection of the incident radiation) onto the distal end of the second fiber optic. The signal was then transmitted onto the photomultiplier tube (PMT). The photoelectrical signal of the PMT was fed to a picoammeter and stripchart recorder.

Sampling and quantitation of the analyte by direct fluoroimmunoassay using the regenerable sensor was accomplished by the following procedure. With all valves closed, the sample was initially drawn through the porous retaining material and into the appropriate column by mild aspiration. This lead was then closed and the excess sample was removed from the sensing microchamber via the outlets as the solution was passed through the microchamber from the inlet columns. This procedure allowed for only a given volume of sample to the retained in the aspiration column. All leads were then closed and a given volume of immunobeads were injected. The immunobeads used were prepared by immobilizing a specific antibody on 7 μm silica beads using previously described procedures *(12)*. Once the beads were in place, the sample was reintroduced into the microchamber with all columns closed in order to retain the solid-phase reagent. The sample was then followed by a rinse solution and the intrinsic fluorescence of the retained antigen was measured as described above. To regenerate the sensor, the entire reagent phase was flushed from the sensing microchamber via the outlet capillaries and the process repeated for the rest of the measurement cycle.

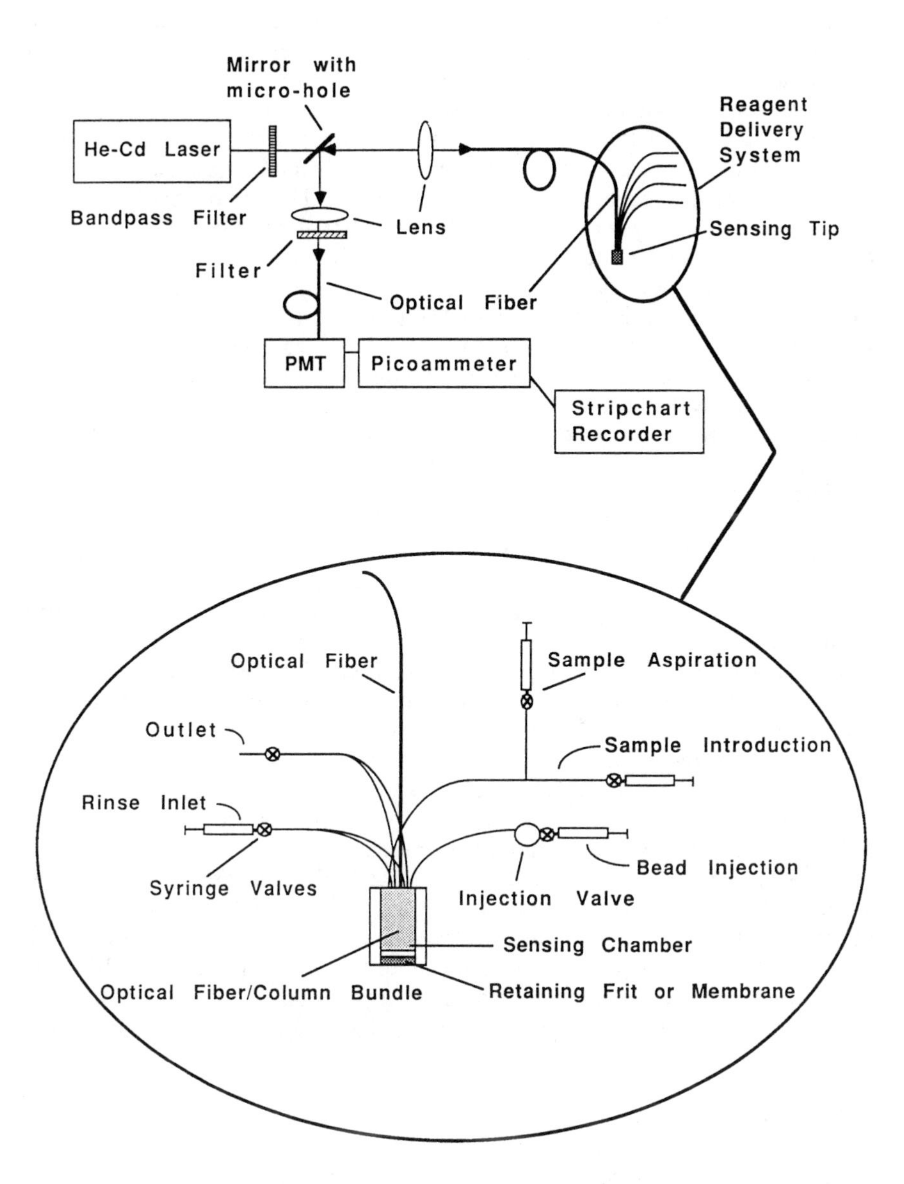

Figure 3: Microregenerable FIS Device

Measurements of Aflatoxins

Preliminary studies have been initiated to determine the feasibility of performing a direct fluoroimmunoassay with the microscale regenerable biosensor (MRB) for aflatoxins (AFT). These initial studies involve determination of the fluorescence signal enhancement of the AFT by a variety of solutions and the effects of addition of the antibody on the AFT fluorescence intensity. These solutions would be used as rinse solutions once the immobilized reagent (anti-aflatoxin) has collected the AFT from the sample. This is possible because of the unique delivery system employed by the MRB (see Figure 3). Figure 4 represents the results of such a study using $AFTB_1$ as an analyte.

The data was obtained using a spectrofluorimeter (Perkin-Elmer MPF43A). Initially, the solution of $AFTB_1$ (1 mL, 2.1×10^{-7}M) is placed in the solution of choice and the fluorescence intensity measured at 430 nm (AFT emission max) using two excitation wavelengths (325 nm and 370 nm). The 325 nm wavelength is the primary line of the He-Cd laser employed in our work and the 370 nm wavelength is the excitation maximum of $AFTB_1$. Once these measurements were made, the anti-$AFTB_1$ (4μL, 6.24 mg/mL) was added and the fluorescence signals again measured.

From Figure 4 one can see the relative fluorescence intensities are better for the bile salt (cholic acid, deoxycholic acid), sodium dodecylsulfate (SDS) micellar solutions and ethanol than for the phosphate buffered saline (PBS) solution. These solutions were initially chosen on the basis of their ability to enhance the fluorescence and secondly, for their interactions with the antibody protein molecule, i.e., denaturing properties thereby decreasing the ability of the antibody to bind the antigen. Cholic acid and deoxycholic acid should not greatly affect the antibody in terms of denaturing the molecule, whereas SDS solutions are know to denature protein molecules. Ethanol is also known to denature the antibody protein and disrupt the antibody antigen binding *(16)*. Preferably the solution chosen as the rinse solution in the actual analysis would help retain the $AFTB_1$ during the measurement step.

The figure also illustrates that the addition of antibody does not significantly decrease the fluorescence intensity of the $AFTB_1$. (The decrease observed is due to the slight dilution of the sample by the addition of antibody). This indicates when binding occurs, the fluorescence is not quenched as has been observed for other immunoassays such as for riboflavin (vitamin B_1) *(17)*. These preliminary results indicate the potential of the device for monitoring aflatoxins.

Conclusion

Immunosensors offer powerful tools for detecting chemicals and studying biological systems. Due to their high sensitivity, immunosensors are well suited to the analysis of trace contaminants in environmental samples. Sensitivity reported for fiberoptics immunosensors are in the 10^{-8} - 10^{-12} M range. In clinical chemistry, the volume of sample available is usually small and the presence of high concentrations of proteins require exquisite specificity only

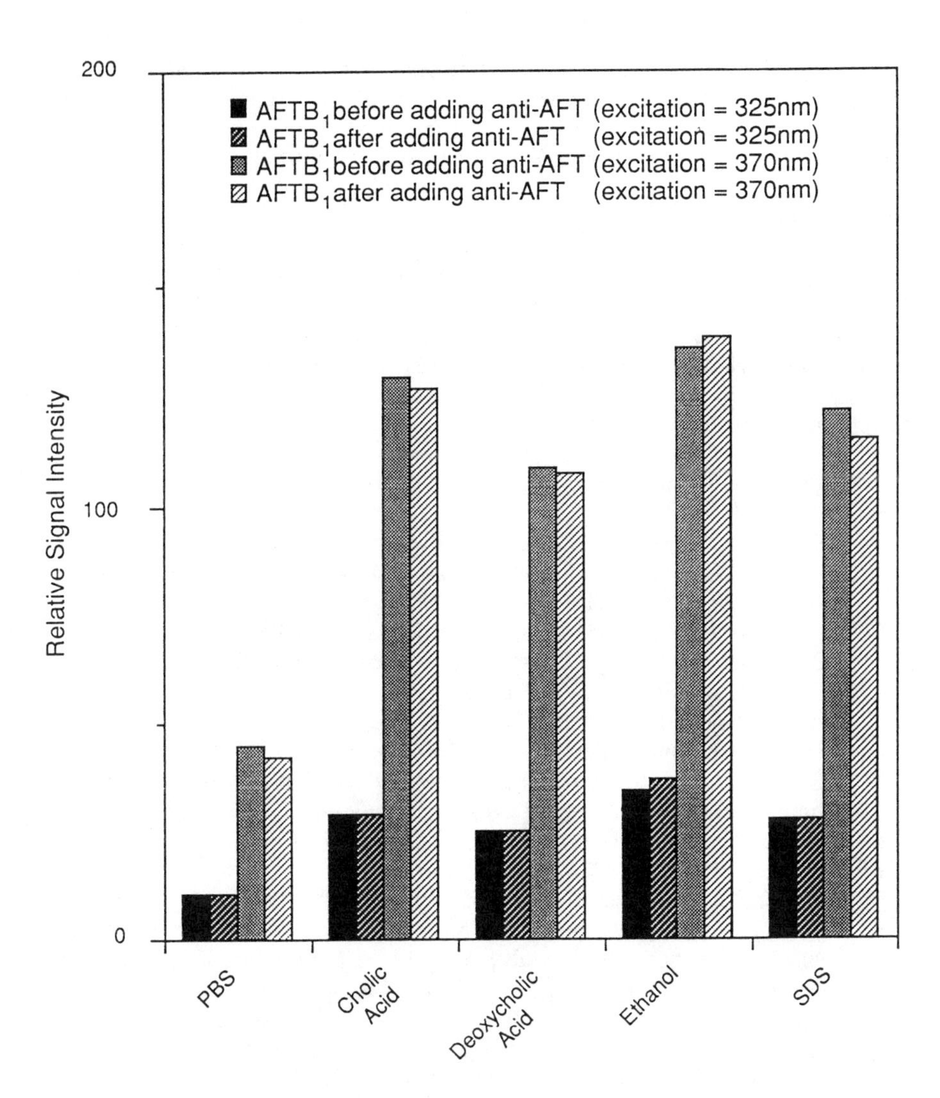

Figure 4: Measurements of Aflatoxins

available in immunological assays. Fiberoptics immunosensors can use very small amounts of sample ranging from 40 nL to a few mL to detect attomole amounts of toxic chemicals and related biomarkers for monitoring environmental exposure and health effects. These devices can also be used to monitor fermentation processes and antibody production in biotechnology-based industries. They can be used in agriculture and the food industry for measuring important chemical, physiological and biological parameters. Finally, immunosensors may be used to detect toxic chemicals in order to protect the health of workers and personnel living near toxic wastes.

Acknowledgments

This work was jointly sponsored by the National Institutes of Health under contract number GM 34730 and the Office of Health and Environmental Research, U. S. Department of Energy, under contract number DE-AC05-84OR21400 with Martin Marietta Energy Systems, Inc.

Literature Cited

1. Vo-Dinh, T., Ed., *Chemical Analysis of Polycyclic Aromatic Compounds*, Wiley, New York, 1990.
2. Beaver, R. W., Determination of Aflatoxins in Corn and Peanuts Using High Performance Liquid Chromatography, *Arch. Environ. Contam. Toxicol.* ***18*****, 1989 p. 315.**
3. Absolom, D. R. and van Oss, C. J., The Nature of the Antigen-Antibody Bond and the Factors Affecting its Association and Dissociation, *CRC Crit. Rev. Immunol.,* ***6*****(1), 1, 1986.**
4. Alzari, P. M., Lascombe, M. B. and Poljak, R. J., Three-Dimensional Structure of Antibodies, *Ann. Rev. Immunol.,* ***6*****, 555, 1988.**
5. Sutton, B. J., Antigen Recognition by B Cells: Antibody-Antigen Interactions at the Atomic Level, *Immunol. Suppl.,* ***1*****, 31, 1988.**
6. Köhler, G., Derivation and Diversification of Monoclonal Antibodies, *Science,* ***233*****, 1281, 1986.**
7. Tijssen, P., Practice and Theory of Enzyme Immunoassays, in Laboratory Techniques in Biochemistry and Molecular Biology Burdon, R. H. and van Knippenberg, P. H. eds., Vol. 15 Elsevier Science Publishers, Amsterdam, Netherlands, 1985, Chaps. 4-8, pp. 39-150.
8. Morrison, S. L., Canfield, S., Porter, S., Tan, L. K., Tao, M.-h., and Wims, L. A., Production and Characterization of Genetically Engineered Antibody Molecules, *Clin. Chem.,* ***34*****(9), 1668, 1988.**
9. DePinho, R. A., Feldman, L. B. and Scharff, M. D., Tailor-Made Monoclonal Antibodies, *Annals Internal Med.,* ***104*****, 225, 1986.**
10. Tromberg, B. J., Sepaniak, M. J., Alarie, J. P., Vo-Dinh, T. and Santella, R. M., Development of Antibody-Based Fiber-Optic Sensors for Detection of a Benzo[a]pyrene Metabolite, *Anal. Chem.,* ***60*****, 1901, 1988.**

11. Vo-Dinh, T., Tromberg, B. J., Sepaniak, M. J., Griffin, G. D., Ambrose, K. R., and Santella, R. M., Immunofluorescence Detection for Fiberoptics Chemical and Biological Sensors, in Fluorescence Detection II, Menzel, E. R.; Ed., Proceedings of SPIE, Bellingham, Washington, Vol. 910, pp. 87-94, 1988.
12. Alarie, J. P., Sepaniak, M. J. and Vo-Dinh, T., Evaluation of Antibody Immobilization Techniques for Fiber Optic-Based Fluorimmunosensing, *Anal. Chim. Acta, 229*, **1990.**
13. Fudenberg, H. H., Stites, D. P., Caldwell, J. L., and Wells, J. V., Basic and Clinical Immunology Lange Medical Publications, Los Altos, CA, 1976.
14. Vo-Dinh, T., Tromberg, B. J., Griffin, G. D., Ambrose, K. R., Sepaniak, M. J., and Gardenhire, E. M., Antibody-based Fiberoptics Biosensor for the Carcinogen Benzo(a)pyrene, *Appl. Spectrosc., 41*, **735, 1987.**
15. Tromberg, B. J., Sepaniak, M. J., Vo-Dinh, T. and Griffin, G. D., Development of Fiberoptics Chemical Sensors, in Optical Fibers in Medicine III, Katzir, A., Ed., SPIE Publishing, Bellingham, Washington, 1988.
16. Cantor, C., Schimmel, P., eds., Biophysical Chemistry Part II: Techniques for the Study of Biological Structure and Function, W. H. Freeman and Company, San Francisco, 1980, p. 680.
17. Wassink, J. H., Mayhew, S., *Anal. Biochem, 68*, **1975, p. 609.**

RECEIVED May 12, 1992

Chapter 23

Sonolysis Transformation of 1,1,1-Trichloroethane in Water and Its Process Analyses

Madeline S. Toy[1], Roger S. Stringham[1], and Thomas O. Passell[2]

[1]Balazs Analytical Laboratory, 1380 Borregas Avenue, Sunnyvale, CA 94089
[2]Electric Power Research Institute, 3412 Hillview Avenue, Palo Alto, CA 94303

The sonolysis transformation of 1,1,1-trichloroethane in water at ambient temperature can be a potential digestion process for CCl_3CH_3. As the volume capacity of the same concentration of CCl_3CH_3 increases, the sonolysis digestion efficiency decreases. The increase of sonolysis transformation increases with sonolysis time. The compound decomposes into gases, volatile organic compounds and ionic species. The ionic species are removable by ion-exchange resins, while the gases and volatiles are easily degassed under sonolysis.

The aqueous digestion of 1,1,1-trichloroethane was studied under sonolysis at ambient temperature. In general the chemical degradation of simple haloalkane under sonolysis is the carbon-halogen bond cleavage in liberating atomic halogen (*1*). The nonaqueous sonochemical decomposition of carbon tetrachloride was reported to form products (Cl_2 and C_2Cl_6) due to the sonolysis effect (*2*). When the same sonochemical decomposition happens in water, the major changes turn to the reactions, which are caused by the degraded water molecule (HO• + H•) and form a number of products (Cl_2, CO_2, HCl, C_2Cl_6, and HOCl) (*3-5*). The nonaqueous sonolysis of chloroform gives a large number of products (notably HCl, and in decreasing amounts of CCl_4, CH_2Cl_2, C_2Cl_5H, C_2Cl_4, C_2Cl_6, and C_2HCl_3), arising from both radical and carbene intermediates (*6*). Recently the extent of sonolysis digestion of CCl_3CH_3 in water was reported to grossly increase in the solution form than in the forms of immiscible pairs (i.e., above the solubility limit of CCl_3CH_3 in H_2O). This phenomenon agrees and confirms that the decreasing vapor pressure of the liquid (e.g., the solution form in this case) increases the oxidative cavitation collapse in decomposing CCl_3CH_3 (*7*). This paper continues to study the sonolysis transformation of aqueous CCl_3CH_3 solution and its process analyses.

0097–6156/92/0508–0284$06.00/0

Experimental

1,1,1-Trichloroethane (99+%) was purchased from Aldrich, confirmed by gas chromatograph and used as received. The commercial distilled water was purified by a deionized unit, whose charcoal filter column (P/N AC5-10-5 micron 10" activated filter), two nuclear grade mixed bed resin columns (P/N DIMB2002 mixed bed 20" ion exchange column, 2 pk) and a particle filter column (P/N CWCFF1492, 0.2 micron capsule filter) were purchased from Continental Water Systems. The plastic water supply tank and a circulation pump were purchased from Cole-Palmer. The unit provided the deionized water at 18 megohm-cm.

The sonicator (Model W-385) was obtained from Heat System Ultrasonics. This ultrasonic liquid processor with 1/2" tapped titanium disruptor horn provided 475 watts output power and 20 kHz (20,000 cycles per second) frequency. The horn was immersed in the aqueous CCl_3CH_3 solution under atmospheric pressure.

The decreasing concentration of CCl_3CH_3 content from the starting solution was measured by EPA Purgeable Halocarbon Method 601 (*8*) on a Hewlett Packard Model 5890 gas chromatograph (GC). A portion of each sample was purged for 10 minutes at a rate of 25 ml per minute in a Tekmar Model LSC-2 liquid sample concentrator. The purged gases were trapped, concentrated, and automatically desorbed onto the GC (*9*). Sample separation was achieved on a packed column of 1% SP-1000 on Carbopack B in a 8' x 1/8" ID stainless steel column with helium as carrier gas. Each eluted component was detected by an electrolytic conductivity detector (ELCD or Hall detector) and the output recorded on a digital plotter/recorder.

The production of chloride anion was determined by ion chromatograph (Dionex 4000i). At the end of sonolysis, the digested aqueous solution was passed through two mixed bed ion exchange columns. The resistance of the inlet solution and the column effluent was measured by a conductivity meter (Continental Water Systems Series 900-Millipore). The CCl_3CH_3 contents of the inlet and column effluent were measured by purge-and-trap GC.

Results and Discussion

As the volume capacity of the same concentration of 1,1,1-trichloroethane increases, the sonolysis digestion efficiency decreases. Table I shows the digestion efficiency of CCl_3CH_3 at 10 ppm based on the production of the total available chloride anion at 8 ppm (Table I, Footnote b). When the volume increases 100X, the digestion process efficiency decreases 5X (i.e. 4% Cl^- produced at 500 ml in Table I divided by 0.8% Cl^- produced at 50,000 ml or 13.21 gallons in Figure 1).

Table II shows the increase of sonolysis transformation with an increase of sonolysis time. At 18 minutes, the percent of chloride anion produced increases 3.1%, whereas the disappearance of CCl_3CH_3 is estimated at 91.4% (i.e., $100\% - \frac{91 \text{ ppb} \times 100\%}{1060 \text{ ppb}} = 100\% - 8.6\% = 91.4$). The data indicate that

Table I. Sonolysis Digestion Efficiency of Aqueous 1,1,1-Trichloroethane (10 ppm) for 18 Minutes Sonolysis Time

Solution Volume	Rel. Ratio	Conc. of Chloride Anion,[a] ppm	Percent Chloride Produced[b]
500 ml	1	0.324	4
7570 ml (2 gal.)	15.1	0.148	1.8
56775 ml (15 gal.)	113.6	0.55	0.7

[a]Measured by ion chromatograph (Dionex 4000i) with reference standard of Cl^- at 300 ppb.

[b]$$\%Cl^- \text{ produced} = \frac{\text{Conc. } Cl^- \text{ ppm} \times 100\%}{10 \text{ ppm} \times \frac{3 \times 35.45}{133.32}} = \frac{\text{Conc. } Cl^- \text{ ppm} \times 100\%}{8 \text{ ppm}}$$

Table II. Sonolysis Effect on 2-Gallon Volume of 1,1,1,-Trichloroethane (1.06 ppm) in Water

Conc. of CCl_3CH_3,[a] ppb	Sonolysis Time min	Corrected Conc. Chloride Anion,[b] ppb	Percent Chloride Anion Produced[c]
1060	0	0.5	0.06
--	8	13.0	1.53
91	18	26.2	3.10
42[d] (133[e])	38	41.4	4.89[d] (3.0[e])

[a]Measured by EPA purgeable halocarbon method 601.
[b]Measured by ion chromatograph (Dionex 4000i) with reference standard of Cl^- at 300 ppb.
[c]$$\%Cl^- \text{ produced} = \frac{\text{Conc. } Cl^- \text{ ppb} \times 100\%}{1060 \text{ ppb} \times \frac{3 \times 35.45}{133.32}} = \frac{\text{Conc. } Cl^- \text{ ppb} \times 100\%}{846 \text{ ppb}}$$

[d]Resistance of the solution was measured at 0.7 megohm-cm, which was increased to 1.5 megohm-cm after passing through the two mixed-bed ion exchange columns.
[e]Another similar run as d, resistance of the solution increased from 0.8 to 2.7 megohm-cm.

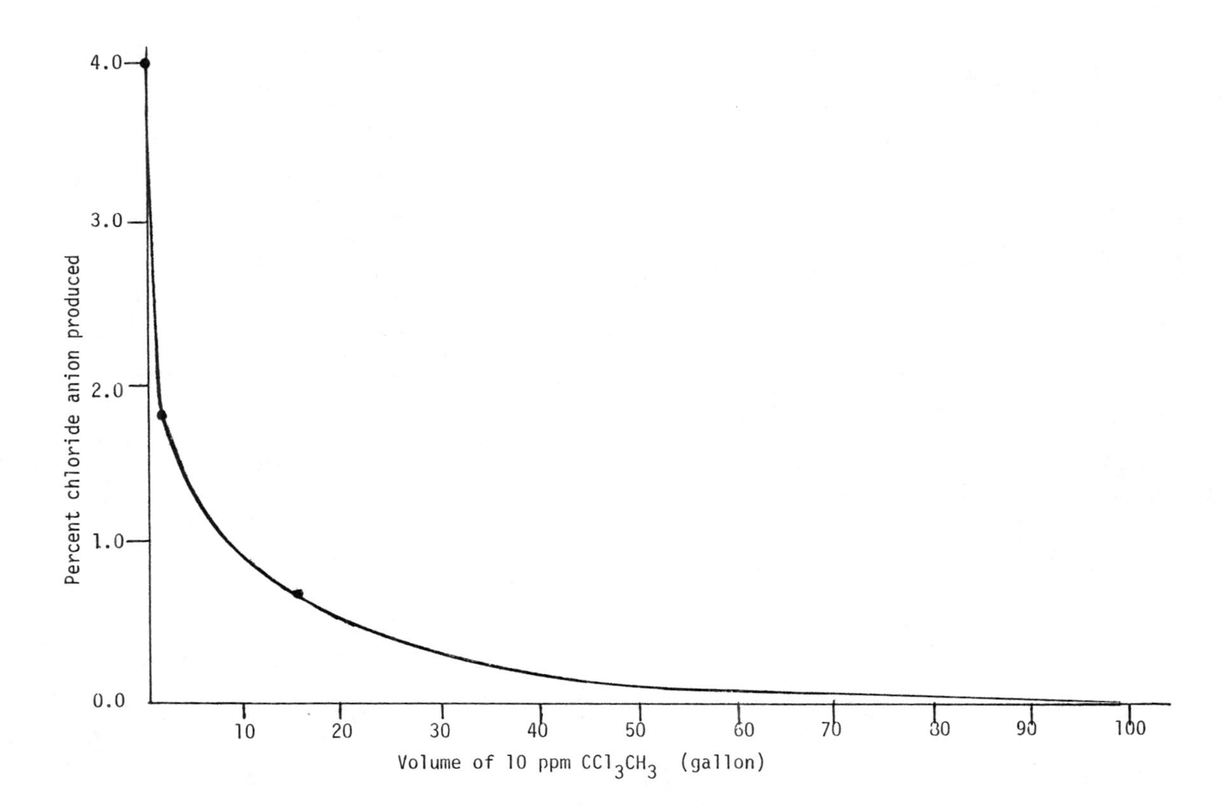

Figure 1. Percent chloride anion produced versus volume of aqueous CCl_3CH_3 at 10 ppm for 18 minutes sonolysis time

a significant amount of the sonolysis digestion products are present besides the chloride anions. They may be ionic as well as gases and volatile covalent compounds. The latter two are easily degassed under sonolysis. At the end of 38 minutes in Table II, the digested aqueous solution was passed through the two mixed-bed ion exchange columns. An increase of resistance from the inlet at 0.7 to 1.5 megohm-cm in the column effluent was measured, while the CCl_3CH_3 content simultaneously decreased from 42 to 3 ppb. Another repeat run was passed through the same mixed-bed ion exchange columns with the resistance increased from 0.8 to 2.7 megohm-cm, while the CCl_3CH_3 content decreased from 133 to 36 ppb from the inlet to the column effluent. The conductivity decrease (or the resistance increase) data support the production of ionic species at ambient temperature. The decrease of CCl_3CH_3 content at the inlet comparing to the column effluent suggests adsorptions also occurring between the mixed-bed resins and CCl_3CH_3.

In conclusion, the sonolysis transformation of CCl_3CH_3 in water at ambient temperature can be a potential digestion process for CCl_3CH_3, which decomposes into gases and volatile covalent compounds and to ionic species, which are well known as removable by ion exchange resins. The gases and volatiles are easily degassed under sonolysis.

Acknowledgment

This work was supported by Electric Power Research Institute under Contract 2997-4. The authors thank Mr. Desmond Kerr for the sample analyses by EPA Purgeable Halocarbon Method 601.

Literature Cited

1. Suslick, K.S. Ultrasound: Its Chemical, Physical, and Biological Effects, VCR Publ., New York, N.Y., 1988, pp 147-150.
2. Weissler, A.; Pecht, I.; Anbar, M. Science, 150, pp 1288-1289 (1965).
3. Mason, T.J.; Lorimer, J.P. Sonochemistry: Theory, Applications and Uses of Ultrasound in Chemistry, John Wiley, New York, N.Y., 1988, pp 72-74.
4. Jennings, B.H.; Townsend, S.N. J. Phys. Chem., 65, pp 1574-1579 (1961).
5. Chendke, P.K.; Fogler, H.S. J. Phys. Chem., 87, pp 1362-1369 (1983).
6. Henglein, A.; Fischer, C.H. Ber. Bunsen-Ges. Phys. Chem., 88, pp 1196-1199 (1984).
7. Toy, M.S.; Carter, M.K.; Passell, T.O. Environ. Tech., 11, pp 837-842 (1990).
8. EPA Test Methods: Methods for Organic Chemical Analysis of Municipal and Industrial Wastewater; Longbottom, J.E.; Lichtenberg, J.J. Ed., EPA-60014-79-020 (Revised), U.S. Environmental Protection Agency, Environmental Monitoring and Support Laboratory, Cincinnati, OH 45268, 1982, 601-1 to 610-10.
9. Grote, J.O.; Westendorf, R.G. Am. Lab., Dec. 1979.

RECEIVED May 18, 1992

Chapter 24

Dependence of Fluoride Accumulations on Aluminum Smelter Emissions

A Combined Fuzzy c-Varieties–Principal Component Regression Analysis

R. W. Gunderson[1] and K. E. Thrane[2,3]

[1]Department of Electrical Engineering, Utah State University, Logan, UT 84322–4120

[2]A/S Miljøplan, Kjørbøveien 23, N–1300, Sandvika, Norway

Samples of grass from locations in Sunndal, Norway, have been collected regularly by technicians of Hydro Aluminum A/S for a period of nearly thirty years. The investigation reported in this paper combined fluoride measurement data from those samples with meteorological information in order to predict the fluoride content in grass grown in the vicinity of the aluminum plant, as a function of daily emissions at the plant site. The methodology made use of the Fuzzy c-Varieties (FCV) family of unsupervised classification algorithms to identify a family of meteorological-dependent fluoride accumulation classes, followed by class-wise application of the Principal Component Regression (PCR) method. Substantial improvement in prediction accuracies were achieved with respect to simply basing the regression on a single, one-class, total data model. This model was used to estimate the extent to which fluoride accumulation levels would be reduced by modernization of plant facilities and introduction of new technologies.

The injurious effects of fluoride accumulations upon plant life and foraging animals has been the subject of a great many investigations, including one recorded nearly 1000 years ago in the Icelandic literature (1). It is well known that relatively large amounts of fluorides are emitted into the atmosphere during the production of aluminum, and that these fluorides are transported to the surrounding regions, where they are deposited and accumulate in the local vegetation. This paper focuses

[3]Current address: M.I.L., Sigurds Syrs Gate 4, N–0273 Oslo 2, Norway

0097–6156/92/0508–0289$06.00/0

upon fluoride emissions from a particular aluminum factory, Sunndal Works, located in the city of Sunndalsøra, at the end of Sunndals Fjord in Western Norway. The problem of fluoride emission is intensified at Sunndalsøra by the unique topography of the Norwegian fjord country. The aluminum plant is located at the end of the fjord and the mouth of the river Driva (see Figure 1), with the valley of Sunndal continuing into the countryside as a spectacularly deep and narrow extension of the fjord. A few equally spectacular side valleys split off from the main valley as it makes its way into the interior. Depending upon meteorological conditions, these valleys create a complicated channelizing of factory emissions which is extremely difficult to model from the standpoint of theoretical atmospheric transport dynamics. A detailed description of the area and weather conditions is given in the report by Thrane (2).

The objective of the investigation was to use the methodology of unsupervised pattern classification to model the dependence of fluoride accumulations in the grassy areas of the valley, as a function of fluoride emissions from the smelter and the meteorological variables believed to play an influential role in fluoride buildup.

The current project is a continuation of several previous studies, including an examination of relevant meteorological and topographical factors (3), and two previous modeling investigations (4,5). The investigation by Thrane was an attempt to model maximal expected fluoride concentrations as a function of fluoride emissions and precipitation variables only. The report by Gunderson and Thrane concerned a preliminary analysis of the data used for the current study.

Data Description

Fluoride Accumulations. Samples of grass have been collected by company technicians for a period of about thirty years. Two independent programs have been conducted. The first provides data required by a monitoring agency of the Norwegian government. These data consist of fluoride concentrations found by analysis of grass samples, collected once each summer month from 28 grazing sites situated in different directions relative to location of the aluminum plant. The report by Thrane and Aldrin (3) provides a detailed description of this measurement program and lists fluoride sample analysis results for the period 1966-1985.

The aluminum company itself sponsors the second program. As opposed to the first program, grass samples are taken every tenth day from eight different sites. All of the sites are up the valley from the installation, where the accumulation of fluorides in plant growth is expected to be higher. Thrane and Aldrin provide fluoride analysis results from these tests for the period 1961-1985 (3).

Sampling for both programs is carried out only during the growing season, roughly from May through September. Because there are fewer sites, samples for the second program can be collected at all of the sites on the same day, rather than the two to three days required for the first program. For this reason, and because of the higher monthly sampling rate, it was decided to use the fluoride concentration records of the second program for this investigation. Locations of the eight sites are shown on the map of Figure 1. Site #1 is at the gate to the plant. Driving distance from the first to the last site (#8) is about 34 km.

Fluoride Emissions. Hydro Aluminum was able to provide monthly averages for gaseous, particulate, and total fluorides emitted from the production facility. The estimates are based upon thrice-monthly measurements at various sites within the facility. Results of the preliminary investigation by Gunderson and Thrane (5) were based partially upon the emission figures provided by the company and listed by Thrane and Aldrin in their report (3). It was later discovered that these data needed to be corrected, for technical reasons. Corrected values are given in Thrane (4), and are the emission averages used in this study.

Meteorological Variables. Hydro Aluminum maintains a twenty-four hour meteorological station at the entrance to the production facility. Records are kept of daily rainfall totals, plus hourly measurements of wind direction and wind force. Wind directions are recorded by dividing the circumference into thirty-six, ten degree, sectors; with the 0-sector indicating a wind from the north, and the 9-sector a wind from the east. Some of these data, including precipitation amounts, are listed for the period 1966 through 1985 by Thrane and Aldrin (3). Hourly wind measurement data was provided to the investigators directly from company records.

Summer winds in Sunndal are generally directed up the valley in the daytime, and down the valley and out over the fjord in the evening. This results in the plant emissions being funneled up the valley during a part of the twenty four hour day, and then back over the valley at night. Precipitation tends to wash deposits of fluorides off the leafy surfaces. This effects the manner in which gaseous fluorides enter the plant system, and the amounts of fluoride particulate left on leaves. Therefore it is important to take rainfall totals into account. Årflot (6) provides a good technical discussion of the processes through which gaseous fluorides are able to enter into the plant leaf.

Methodology

Data Preparation. After extensive preliminary analysis, including that reported in Gunderson and Thrane (5), an input data set **X** was constructed which consisted of 72 data vectors, each corresponding to a measurement day from the period 1978 through 1985. Data vectors with missing measurements were not allowed. In all, the data **X** consisted of n = 72 rows, corresponding to the data vectors, and d = 6 columns, corresponding to the six selected measurement, or feature, variables shown in Table I.

The first feature was obtained by summing the fluoride concentrations measured in leaf samples collected at the eight sites on a particular day, then normalizing the sum to the average fluoride emission (kg/hr) over the reporting month which contained that sampling day. A "reporting month" spanned the period from the 25th of one month to the 25th of the next. Since none of the samples were taken any later than the 25th day of the month, the five-day period consisting of the sampling day and the four days preceding it always fell into the same reporting month. Measurement values for the remaining five, meteorological, features were all computed over these five day periods.

The second feature corresponds to the total five-day rainfall amount recorded at the meteorological station. The third and fifth features correspond to the percentage of hourly measurements during the five days when the wind was blowing up the valley, and the average wind force during those hours, respectively.The fourth and sixth features correspond to a wind direction down the valley, and out into the fjord. Winds in Sunndal are assumed to be "up" the valley when coming out of sectors 21 through 36 at the observation station. A "down" valley wind corresponds to winds coming out of the remaining sectors (1-20).

FCV Disjoint Principal Component Modeling. There are numerous sources available which explain the ideas and algorithms of FCV modeling in detail (for example; Bezdek et. al. (7), Gunderson and Jacobsen (8), Gunderson and Thrane (9)). We shall therefore only provide a brief discussion of the c-means form of the algorithms used in this investigation.

The FCV algorithms are initiated by specifying the number the number of classes, c, which are assumed to be modeled by the data **X**, and a first guess at centers $\mathbf{v}_1$, $\mathbf{v}_2$, .. $\mathbf{v}_c$ for those classes as viewed in data space $\mathbf{R}_d$. The starting guess is used to iteratively solve the two nonlinear equations

$$i) \quad u_{ik} = \frac{1}{\sum_{j=1,c}(\|\mathbf{x}_i-\mathbf{v}_k\|/\|\mathbf{x}_i-\mathbf{v}_j\|)^2}$$

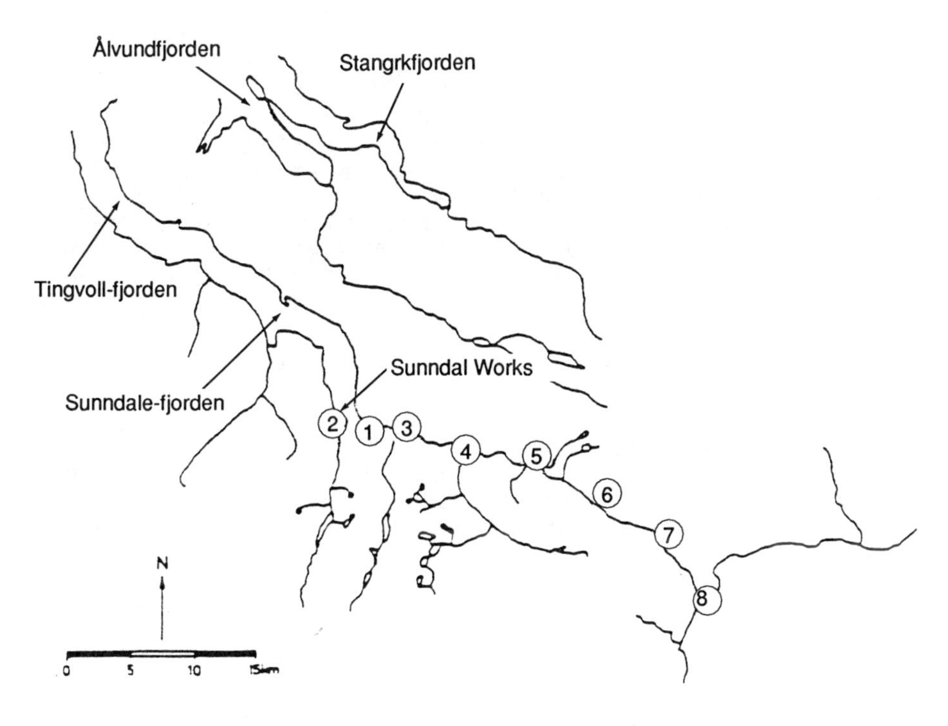

Figure 1. Map illustrating the locations of the eight stations along the valley Sunndal and the Sunndal Works.

Table I. Feature (measurement) definitions

Variable	Description
y - response	normalized sum of 8-site F accumulations
x_1	5-day precipitation total
x_2	% up-valley wind
x_3	% down-valley wind
x_4	5-day average up-valley wind velocity
x_5	5-day average down-valley wind velocity

and

$$ii)\quad \mathbf{v}_i = \frac{\sum_{k=1,n}(u_{ik})^2\mathbf{x}_{ik}}{\sum_{k=1,n}(u_{ik})^2}$$

over all the classes i=1,2,..,c and data vectors k=1,2,..n. The iterative solution is stopped when the maximal change in the coefficients u_{ik} at the end of any iteration is less than a pre-specified threshold value. The final u_{ik} values from i) are then used, together with the final $\mathbf{v}_i$ vectors from ii), to compute a weighted scatter matrix

$$iii)\quad S_i = \sum_{k=1,n}(u_{ik})^2(\mathbf{x}_k-\mathbf{v}_i)(\mathbf{x}_k-\mathbf{v}_i)^T$$

for each class i=1,2,..c.

The weights u_{ik} measure the extent to which data vectors $\mathbf{x}_k$ influence computations for the ith class model. Because they are constrained to satisfy the inequalities

$$iv)\quad u_{ik} \in [0,1]$$

and

$$v)\quad \sum_{i=1,c}u_{ik} = 1$$

for all i=1,2,..c and k=1,2,..n, they can also be interpreted as defining the "fuzzy" membership of the data vectors in c "fuzzy" sets (7). Rather than thinking of the classes as being defined by these c "fuzzy" sets, however, we prefer to view each class as being <u>modeled</u> by one of the c vectors vi obtained from equation ii). It is clear from equation i) that a weight u_{ik} will take on the value unity if, and only if, the vector $\mathbf{x}_k$ exactly coincides with a vector $\mathbf{v}_i$ for some i and k, and, for that k, all of the remaining u_{jk} (j=i) will be zero. Thus, each of the vectors $\mathbf{v}_i$ (i=1,2,..,c) can be considered to be a "prototypical", or canonical, representative of its class (the best apple in the barrel, so to speak). Values of weighting coefficients lying between 0 and 1 can then be interpreted as measuring the extent of similarity a vector $\mathbf{x}_k$, associated with a particular coefficient u_{ik}, exhibits relative to each of the class prototypical vectors $\mathbf{v}_i$. Computation of the eigenvalues and eigenvectors for the c scatter matrices of equation iii) provides additional geometric information, relative to the orientation (eigenvectors) and shapes (eigenvalues) of the classes as viewed in d-dimensional data space. Similarly, the vectors $\mathbf{v}_i$ can be thought of as locating the position of each class in data space, by virtue of defining the class centers.

The FCV acronym for the algorithms is derived from their description in (7) as the "fuzzy c-varieties" clustering algorithms. (The "c-varieties" refers to the algebraic linear varieties defined by the eigenvectors of the weighted scatter matrices S_i.) In the case of the c-means form of the algorithms used in this paper, the dimension of the linear varieties is taken to be the same as that of the data space $\mathbf{R}_d$. Although several figures of merit exist for evaluating whether this assumption is a good one, none of them turn out to be entirely satisfactory. The authors have found that the false-color data imaging technique described in Gunderson and Thrane (9) is generally a reliable approach to the problem, and it was therefore employed in this investigation.

Principal Component Regression. Use of principal component techniques for purposes of multilinear regression (PCR) prediction purposes is very well known (cf.(10), for example) and further elaboration is not required here. The pseudo-inverse function of the program MATLAB was used to implement the computations of PCR. MATLAB employs the numerically accurate and reliable technique of singular value decomposition to compute the pseudo-inverse. Before application of PCR, each input data set was first normalized by subtracting the mean of a column from each of the columns elements, and than dividing the column by its standard deviation.

Data Analysis

Class Structure. The first step in the data analysis was to employ the FCV algorithms to establish the existence, if any, of a class structure for the data $\mathbf{X}$. Figure 2 is the appearance of the data when projected from d=6 dimensional data space $\mathbf{R}_d$ onto the 3-dimensional subspace of $\mathbf{R}_d$ formed by the first three principal component vectors of the $\mathbf{X}$. This figure suggests that the data suffers badly from the standpoint of class-separation. For this reason the FCV algorithms, which were written specifically for cases such as this, were selected as the unsupervised classification methodology.

Because of the known influence of weather conditions upon accumulation of fluoride concentrations, the principal criteria used in establishing a valid class structure was that it should reflect the dependence of the normalized concentration values of measurement #1 on the five remaining meteorological variables. The desired dependency began to clearly emerge with the assumptions of c = 4 classes, and led to a final selection of c=6 class models. The assumption of c > 6 models provided only minor subpartitions of the classes obtained under the assumption c = 6.

Figures 3 through 6 provide bar graphs of the centers of four out of the six classes. Figure 3 is interesting because of the relatively low level of normalized fluoride concentrations, feature #1, and an unusually high level of precipitation, the second feature. This correspondence is in good agreement with field experience and observations, since, as mentioned earlier in this paper, precipitation tends to wash particulates from the leaves onto the ground, where they are less likely to be absorbed into the plant system (Årflot, 1981). Figure 4 was included because it is typical of classes with lower normalized fluoride concentrations. In the case of this class, the lower concentrations were explained by average precipitation amounts, with a significantly higher percentage of time when the wind was blowing down the valley. This dependency again agrees with field observations. Figures 5 and 6 correspond to the classes of greatest interest to this investigation, since they show how the meteorological variables can combine to create conditions most favorable for accumulation of higher concentrations of fluorides, and the levels to be expected under those conditions. Figure 5 shows the accumulations expected with wind conditions opposite to those of Figure 4; that is, average precipitation with winds this time predominantly up the valley. Figure 6 is interesting because neither wind direction appears to be dominant. This is often the case if the winds are either calm or shifting. The conditions and accumulation levels of Figures 5 and 6 also conform to field experience.

Figure 7 shows the result of projecting only those data vectors whose maximum influence occurs in the formation of the classes corresponding to Figures 4 and 5, the down-wind and up-wind classes described above. The two classes shown can be seen to exhibit good separation, further validating the FCV unsupervised classification results. Figure 8 shows the centers of all six classes as projected onto the same 3-dimensional subspace of $\mathbf{R}_d$.

Multilinear Regression Models. The objective of partitioning the data **X** into class models was to be able to improve prediction accuracies through replacing the single regression equation obtainable from the total data by a family of regression equations, one holding for each class within **X**. Data for calibrating a class-defined regression equation can be easily obtained from the FCV principal component model for the class by assigning a data vector to the training set for that class only if its influence in forming the model for the class (i.e. class membership) is greater than a pre-specified threshold value. A threshold of 0.40 was used in this study.

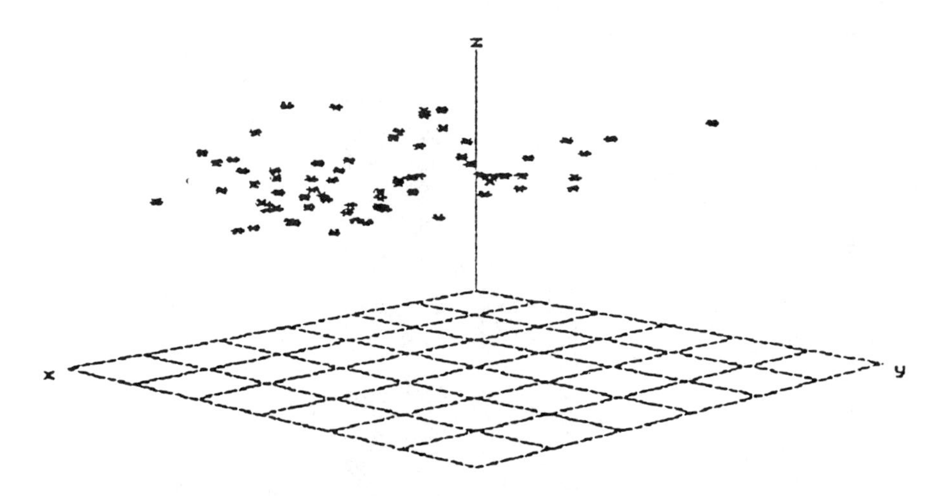

Figure 2. Data vectors of X projected onto 3-dimensional principal component subspace.

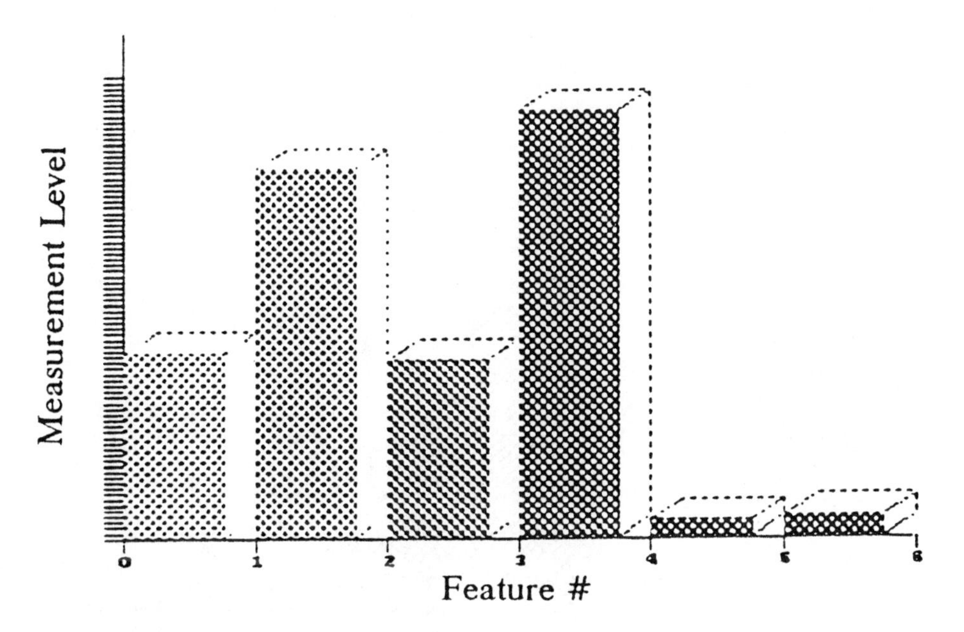

Figure 3. Bar graph for class #1. Class assignment by maximum membership.

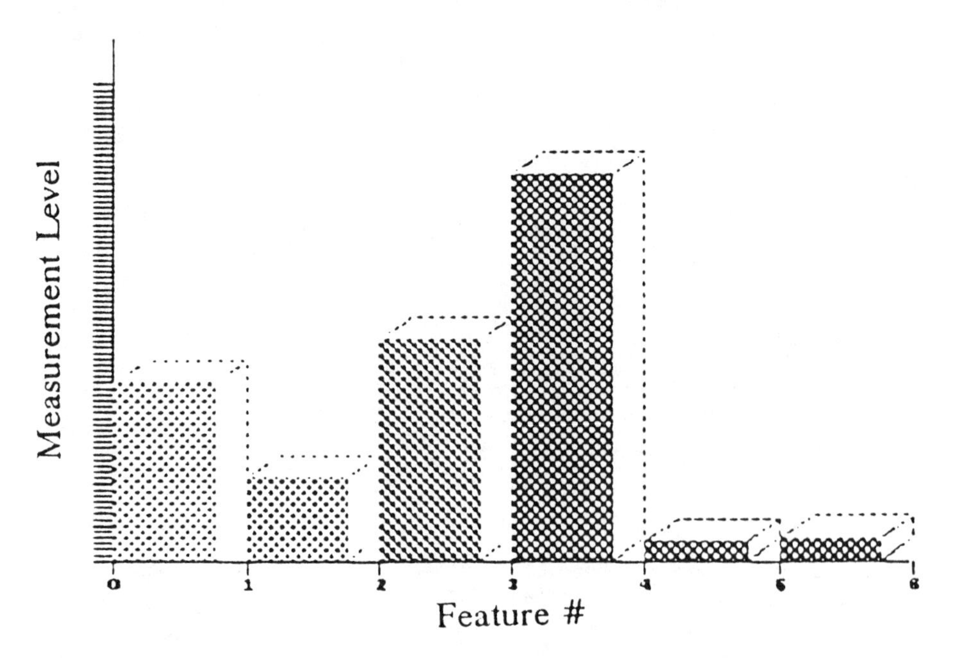

Figure 4. Bar graph for class #3. Class assignment by maximum membership.

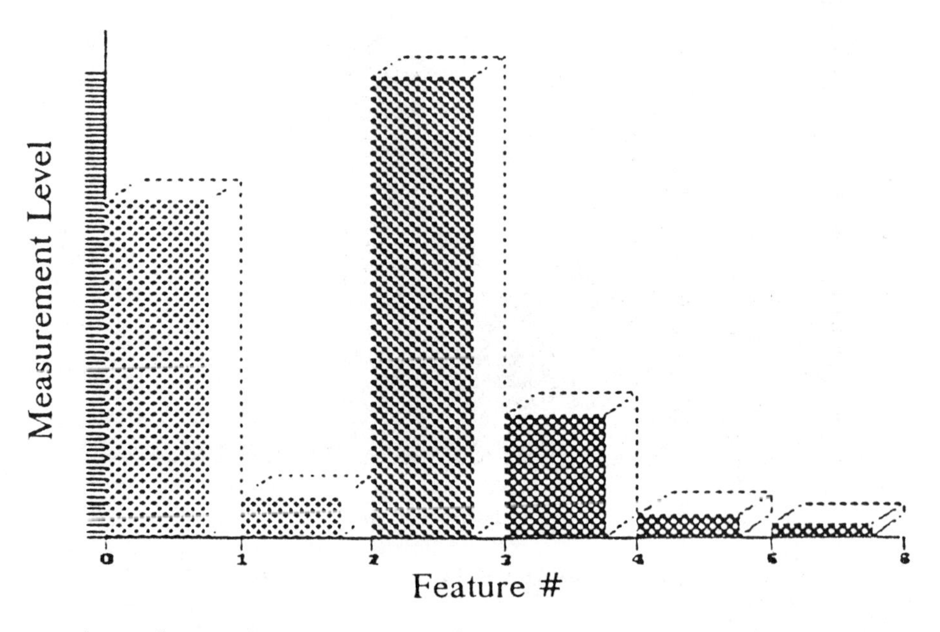

Figure 5. Bar graph for class #2. Class membership by maximum membership.

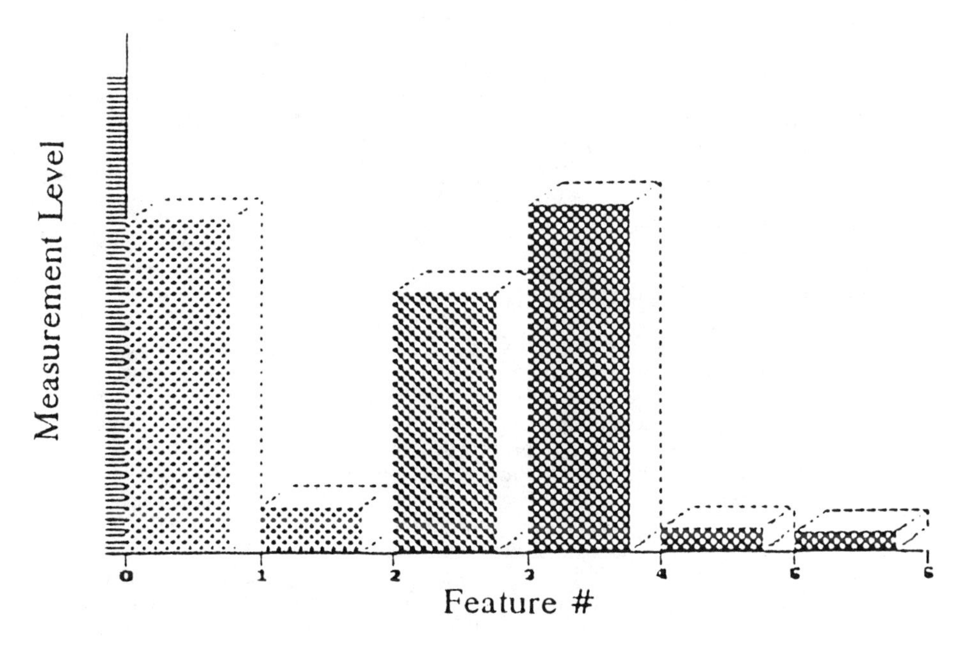

Figure 6. Bar graph for class #5. Class assignment by maximum membership

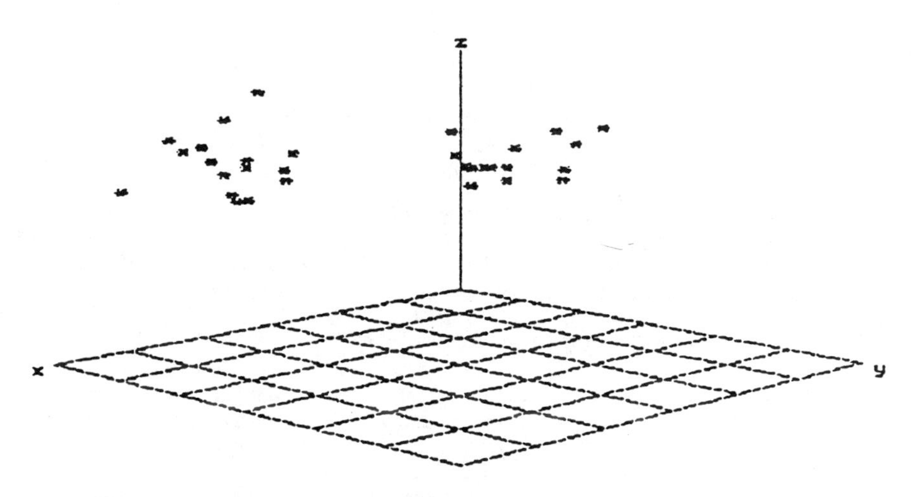

Figure 7. Projection of classes #1, #2, #3 onto 3-dimensional principal component subspace.

Table II summarizes the regression equation constant terms and coefficients obtained for several cases of interest. The first row corresponds to the use of all the data X for calibration purposes. The remaining rows correspond to classes #2, #3, and #5, whose centers are shown in the bar graphs of Figures 4 through 6. Classes #2 and #5 are of primary interest to this investigation, since they correspond to the classes with highest fluoride concentrations. Class #3 is included for comparison purposes.

Given the paucity of training data, it is difficult to get a good indication of the prediction accuracy of the different equations. The results of one approach are summarized by Table III. The first entry in a row shows the ratio of data vectors in the training set for the class in that row, with respect to the total number of vectors available from the original data **X**. The next entry provides a measure of the resubstitution prediction accuracy for the equation calibrated with the training data of that row. As suggested by the heading for the column, its values were obtained by i) computing the average sum of the squared residuals between the response values predicted using data vectors of the training data and the actual recorded values from the first column of **X**; and ii) then normalizing the result by the variance of the recorded y-vector for that row. The values obtained suggest the improvement in prediction accuracies which may be expected by using class-wise regression.

The remaining three columns of Table III list i) the maximum field-test (i.e., measured) value for the response variable from the training data for the row; ii) the maximum value of the response variable predicted by the regression equation with respect to all of the training vectors defining the class; and iii) the value of the response variable for that particular vector with maximum u_{ik} membership in the class represented by the row.

Effect of a Reduction in Emissions. As stated in the introduction to this paper, a major objective of the investigation was to use the models developed in the preceding section to estimate the reduction in the accumulation of fluorides in the leaves of surrounding vegetation resulting from lowering the average monthly emission level. Table IV summarizes the results. The computations were based upon the assumption that reduction of emissions would result in average hydrogen fluoride content in installation emissions of about 9.5 kg/hr (see (2)). Predictions are listed only for two classes, with class #2 chosen because it presented a worst case, and class #3, because it represented levels resulting from more benign meteorological conditions. Predicted concentrations at the individual site were computed under the two percentage-of-total accumulation assumptions shown at the bottom of the table.

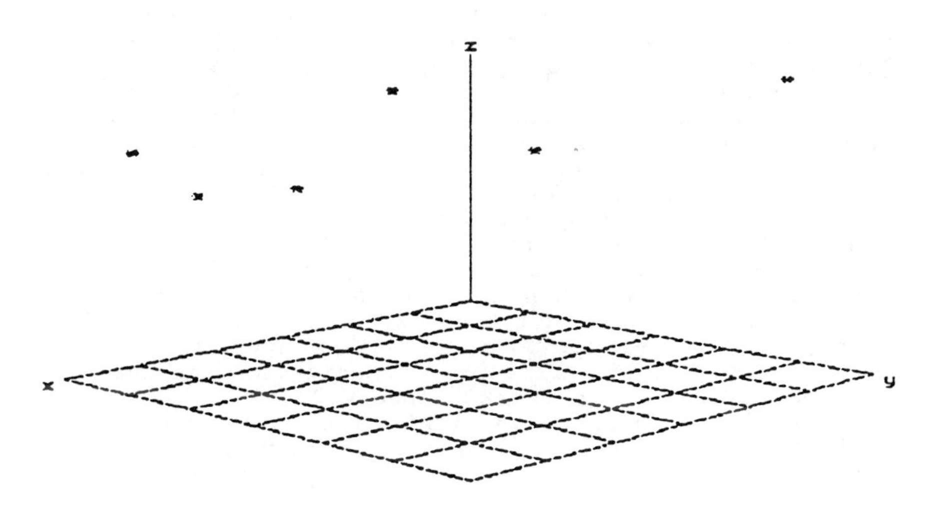

Figure 8. Projection of class centers onto 3-dimensional principal component subspace.

Table II. Regression equation constants and coefficients

Training Data	y_o	a_1	a_2	a_3	a_4	a_5
X	49.23	-0.11	-0.10	-0.27	2.94	-0.76
class #2	98.75	0.47	-0.41	-0.27	-3.10	-0.39
class #3	87.11	-0.35	-0.70	-0.52	0.27	-0.25
class #5	-255.37	0.15	2.54	2.00	23.64	-6.80

When inspecting the results of Table IV, it might be kept in mind that the accepted "safe" level for fluoride in grass used for grazing animals is about 30 mg/kg. However, it might also be pointed out that one could put forth a reasonable hypotheses that fluoride concentration levels may accumulate in a sort of bi-level mode. The first mode could be called a "steady-state", or "plateau" mode; that is, a gradual adjustment to a certain, more or less constant, long-term average of fluoride emissions. The second mode may be a "transient" response, which is superimposed upon the steady state level. This mode is a "fast"' mode, responding to short term (five day) variations in emissions and meteorological variables. Although suggested by field observations, such a hypothesis would certainly not conflict with the model one would expect from theoretical modeling based upon the laws of mathematical physics. In fact, it conforms nicely with the form of the general solution to a system of non-homogeneous, linear, differential equations. At this point, such a bi-level model is only a hypothesis, but certainly one worthy of future investigation. As far as the results of Table IV are concerned, given that the hypothesis is true, one should take into consideration that it may be the sum of the two levels which is being represented.

Other Model Applications. It may be reasonable to use the class-wise calibrated regression models as means to optimally regulate plant production for maximum production within a safe level of emissions. In particular, one could use the x-variable measurements and the results of the FCV class modeling to label the data vectors of **X** relative to the six classes. Using any of several techniques, it would then be possible to build a classifier which could be used to indicate which of the six class regression models to use for estimating the maximum output leading to a safe level of emissions, given a forecast of meteorological variables for the period.

Summary

After application of the FCV method to a combined data set of meteorological and emission variables, it was concluded that the data could reasonably be viewed as consisting of six classes. Calibration of multilinear regression models by data defined from the six class models does lead to an apparent improvement in prediction accuracies. Various applications can be made of the regression models, including worst-case levels of fluoride concentrations to be found in the grass of grazing regions surrounding the aluminum production facility, and regulating plant operation for maximum production under designated emission levels.

TABLE III. Summary of prediction accuracy and predicted values of total fluoride accumulations normalized to average emissions

Training Data	n_i/n	$\frac{ave(r^2)}{var(y)}$	y_{meas} (max)	y_{pred} (max)	y_{pred} (max-memb)
X	72/72	0.83	76.5	49.5	n/a
Class #2	16/72	0.43	76.5	69.8	53.7
Class #3	15/72	0.16	44.0	43.6	26.3
Class #5	9/72	0.15	71.3	67.4	47.4

Table IV. Predicted accumulation levels at the eight sites assuming an emission rate of 9.5 kg/h

	Predicted Levels (mg/kg) Class#3		Predicted Levels (mg/kg) Class#2	
Station #	distrib*1	distrib*2	distrib 1	distrib 2
1. Entrance	214	147	516	240
2. Beitet	15	24	13	21
3. Hoel	63	16	70	75
4. Hoås	37	18	22	36
5. Fahle	24	11	42	44
6. Romfo	30	9	29	45
7. Ottem	13	14	10	33
8. Gjøra	11	9	7	11

* Distribution 1 - Based on % contribution of each site to 8-site total observed on sampling day corresponding to maximum prediction in Table 1.

* Distribution 2 - Based on % contribution observed on sampling day corresponding to maximum membership prediction of Table 1.

Acknowledgments

The authors would like to extend their thanks to Mr. Magne Leinum, and the several other employees of Hydro Aluminum A/S for use of data and extensive assistance provided in the preparation of this report. The authors would also like to express their gratitude for the sponsorship of NATO grant 5-2-05/RG no. 0-33-88.

Literature Cited

1. Weinstein, L.H. J. of Occ. Med. **1977**, vol. 19, No.1, pp 49-77.
2. Thrane, K.E. Atmosph. Env. **1988**, vol.22, pp 587-594.
3. Thrane, K.E.; Aldrin, M. NILU report no. 77/88, Norsk Insitutt for Luftforskning, Lillestrøm, Norway, 1988.
4. Thrane, K.E. A/S Miljøplan report no. P89-010, Oslo, Norway,1989.
5. Gunderson, R.W; Thrane, K.E. In Proc. of the Third IFSA Congress; Editor, J.C.Bezdek, Seattle, Washington, 1989.
6. Årflot, O. Landsbruksforlaget, Oslo, Norway, 1981.
7. Bezdek, J.C.; Coray, C.; Gunderson, R.W.; Watson J.D. SIAM J. of Appl. Math. **1981,** vol.40, No. 2, Part I pp 339-357, Part II pp 358-372.
8. Gunderson, R.W.;Jacobsen, T. In Proc. of the Third Nordic Symposium on Appl. Stat.; Editor, O.J.H. Christy; Stokkand Forlag Publ., N-4000 Stavanger, Norway, 1983.
9. Gunderson, R.W.;Thrane, K.E. J. of Chemomet. and Instr. **1987,** vol.2, No.1, pp 88-98.
10. Beebe, K.R.;Kowalski, B.R. Anal. Chem. **1987**, vol. 57, pp 1007A-1017A.

RECEIVED January 15, 1992

Author Index

Affiliation Index

Subject Index

N

O

W

Production: C. Buzzell-Martin
Indexing: Deborah H. Steiner
Acquisition: Rhonda Bitterli
Cover Design: Pat Cunningham

Printed and bound by Maple Press, York, PA